Lecture Notes in Computer Science 6124

Bo Chen (Ed.)

Algorithmic Aspects in Information and Management

6th International Conference, AAIM 2010
Weihai, China, July 19-21, 2010
Proceedings

Volume Editor

Bo Chen
Warwick Business School
University of Warwick
Coventry, UK
E-mail: B.Chen@warwick.ac.uk

Library of Congress Control Number: 2010929854

CR Subject Classification (1998): F.2, G.2, J.1, I.2, G.2.2, G.2.1

LNCS Sublibrary: SL 3 – Information Systems and Application, incl. Internet/Web and HCI

ISSN 0302-9743
ISBN-10 3-642-14354-7 Springer Berlin Heidelberg New York
ISBN-13 978-3-642-14354-0 Springer Berlin Heidelberg New York

springer.com

Printed in Germany

Typesetting: Camera-ready by author, data conversion by Scientific Publishing Services, Chennai, India
Printed on acid-free paper 06/3180

Preface

While the areas of information management and management science are full of algorithmic challenges, the proliferation of data has called for the design of efficient and effective algorithms and data structures for their management and processing.

The International Conference on Algorithmic Aspects in Information and Management (AAIM) is intended for original algorithmic research on immediate applications and/or fundamental problems pertinent to information management and management science to be broadly construed. The conference aims at bringing together researchers in computer science, operations research, applied mathematics, economics, and related disciplines.

This volume contains papers presented at AAIM 2010: the 6th International Conference on Algorithmic Aspects in Information and Management, which was held during July 19-21, 2010, in Weihai, China. We received a total of 50 submissions. Each submission was reviewed by three members of the Program Committee or their deputies on the quality, originality, soundness, and significance of its contribution. The committee decided to accept 31 papers. The program also included two invited keynote talks.

The success of the conference resulted from the input of many people. We would like first of all to thank all the members of the Program Committee for their expert evaluation of the submissions. The local organizers in the School of Computer Science and Technology, Shandong University, did an extraordinary job, for which we are very grateful. We thank the National Natural Science Foundation of China, Montana State University (USA), University of Warwick (UK), and Shandong University (China) for their sponsorship.

July 2010 Bo Chen

Preface

Conference Organization

General Conference Chairs

Xiangxu Meng	Shandong University, China
Binhai Zhu	Montana State University, USA

Program Committee

Peter Brucker	Universität Osnabrück, Germany
Bo Chen (Chair)	University of Warwick, UK
Jianer Chen	Texas A&M University, USA
Zhi-Long Chen	University of Maryland, USA
Chengbin Chu	University of Troyes, France
Van-Dat Cung	Institut polytechnique de Grenoble, France
Ovidiu Daescu	University of Texas at Dallas, USA
Shijie Deng	Georgia Institute of Technology, USA
Donglei Du	University of New Brunswick, Canada
Paul Goldberg	University of Liverpool, UK
Xiaodong Hu	Chinese Academy of Sciences, China
Jianhua Ji	Shanghai Jiaotong University, China
Ellis Johnson (Co-chair)	Georgia Institute of Technology, USA
Ming-Yang Kao	Northwestern University, USA
Han Kellerer	University of Graz, Austria
Adam Letchford	Lancaster University, UK
Mohammad Mahdian	Yahoo! Research, USA
Silvano Martello	University of Bologna, Italy
Peter Bro Miltersen	University of Aarhus, Denmark
Yoshiaki Oda	Keio University, Japan
Chris Potts	University of Southampton, UK
Tomasz Radzik	King's College London, UK
Rajeev Raman	University of Leicester, UK
Frits Spieksma	Katholieke Universiteit Leuven, Belgium
Paul Spirakis	University of Patras, Greece
Defeng Sun	National University of Singapore, Singapore
Jie Sun	National University of Singapore, Singapore
Tami Tamir	The Interdisciplinary Center Herzliya, Israel
Jens Vygen	University of Bonn, Germany
Shouyang Wang	Chinese Academy of Sciences, China
Wenxun Xing	Tsinghua University, China
Yinfeng Xu	Xi'an Jiaotong University, China
Yinyu Ye	Stanford University, USA
Guochuan Zhang	Zhejiang University, China

Jiawei Zhang	New York University, USA
Yuzhong Zhang	Qufu Normal University, China
Daming Zhu	Shandong University, China

Local Organizing Committee

Shandong University: Haodi Feng, Hong Liu, Junfeng Luan, Hongze Qiu, Peng Zhang, Daming Zhu (Chair)

Table of Contents

Comparison of Two Algorithms for Computing Page Importance

Yuting Liu[1] and Zhi-Ming Ma[2]

[1] Beijing Jiaotong University,
School of Science, Beijing, China, 100044
ytliu@bjtu.edu.cn
[2] Academy of Mathematics and Systems Science,
CAS, Beijing, China, 100190
mazm@amt.ac.cn

Abstract. In this paper we discuss the relation and the difference between two algorithms BrowseRank and PageRank. We analyze their stationary distributions by the ergodic theory of Markov processes. We compare in detail the link graph used in PageRank and the user browsing graph used in BrowseRank. Along with the comparison, the importance of the metadata contained in the user browsing graph is explored.

Keywords: PageRank, BrowseRank, Continuous-time Makrov process, Stationary distribution, Ergodic theorem, link graph, user browsing graph.

1 Introduction

Page importance is a critical factor for web search. Currently, the link analysis algorithms are successfully used to calculate the page importance from the hyperlink graph of the Web. Link analysis algorithms take the link from one webpage to another as an endorsement of the linking page, and assume that the more links point to a page, the more likely the page being pointed is important. Well known link analysis algorithms include PageRank [1,10], HITS [6], TrustRank [5], and many others. In this paper we take PageRank as a typical example of link analysis. PageRank is usually modeled as a discrete-time Markov process on the web link graph [7]. Actually the process simulates a random walk of a surfer on the Web along hyperlinks. Although PageRank has many advantages, recently people have realized that it has also certain limitations as a model for representing page importance. For example, the link graph, which PageRank relies on, is not a very reliable data source, because hyperlinks on the Web can be easily added or deleted by web content creators. It is clear that those purposely created hyperlinks (e.g. created by link farms[4]) are not suitable for calculating page importance.

To tackle the limitations of PageRank, recently a new algorithm called BrowseRank has been proposed ([8,9]). The new algorithm collects the user behavior data in web surfing and builds a user browsing graph, which contains both user transition information and user staying time information. A continuous-time Markov process

B. Chen (Ed.): AAIM 2010, LNCS 6124, pp. 1–11, 2010.

is employed in BrowseRank to model the browsing behavior of a web surfer, and the stationary distribution of the process is regarded as the page importance scores. The experimental results conducted in [8,9] testified that BrowseRank has better performance than PageRank in finding top websites and filtering spam websites.

In this paper we shall further analyze and compare the two algorithms PageRank and BrowseRank. In the next section we briefly review the probabilistic meaning of PageRank and explain it by ergodic theorem of Markov chains. We then give a description of BrowseRank in Section 3, which include a description of the model by time-continuous Markov process, and a further analysis of BrowseRank algorithm with ergodic theory. A detail comparison between BrowseRank and PageRank is given in Section 4. In Subsection 4.1 we explain that a significant difference between the two algorithms is that they rely on different data bases and take different attitudes in evaluating the importance of web pages. In short, PageRank is "voted" by web content creators , while BrowseRank is "voted" by users of Web pages. In Subsection 4.2 we make a comparison of the two stationary distributions. We show that if we ignore the difference between the link graph and the user browsing graph, then BrowseRank could be regarded as a generalization of PageRank. However, the difference between the user browsing graph and the link graph should not be ignored. We then present a detail comparison between the link graph and the user browsing graph in Subsection 4.3. Along with the comparison, the importance of the metedata contained in the user browsing graph is explored.

2 Probabilistic View of PageRank Algorithm

PageRank is one of the most famous link analysis algorithms. It was proposed by Brin and Page in 1998 [1,10], and has been successfully used by Google search engine. The basic assumptions of PageRank algorithm are: (1) the hyperlink from one page to another page is considered as an endorsement to the in-linked one; (2) if many important pages link to a page on the link graph, then the page is also likely to be important. The algorithm employs a power method to calculate the nonnegative eigenvector of the PageRank matrix, accordingly the eigenvector is interpreted as the PageRank values of webpages.

The probabilistic meaning of PageRank has been explained by Langville and Meyer in [7]. They modeled the surfing on the Web as a random walk on the link graph, and found that the nonnegative eigenvector of PageRank matrix is nothing but the stationary distribution (up to a constant) of the random walk. For the purpose of our further discussion and comparison, in what follows we briefly review the probabilistic meaning of PageRank and explain it by ergodic theorem of Markov chains.

We regard the hyperlink structure of webpages on a network as a directed graph $\tilde{G} = (\tilde{V}, E)$. A vertex $i \in \tilde{V}$ of the graph represents a webpage, and a directed edge $\overrightarrow{ij} \in E$ represents a hyperlink from page i to page j. Suppose that $|\tilde{V}| = N$. Let $B = (B_{ij})_{N \times N}$ be the adjacent matrix of $\tilde{G}$ and b_i be the sum of the ith row of B, i.e. $b_i = \sum_{j=1}^{N} B_{ij}$. Let D be the diagonal matrix with diagonal entry b_i (if $b_i = 0$, then we set $b_i = N$, and change all entries of the ith row of B to 1). Now, we construct a stochastic matrix $\bar{P} = D^{-1}B$.

When a surfer browses on the Web, he may choose the next page by randomly clicking one of the hyperlinks in the current page with a large probability α (in practice it is often set $\alpha = 0.85$), which means that with probability α, the surfer may randomly walk on $\tilde{G}$ with transition probability $\bar{P}$; while with a probability of $(1-\alpha)$, the surfer may also open a new page from the Web, the new page might be selected randomly according to his personal preference φ, which means that he walks randomly on $\tilde{G}$ with transition probability $e^T\varphi$, where e is a row vector of all ones, and φ is an N-dimensional probability vector (in practice it is often set $\varphi = \frac{1}{N}e$ for simplicity), which is called the personalized vector. Combining the above two random walks, a transition matrix describing the browsing behavior of a random surfer may be formulated as

$$P = \alpha\bar{P} + (1-\alpha)e^T\varphi. \tag{1}$$

Let $\{Y_n\}_{n\geq 0}$ be a discrete-time Markov chain with transition matrix P. Then the evaluation of the Markov chain at time n represents the webpage that the random surfer is browsing on his n-th surfing. Since P is an irreducible stochastic matrix with a finite state space, hence it admits a unique stationary distribution, denoted by $\pi = (\pi_i)_{i=1,\ldots,N}$. π is then called the PageRank, which is indeed the unique positive (left) eigenvector of P satisfying

$$\pi = \pi P, \quad \pi e^T = 1, \tag{2}$$

and can be calculated iteratively. Furthermore, by the ergodic theorem (cf. e.g. [11]) we have:

$$\pi_i = \lim_{n\to\infty}\frac{1}{n}\sum_{k=0}^{n-1} p_{ii}^{(k)} = \Big(\sum_{n=1}^{\infty} n f_{ii}(n)\Big)^{-1} = \frac{1}{\mu_{ii}}, \tag{3}$$

where π_i is the ith entry of π, $p_{ii}^{(k)}$ is the ii^{th} entry of the k-step transition matrix P^k, $f_{ii}(n)$ is the probability that the random surfer starts from the page i and returns for the first time back to the page i exactly at the n-step, and μ_{ii} is the so called mean re-visiting time (or mean re-visiting steps). Therefore, the right hand side of Equation (3) is equal to the mean frequency of visiting webpage i. Thus the more often a webpage is visited, the higher value of its PageRank will be. This reveals that PageRank defined as the stationary distribution of the Markov chain is a very suitable measure for page importance.

3 Probabilistic View of BrowseRank Algorithm

As mentioned at the beginning of this paper, to overcome some limitations of PageRank, recently a new algorithm called BrowseRank has been proposed ([8,9]). We now give a detail description of the algorithm BrowseRank.

3.1 Continuous Time Markov Process Model

In the algorithm of BrowseRank, the browsing behavior of a random surfer is modeled as a continuous-time Markov process $\{X_t, t \geq 0\}$, which is called the user browsing process in [8,9]. By the algorithm one should collects the user behavior data in web surfing and builds a user browsing graph. The process takes its value in the state space V consisting of all the webpages in the user browsing graph. Thus the user browsing process X contains both user transition information and user staying time information. The evaluation of X_t at time t represents the webpage that the random surfer is browsing at the time point t, here t may take value in the set $\mathbb{R}^+$ of all the nonnegative real numbers. The following two basic assumptions are imposed in [8,9].

Firstly, the time and the page that a surfer will visit in the next step depends only on the current situation, and is independent of his browsing history. This assumption is also a basic assumption in PageRank. It is so called the one-step Markov property, that is, $\forall\, t_1 \leq \cdots \leq t_n \leq t_{n+1} \in \mathbb{R}^+$, and $\forall\, i_1, \ldots, i_{n+1} \in V$,

$$P(X_{t_{n+1}} = i_{n+1}|X_{t_1} = i_1, \ldots, X_{t_n} = i_n) = P(X_{t_{n+1}} = i_{n+1}|X_{t_n} = i_n). \quad (4)$$

Secondly, the process $\{X_t, t \geq 0\}$ is time-homogeneous, that is, the transition probability of the process depends only on the time difference, and is independent of the current time point. More precisely, $\forall\, i, j \in V, s \geq 0, t \geq 0$,

$$P(X_{s+t} = j|X_s = i) = P(X_t = j|X_0 = i) \triangleq p_{ij}(t). \quad (5)$$

With the above assumptions, the user browsing process $\{X_t, t \geq 0\}$ is a continuous-time time-homogeneous Markov process. Denote by $P(t)$ its transition probability matrix in time slot t, that is, $P(t) = (p_{ij}(t))_{i,j \in V}$ (cf. Equation (5)). Let τ_k be the kth jumping time of the process X, i.e., $\tau_0 = 0$ and $\tau_k = \inf\{t : t > \tau_{k-1}, X_t \neq X_{\tau_{k-1}}\}, k \geq 1$. Then $\{\tau_k\}_{k \geq 0}$ forms a sequence of stopping times. By the property of time-homogeneous Markov processes (cf. e.g. [13]), for any $k \geq 1$ we have

$$\begin{aligned} P(\tau_{k+1} - \tau_k > t, X_{\tau_{k+1}} = j|X_k = i) &= P(\tau_1 > t, X_{\tau_1} = j|X_0 = i) \\ &= P(X_{\tau_1} = j|X_0 = i) \times P(\tau_1 > t|X_0 = i) \\ &= \tilde{p}_{ij} \times P(\tau_1 > t|X_0 = i), \end{aligned} \quad (6)$$

where $\tilde{p}_{ij} \triangleq P(X_{\tau_1} = j|X_0 = i)$ represents the probability that the random surfer jumps from page i to page j, which is independent of the staying time of the process X.

Write $\tilde{P} = (\tilde{p}_{ij})_{i,j \in V}$ and define $Z_n = X_{\tau_n}$ for $n \geq 0$, then $\{Z_n, n \geq 0\}$ is a discrete-time Markov chain with state space V and transition probability matrix $\tilde{P}$. Z is called the embedded Markov chain of the process X [12]. It was shown in [8,9] that the embedded Markov chain $\{Z_n, n \geq 0\}$ is irreducible and aperiodic, which implies that the continuous-time process $\{X_t, t \geq 0\}$ is also irreducible and aperiodic, and hence X admits a unique stationary distribution, denoted by $\pi = (\pi_i)_{i \in V}$, which is independent of t. BrowseRank is then defined in [8,9] as the stationary distribution π of X. By the definition, π is the unique nonnegative vector satisfying

$$\pi = \pi P(t), \quad \pi e^T = 1. \tag{7}$$

We shall see that π_i stands for the ratio of the time that the surfer spends on the page i over the time he spends on all the pages when time interval t goes to infinity (cf. Equation (9) below). Hence, π is a suitable measure of page importance.

3.2 Analyze BrowseRank with Ergodic Theory

Let $\{X_t, t \geq 0\}$ be the user browsing process described as above. Suppose that a user starts from a page $i \in V$ and randomly surfs on webpages according to the process $\{X_t, t \geq 0\}$. Let T_i be the staying time of the random surfer on page i. That is, T_i stands for the first jumping time τ_1 of X given the condition that $X_0 = i$. Let T_{ii} be the length of the time period that the random surfer walks a circle from page i to page i, that is, the time period that the random surfer starts from page i and returns for the first time back to page i. We denote the expectation of T_i as μ_i, and the expectation of T_{ii} as μ_{ii}.

The following results can be found e.g. in [2,14]. For the convenience of the reader we provide a proof here.

Theorem 1. *Let $\{X_t, t \geq 0\}$ be a time homogeneous Markov process. Suppose that X is irreducible and aperiodic, and $\pi = (\pi_i)_{i \in V}$ is its stationary distribution. Then with the above notations we have:*

(1) $\pi_i = \frac{\mu_i}{\mu_{ii}}$, a.s..
(2) Suppose $\hat{\pi} = (\hat{\pi}_i)_{i \in V}$ is the stationary distribution of the embedded Markov chain $\{Z_n, n \geq 0\}$ of X, then we have

$$\pi_i = \frac{\hat{\pi}_i \mu_i}{\sum_{j \in V} \hat{\pi}_j \mu_j}. \tag{8}$$

Proof. By the definition of stationary distribution, we have

$$\pi_i = \lim_{t \to \infty} P(X_t = i | X_0 = j),$$

where j is an arbitrary page of V. Without loss of generality, we may take $X_0 = i$. By this way we can regard the process $\{X_t, t \geq 0\}$ as a renewal process which starts from page i and returns to page i when the next renewal occurs. The lengths of time period between two successive renewals form a family of i.i.d random variables with the same distribution as T_{ii}. Let N_t denotes the number of renewals occurred before time t. By a property of renewal processes (cf. e.g. Theorem 3-3-2 in [2]), we have

$$\lim_{t \to \infty} \frac{N_t}{t} = \frac{1}{\mu_{ii}} \ a.s. \ ,$$

Moreover, by the ergodic theorem of continuous-time Markov processes we know that

$$\pi_i = \lim_{t\to\infty} \frac{\text{The time spent on page } i \text{ in interval [0,t]}}{t}, \ a.s. \ . \tag{9}$$

Let $T_i(k)$ be the staying time of the kth visit on page i. Then $T_i(k), k \geq 1$, form a family of i.i.d random variables with the same distribution as T_i. Consequently by the strong law of large numbers,

$$\begin{aligned}
\pi_i &= \lim_{t\to\infty} \frac{\text{The time spent on page } i \text{ in interval [0,t]}}{t} \\
&= \lim_{t\to\infty} \frac{\sum_{k=1}^{N_t} T_i(k)}{t} = \lim_{t\to\infty} \frac{N_t}{t} \frac{\sum_{k=1}^{N_t} T_i(k)}{N_t} \\
&= \frac{\mu_i}{\mu_{ii}}, \ a.s. \ .
\end{aligned}$$

So we get assertion (1).

We now prove assertion (2). Let $N_i(m)$ be the number of times visiting page i in past m transitions. Based on the fact that the embedded Markov chain $\{Z_n, n \geq 0\}$ is ergodic, by the ergodic theorem we have

$$\lim_{m\to\infty} \frac{N_i(m)}{m} = \hat{\pi}_i \ a.s. \ . \tag{10}$$

By the strong law of large number, the definition of μ_i, and Equation (10), we have

$$\lim_{m\to\infty} \frac{1}{m} \sum_{k=1}^{N_i(m)} T_i(k) = \lim_{m\to\infty} \frac{N_i(m)}{m} \frac{1}{N_i(m)} \sum_{k=1}^{N_i(m)} T_i(k) = \hat{\pi}_i \mu_i \ a.s. \ . \tag{11}$$

Combining Equation (9) and Equation (11), we get

$$\begin{aligned}
\pi_i &= \lim_{t\to\infty} \frac{\text{The time spent on page } i \text{ in interval [0,t]}}{t} \\
&= \lim_{m\to\infty} \frac{\sum_{k=1}^{N_i(m)} T_i(k)}{\tau_m} \\
&= \lim_{m\to\infty} \frac{\sum_{k=1}^{N_i(m)} T_i(k)}{\sum_{j\in V} \sum_{k=1}^{N_j(m)} T_j(k)} \\
&= \lim_{m\to\infty} \frac{\frac{1}{m}\sum_{k=1}^{N_i(m)} T_i(k)}{\sum_{j\in V} \frac{1}{m}\sum_{k=1}^{N_j(m)} T_j(k)} \\
&= \frac{\hat{\pi}_i \mu_i}{\sum_{j\in V} \hat{\pi}_j \mu_j} \ a.s. \ .
\end{aligned}$$

So we get assertion (2).

From Theorem 1 we can draw the following conclusions.

(i) By assertion (1), we know that the evaluation of the stationary distribution at page i is determined by the fraction $\frac{\mu_i}{\mu_{ii}}$, where the numerator μ_i is the mean staying time on page i, while the denominator μ_{ii} stands for the mean re-visit time on page i. This means that, the more important a page is, the longer time the user would like to stay on it, and hence the higher evaluation of the stationary distribution at this page will be. Also the more important a page is, the shorter the mean re-visit time on this page will be, the more often this page will be visited, and hence the higher evaluation of the stationary distribution at this page will be. Therefore, the stationary distribution of the user browsing process is a suitable measure of page importance.

(ii) Assertion (2) provides us feasible algorithms to calculate BrowseRank π. Indeed, μ_i is the mean staying time which can be estimated from a large number of observation data extracted from users' browsing log data. The stationary distribution $\tilde{\pi}_i$ of the discrete-time embedded Markov chain can be computed similar to PageRank by Power Method[3]. Based on assertion (2), an efficient and feasible flow chart of BrowseRank algorithm has been devised in [8,9].

4 Comparison between BrowseRank and PageRank

In this section we make further analysis and comparison of the two algorithms BrowseRank and PageRank.

4.1 Different Data Bases and Different Votes

We note first that a significant difference between BrowseRank and PageRank is that they make use of different data bases and take different attitudes in evaluating the importance of web pages.

PageRank algorithm relies solely on the data of the link graph of the Web. It interprets a link from page i to page j as a vote by page i for page j, and analyzes the page that casts the vote. Votes cast by pages that are themselves "important" weigh more heavily and help to make other pages "important". In other words, the importance of pages is evaluated by the votes of web content creators who create or delete hyperlinks.

BrowseRank algorithm collects the user behavior data in web surfing and builds a user browsing graph, which contains both user transition information and user staying time information. In this way, the importance of pages relies on the browsing behavior of a large number of users, it is evaluated by hundreds of millions of users' implicit votes.

4.2 Comparison of Stationary Distributions

BrowseRank algorithm employs a continuous-time Markov process $\{X_t, t \geq 0\}$ to model the browsing behavior of a web surfer, and the BrowseRank is defined as the stationary distribution of the browsing process X. To make comparison, in this subsection we denote the stationary distribution of X as $\pi^B = (\pi_i^B)_{i \in V}$. By (8) we know that

$$\pi_i^B = \frac{\hat{\pi}_i \mu_i}{\sum_{j \in V} \hat{\pi}_j \mu_j}. \tag{12}$$

Thus the BrowseRank value of a page i is proportional to the product of two factors $\hat{\pi}_i$ and μ_i, where $\hat{\pi}_i$ is the stationary distribution of the embedded Markov chain Z on the page i, and μ_i is the mean staying time of the surfer on the page i. On the other hand, PageRank is modeled as a discrete-time Markov process $\{Y_n\}_{n \geq 0}$ on the web link graph, and PageRank value is defined as the stationary distribution π of Y (cf. (2)).

If we assume that the staying times on the pages are all the same (this is indeed implicitly assumed in PageRank algorithm), say, $P(T_j = 1) = 1$ for all $j \in V$, then (12) will be reduced to $\pi_i^B = \hat{\pi}_i$. In this case the BrowseRank algorithm would be reduced to the PageRank algorithm. Assume further that the user browsing graph was the same as the link graph, then π^B would be equal to π. In other words, if we ignore the difference between the user browsing graph and the link graph, then BrowseRank could be regarded as a generalization of PageRank.

However, we emphasize that the difference between the user browsing graph and the link graph should not be ignored. See the discussion in the next subsection.

4.3 User Browsing Graph vs. Link Graph

We start with a brief description of the user browsing graph. For more details the reader is refered to [8,9].

The user browsing graph is a weighted graph with vertices containing metadata and edges containing weights. We denote it as $G = \langle V, W, C, R, Z, \gamma \rangle$, where $V = \{v_i\}$, with size $|V| = M$, is the state space consisting of all the webpages browsed by users during a long period of observation; an element w_{ij} in $W = \{w_{ij} | i, j \in V\}$ represents the number of transitions from page i to page j conducted by users during the period; $C = \{c_i | i \in V\}$ collects the numbers of visiting times on each page observed during the period; The entry r_i in $R = \{r_i | i \in V\}$ is the number of resettings[1] occurred at page i; $Z = \{Z_i | Z_i = \{z_i^1, z_i^2, \ldots, z_i^{m_i}\}, i \in V\}$, each Z_i collects all the observed staying times on page i; and $\gamma = \{\gamma_i | i \in V\}$, γ_i stands for the resetting probability on page i.

Below we list some major differences and relations between the user browsing graph (UBG for short) employed by BrowseRank and the link graph employed by PageRank:

1. From the above description we know that UBG is a multi-graph. Note that the vertex set of UBG is a subset of that of the link graph, and the edge set

[1] If user dose not click any hyperlinks on some page, but choose a new page to start another round of browsing, we say it occurs a resetting at such page.

of UBG is also a subset of that of the link graph. Hence if we condense UBG into a simple graph, i.e. ignore the metadata of each vertex and ignore the weight of each edge, then UBG is a subgraph of the link graph. However, those vertices of link graph which are not contained in UBG are exactly low qualified and negligible pages. Because those pages have never been visited by any user in the long period of observation. Similarly, those edges in the link graph which are not contained in UBG are exactly low qualified and negligible hyperlinks. Because those links have never been clicked by any user in the long period of observation. But UBG can not be simply described as a subgraph of the link graph. The metadata and weights contained in UBG play important roles in BrowseRank algorithm and influence efficiently the evaluation of the importance of webpages.

2. Being a multi-graph, each edge in UBG has a weight $w_{ij} \in W$ which records the number of transitions from one page to another page. Being a simple graph, the weights of edges in the link graph are assumed implicitly all equal to one. In BrowseRank algorithm, the weights $W = \{w_{ij}|i,j \in V\}$ of edges together with the resetting probabilities $\gamma = \{\gamma_i|i \in V\}$ of pages influence efficiently the transition probabilities of the embedded Markov chain. Ignore the detail techniques, in BrowseRank algorithm[8] the transition probability from page i to page j is computed by the formula

$$\tilde{p}_{ij} = \frac{\alpha w_{ij}}{\sum_{k \in V} w_{ik}} + (1-\alpha)\gamma_j, \tag{13}$$

when $\sum_{k \in V} w_{ik} > 0$, where α stands for the probability that the user surfs on Web along hyperlinks. In other words, when the user surfs along hyperlinks, the probability that he jumps from page i to page j is proportional to the weight w_{ij}; when the user wants to start another round of browsing, with probability γ_j he will choose page j. Comparing with PageRank algorithm based on the link graph, the transition probability from page i to page j is in practice computed by the formula

$$\bar{p}_{ij} = \frac{\alpha}{\text{out degree of page } i} I_{[i \to j]} + (1-\alpha)\frac{1}{N}, \tag{14}$$

when out degree of page i is not zero, where $I_{[i \to j]} = 1$ if there is a hyperlink from page i to page j and $I_{[i \to j]} = 0$ otherwise. That is, when the user surfs along hyperlinks, he will randomly (i.e. with equal probability) click a link; when he wants to start another round of browsing, he will also randomly choose a page.

3. In PageRank algorithm some authors suggested to employ personalized probability vector φ in place of the uniform probability in the Equation (14). In practice there is no algorithm to compute φ. In BrowseRank the resetting probability γ employed in the Equation (13) plays exactly the role of a personalized probability. But γ is estimated from the real users' behaviors data by a feasible algorithm.

4. As we have mentioned in the last subsection, all the staying times in PageRank algorithm are implicitly assumed to be equal to one. While in BrowseRank

algorithm, the observed staying times Z contained in UBG are essential in estimating the probability distribution of the continuous-time Markov process. They are also used to estimate the mean staying time of pages, which compose a factor of the stationary distribution of X (cf. Equation (8)).

5. Other metadata contained in UBG are also very useful in BrowseRank algorithm. For example, the number of visiting times $C = \{c_i|i \in V\}$ and the number of resettings $R = \{r_i|i \in V\}$ are used to improve the transition probability $\tilde{p}_{ij}$ from page i to page j by the formula:

$$\tilde{p}_{ij} = \alpha \frac{w_{ij} + r_i \, \gamma_j}{c_i} + (1 - \alpha) \, \gamma_j. \tag{15}$$

Theoretically, the improved formula (15) is better than the formula (13), because (15) contains more information. Indeed, the experimental results in [9] have shown that the algorithm besed on (15) outperforms the algorithm based on (13).

Acknowledgments

We are grateful to Bin Gao, Tie-Yan Liu and Hang Li from MSRA for the stimulating discussions and valuable suggestions during our pleasant collaborations. We thank the financial support of the Fundamental Research Funds for the Central Universities (2009JBM106) and the key lab project of CAS. Z. M. Ma would like to thank the organizers of AAIM 2010 for inviting him to give a keynote speech.

References

1. Brin, S., Page, L.: The anatomy of a large-scale hypertextual Web search engine. Computer Networks and ISDN Systems 30(1-7), 107–117 (1998)
2. Deng, Y., Liang, Z.: Stochastic Point Processes and Their Applications. Science Press, Beijing (1998) (in Chinese)
3. Golub, G.H., Loan, C.F.V.: Matrix computations, 3rd edn. Johns Hopkins University Press, Baltimore (1996)
4. Gyöngyi, Z., Garcia-Molina, H.: Web spam taxonomy. Technical Report TR 2004-25, Stanford University (2004)
5. Gyöngyi, Z., Garcia-Molina, H., Pedersen, J.: Combating web spam with trustrank. In: VLDB 2004: Proceedings of the Thirtieth International Conference on Very Large Data Bases, pp. 576–587, VLDB Endowment (2004)
6. Kleinberg, J.M.: Authoritative sources in a hyperlinked environment. In: SODA 1998, Philadelphia, PA, USA, pp. 668–677 (1998)
7. Langville, A.N., Meyer, C.D.: Deeper inside pagerank. Internet Mathematics 1(3), 335–400 (2004)
8. Liu, Y., Gao, B., Liu, T.-Y., Zhang, Y., Ma, Z., He, S., Li, H.: BrowseRank: letting web users vote for page importance. In: SIGIR 2008: Proceedings of the 31st Annual International ACM SIGIR Conference on Research and Development in Information Retrieval, pp. 451–458. ACM, New York (2008)

9. Liu, Y., Liu, T.-Y., Gao, B., Ma, Z., Li, H.: A framework to compute page importance based on user behaviors. Information Retrieval 13(1), 22–45 (2010)
10. Page, L., Brin, S., Motwani, R., Winograd, T.: The pagerank citation ranking: Bringing order to the web. Technical Report 1999-66, Stanford InfoLab, Previous number = SIDL-WP-1999-0120 (November 1999)
11. Qian, M.P., Gong, G.L.: Stochastic Process. Second version. Peking University Press (1997) (in Chinese)
12. Stewart, W.J.: Introduction to the Numerical Solution of Markov Chains. Princeton University Press, Princeton (1994)
13. Stroock, D.W.: An Introduction to Markov Processes, Graduate Texts in Mathematics. Springer, Heidelberg (2005)
14. Wang, Z.K., Yang, X.Q.: Birth and Death Processes and Markov Chains. Springer, New York (1992)

The Invisible Hand for Risk Averse Investment in Electricity Generation

Daniel Ralph[1] and Yves Smeers[2]

[1] University of Cambridge, Judge Business School
d.ralph@jbs.cam.ac.uk
[2] Université catholique de Louvain, Centre for Operations Research & Econometrics
smeers@core.ucl.ac.be

Abstract. We consider a perfectly competitive situation consisting of an electricity market (2nd stage) preceded by investment in generating plant capacity (1st stage). The second stage environment is uncertain at the time of investment, hence the first stage also involves trading in financial instruments, eg, hedges against high generation costs due to rising fuel costs.

The classical Invisible Hand says that if generators and consumers act in their own best interests, the result will be to minimize the net cost (or max net welfare) of the system. This holds true in the stochastic risk neutral case, when a probability distribution of future events is known and used by all generators to evaluate their investment strategies (via two stage stochastic programming with recourse).

Motivated by energy developments in the European Union, our interest is the case when electricity generators are risk averse, and the cost of future production is assessed via coherent risk measures instead of expectations. This results in a new kind of stochastic equilibrium framework in which (risk neutral) probability distributions are endogenous can only be found at equilibrium.

Our main result is that if there are enough financial products to cover every future situation, ie, the financial market is complete, then the Invisible Hand remains in force: system equilibrium is equivalent to system optimization in risk averse investment equilibria. Some practical implications will be discussed.

B. Chen (Ed.): AAIM 2010, LNCS 6124, p. 12, 2010.

Efficient Algorithms for the Prize Collecting Steiner Tree Problems with Interval Data*

E. Álvarez-Miranda[1], A. Candia[1], X. Chen[2], X. Hu[2], and B. Li[2,**]

[1] Industrial Management Department, Universidad de Talca, Chile
[2] Institute of Applied Mathematics, Chinese Academy of Sciences
P. O. Box 2734, Beijing 100190, China
libi@amss.ac.cn

Abstract. Given a graph $G = (V, E)$ with a cost on each edge in E and a prize at each vertex in V, and a target set $V' \subseteq V$, the Prize Collecting Steiner Tree (PCST) problem is to find a tree T interconnecting vertices in V' that has minimum total costs on edges and maximum total prizes at vertices in T. This problem is NP-hard in general, and it is polynomial-time solvable when graphs G are restricted to 2-trees. In this paper, we study how to deal with PCST problem with uncertain costs and prizes. We assume that edge e could be included in T by paying cost $x_e \in [c_e^-, c_e^+]$ while taking risk $\frac{c_e^+ - x_e}{c_e^+ - c_e^-}$ of losing e, and vertex v could be awarded prize $p_v \in [p_v^-, p_v^+]$ while taking risk $\frac{y_v - p_v^-}{p_v^+ - p_v^-}$ of losing the prize. We establish two risk models for the PCST problem, one minimizing the maximum risk over edges and vertices in T and the other minimizing the sum of risks. Both models are subject to upper bounds on the budget for constructing a tree. We propose two polynomial-time algorithms for these problems on 2-trees, respectively. Our study shows that the risk models have advantages over the tradional robust optimization model, which yields NP-hard problems even if the original optimization problems are polynomial-time solvable.

Keywords: Prize collecting Steiner tree, interval data, 2-trees.

1 Introduction

The Prize Collecting Steiner Tree (PCST) problem has been extensively studied in the areas of computer science and operation research due to its wide range of applications [2,9,10]. A typical application occurs when a natural gas provider wants to build a most profitable transportation system to send natural gas from a station to some customers on scattered locations, where each link (segment

* Supported in part by NNSF of China under Grant No. 10531070, 10771209, 10721101,10928102 and Chinese Academy of Sciences under Grant No. kjcx-yw-s7. P^4 Project Grant Center for Research and Applications in Plasma Physics and Pulsed Power Technology, PBCT-Chile-ACT 26, CONICYT; and Dirección de Programas de Investigación, Universidad de Talca, Chile.
** Corresponding author.

B. Chen (Ed.): AAIM 2010, LNCS 6124, pp. 13–24, 2010.

of pipleline) is associated with a cost which is incurred if the link is installed, and each location is associated with a profit which is obtained if the location is connected to the station by links installed. Moreover, the transportation system is required to contain some specified customers. One of the most important special cases of PCST problem is the Steiner Minimum Tree (SMT) problem [7] where the profits associated with all locations are zero.

Since the SMT problem is NP-complete in general [8], so is the PCST problem, and the latter admits a 2-approximation polynomial-time algorithm [5]. When restricted to 2-trees, Wald and Colbourn [12] proved that the SMT problem is polynomial-time solvable; In this paper, we will extend their algorithm to the PCST problem for this special class of graphs.

In contrast to the above deterministic setting, it is often necessary to take uncertainty into account in the real application of algorithms for PCST and SMT problems. For example, in the above application, the gas provider may not be able to know exactly how much cost he/she needs to pay for installing a link and how much profit he/she can obtain from connecting a location. Instead, the provider may manage to estimate the highest and lowest costs or profits he/she needs to pay or can obtain. In such a situation, the gas provider could get a more reliable link if he/she would like to pay more, and the possibility for he/she to obtain a small profit is higher than that to obtain a large profit.

Robust optimization is one of most frequently used approaches to dealing with problems under nondeterministic setting. However, many robust optimization problems (such as the robust shortest path [13] and robust spanning tree problems [1]) are NP-hard even though their deterministic counterparts are polynomial-time solvable. Recently, Chen et. al. [4] and Hu [6] proposed two novel models for network optimization with interval data on network links under which the corresponding problems are polynomial-time solvable, preserving the polynomial-time solvability of the original optimization problems with deterministic data. In this paper, we will extend their approaches to the PCST problem in 2-trees by considering not only uncertain costs on edges but also uncertain profits at vertices.

The reminder of the paper is organized as follows: In Section 2, we present a linear-time algorithm for the PCST problem on 2-trees. In Section 3, we first establish min-max and min-sum risk models for the PCST problem, respectively, and then propose two polynomial-time algorithms for the PCST problem on 2-trees under these two models. In Section 4, we conclude the paper with discussions on the obtained results and future work.

2 Efficient Algorithm for PCST Problem on 2-Trees

In the PCST problem, we are given a undirected graph $G = (V, E)$ with vertex-set V of size n and edge-set E of size m, where each vertex $v \in V$ is assigned a nonnegative prize $p_v \in \mathbb{R}_+$, and each edge $e \in E$ is assigned a nonnegative cost $c_e \in \mathbb{R}_+$. Conventionally, a *target set* $V' \subseteq V$, which is also called a *terminal set*, is given. The goal of the PCST problem is to find a tree T in G such that $V' \subseteq V(T)$ and its value

$$\nu(T,c,p) \equiv \sum_{e\in E(T)} c_e - \sum_{v\in V(T)} p_v \tag{1}$$

is minimum among all trees in G spanning V'. Such a tree is called an *optimal* PCST in G and denoted by $T_{opt}(G,c,p)$. We assume that G is connected, as otherwise we can consider the connected component of G that contains V'.

For easy discussion, we remove target set V' from the input of the PCST problem. In fact, we could force all vertices in V' to be included in $T_{opt}(G,c,p)$ by assigning each of them a sufficiently large prize. The following lemma gives the details. Its proof, along with the proofs of all other lemmas and theorems, is ommited due to space limitation.

Lemma 1. *Given* $G=(V,E)$, $c\in\mathbb{R}^E_+, p\in\mathbb{R}^V_+$, $M=\sum_{e\in E} c_e + \sum_{v\in V} p_v + 100$ *and target set* $V'\subseteq V$, *let* $p'\in\mathbb{R}^V_+$ *with* $p'_v = M$ *for every* $v\in V'$, $p'_v = p_v$ *for every* $v\in V\setminus V'$. *Let* T_{opt} *denote an optimal PCST* $T_{opt}(G,c,p')$, *then* $V'\subseteq V(T_{opt})$, *and* $\nu(T_{opt},c,p)\le\nu(T,c,p)$ *for any tree* T *in* G *with* $V(T)\supseteq V'$.

The graph class of 2-*trees* can be defined recursively as follows: A *triangle* (i.e., a complete graph of three vertices) is a 2-tree; given a 2-tree T with an edge uv in T, adding a new vertex z adjacent with both u and v yields a 2-tree.

Wald and Colbourn [12] gave an $O(n)$-time algorithm for finding a SMT in a given 2-tree of n vertices with target set. Their algorithm is a dynamic programming method that finds a Steiner tree on 2-tree by repeatedly eliminating vertices of degree 2 until the remaining graph is a single edge. During this vertex elimination procedure, they record information associated with triangle $\{u,v,z\}$, where vertex v has degree 2 in the current 2-tree, on the ordered pairs (u,z) and (z,u) corresponding to the edge uz in G, when considering and deleting v.

In the following we will describe how to extend the above method for the SMT problem to the PCST problem on 2-trees. Assume that $G=(V,E)$ is a 2-tree. As target set is not presented (explicitly) in our PCST problem, our task can be accomplished by introducing five measures, instead of six as in [12] for the SMT problem on 2-tree. For each ordered pair (u,v) that corresponds to edge $uv\in E$, we introduce five measures $st(u,v), dt(u,v), un(u,v)$, $nv(u,v)$, $nn(u,v)$ to record the values computed so far for the subgraph S of G which has been reduced onto edge uv.

$st(u,v)$ is the min value of a tree T in S with $u,v\in V(T)$;
$dt(u,v)$ is the min value of two disjoint trees T_1,T_2 in S with $u\in V(T_1)$, $v\in V(T_2)$;
$un(u,v)$ is the min value of a tree T in S with $u\in V(T)$ but $v\notin V(T)$;
$nv(u,v)$ is the min value of a tree T in S with $v\in V(T)$ but $u\notin V(T)$;
$nn(u,v)$ is the min value of a tree T in S with $u,v\notin V(T)$.

In the above, the value of two disjoint trees T_1 and T_2 is defined as

$$\nu(T_1,c,p)+\nu(T_2,c,p)\equiv\Big(\sum_{e\in E(T_1)} c_e - \sum_{v\in V(T_1)} p_v\Big)+\Big(\sum_{e\in E(T_2)} c_e - \sum_{v\in V(T_2)} p_v\Big).$$

Using the five measures, we design a dynamic programming ALG_PCST, whose pseudo-code is omitted, for solving the PCST problem on 2-*tree*.

Theorem 1. *Algorithm* ALG_PCST *outputs the optimal value* ν^* *of the PCST problem on 2-trees of* n *vertices in* $O(n)$ *time.*

3 Algorithms for PCST Problem with Interval Data

In this section, we consider the PCST problem on 2-trees with interval data. Given a undirected graph $G = (V, E)$, each edge $e \in E$ is associated with a cost interval $[c_e^-, c_e^+]$, and each vertex $v \in V$ is associated with a prize interval $[p_v^-, p_v^+]$. These intervals indicate possible ranges of construction cost of edge e and prize of vertex v, respectively. We define the *risk* at edge e as $r(x_e) \equiv \frac{c_e^+ - x_e}{c_e^+ - c_e^-}$ when charging cost $x_e \in [c_e^-, c_e^+]$, and the *risk* at vertex v as $r(y_v) \equiv \frac{y_v - p_v^-}{p_v^+ - p_v^-}$ when collecting prize $y_v \in [p_v^-, p_v^+]$. For ease of description, we make the notational convention that $\frac{0}{0} = 0$. With these definitions, risks $r(x_e)$ and $r(y_v)$ both range from 0 to 1. In particular, $r(x_e) = 0$ when $x_e = c_e^+$ ($r(y_v) = 0$ when $y_v = p_v^-$), meaning no risk occurs if the payment is high enough (the expected prize is low enough). On the other hand, $r(x_e) = 1$ when $x_e = c_e^-$ ($r(y_v) = 1$ when $y_v = p_v^+$), meaning a full risk is doomed at the lowest payment (the highest prize). Let B be a given budget bound on constructing a PCST and $\mathscr{T}$ the set of trees in G. We define the value of tree T with charged payment x and collected prize y as

$$\nu(T, x, y) \equiv \sum_{e \in E(T)} x_e - \sum_{v \in V(T)} y_v. \tag{2}$$

In the following two subsections, we will study two risk models that adopt distinct objective functions, respectively.

3.1 PCST Problem under Min-Max Risk Model

In this subsection, we first formulate the PCST problem under min-max risk model denoted by MMR_PCST, and present the property of its optimal solutions, and then present a polynomial-time algorithm MMR_PCST on 2-trees.

The goal of the MMR_PCST problem is to find a tree T along with payment x and prize y such that the maximum risk at edges and vertices in T is minimized and the value $\nu(T, x, y)$ is less than or equal to the given budget B. This problem can be formulated as follows:

$$\text{(MMR\_PCST)} \qquad \min_{T \in \mathscr{T}, \nu(T,x,y) \le \mathsf{B}} \ \max_{e \in E(T), v \in V(T)} \left\{ \frac{c_e^+ - x_e}{c_e^+ - c_e^-}, \frac{y_v - p_v^-}{p_v^+ - p_v^-} \right\}$$
$$\text{s.t.} \quad \begin{cases} x_e \in [c_e^-, c_e^+], \ \forall\ e \in E; \\ y_v \in [p_v^-, p_v^+], \ \forall\ v \in V. \end{cases}$$

The following lemma shows that the optimal solution (T^*, x^*, y^*) to the above problem possesses an evenness property, which will play an important role in our algorithm design. We reserve symbol r^* for the value of the optimal solution to the MMR_PCST problem, i.e.

$$r^* \equiv r_m(T^*, x^*, y^*) \equiv \max_{e \in E(T^*), v \in V(T^*)} \left\{ \frac{c_e^+ - x_e^*}{c_e^+ - c_e^-}, \frac{y_v^* - p_v^-}{p_v^+ - p_v^-} \right\}.$$

Lemma 2. *For every edge e and vertex v in T^*, it holds that* $\frac{c_e^+ - x_e^*}{c_e^+ - c_e^-} = \frac{y_v^* - p_v^-}{p_v^+ - p_v^-} = r^*$.

To present our analysis for determining the optimal value r^* in an easy way, we need to introduce more notations. For any $r \in [0,1]$, let $x_e^r = c_e^+ - r(c_e^+ - c_e^-)$, $y_v^r = p_v^- + r(p_v^+ - p_v^-)$ for every edge $e \in E$ and every vertex $v \in V$. In addition, let $T^r \in \mathscr{T}$ be an optimal PCST $T_{opt}(G, x^r, y^r)$, i.e.

$$\nu(T^r, x^r, y^r) = \min_{T \in \mathscr{T}} \nu(T, x^r, y^r). \tag{3}$$

From the above lemma we have $\nu(T^{r^*}, x^{r^*}, y^{r^*}) = \nu(T^*, x^*, y^*)$.

Lemma 3. *Suppose* B *is a given bound on the budget for constructing PCSTs. If* $\nu(T^r, x^r, y^r) > \mathsf{B}$*, then* $r < r^*$*; otherwise* $r \geq r^*$.

From the definitions of x^r, y^r, T^r it is easy to see that if $\nu(T^0, x^0, y^0) \leq \mathsf{B}$, then (T^0, x^0, y^0) is an optimal solution to the MMR_PCST problem. Hence we assume $\nu(T^0, x^0, y^0) > \mathsf{B}$. By Lemma 3, we know $r^* > 0$ implies $\nu(T^*, x^*, y^*) = \mathsf{B}$, as otherwise we could increase x_e^* for every edge $e \in E(T^*)$ and decrease y_v^* a little bit for every vertex $v \in V(T^*)$ and get a smaller r^*. Furthermore, by virtue of Lemmas 2 and 3, we can apply Megiddo's parametric search method [11] to determine r^* in polynomial time.

In the following pseudo-code, r^* is always kept being contained in $[r_l, r_u]$. We will narrow the interval $[r_l, r_u]$ by comparing value $\nu(T^r, x^r, y^r)$ with bound B. By Lemma 3, we set $r_l = r$ if $\nu(T^r, x^r, y^r) > \mathsf{B}$ and set $r_u = r$ otherwise. Algorithm ALG_PCST is used as a searching tool for computing $\nu(T^r, x^r, y^r)$. In the end, we are able to locate a unique $r^* \in [r_l, r_u]$ by solving the equation $\nu(T^r, x^r, y^r) = \mathsf{B}$.

Algorithm for MMR_PCST on 2-tree (ALG_MMR)

Input 2-tree $G = (V, E)$ with $c^- \in \mathbb{R}_+^E$, $c^+ \in \mathbb{R}_+^E$, $p^- \in \mathbb{R}_+^V$, $p^+ \in \mathbb{R}_+^V$, $\mathsf{B} \in \mathbb{R}_+$
Output optimal value r^* of the MMR_PCST problem

1. $F \leftarrow G$, $D \leftarrow \emptyset$, $r_l \leftarrow 0$, $r_u \leftarrow 1$
2. **for** $uv \in F$ **do begin**
3. $\quad st^r(u,v) = st^r(v,u) \leftarrow c_{uv}^r - p_u^r - p_v^r$, $\quad dt^r(u,v) = dt^r(v,u) \leftarrow -p_u^r - p_v^r$;
4. $\quad un^r(u,v) = nu^r(v,u) \leftarrow -p_u^r$, $\quad nv^r(u,v) = vn^r(v,u) \leftarrow -p_v^r$;
5. $\quad nn^r(u,v) = nn^r(v,u) \leftarrow 0$
6. **end-for**
7. $D \leftarrow \{$degree-2 vertices in $G\}$
8. **while** $D \neq \emptyset$ **do begin**
9. $\quad$ Take vertex $v \in D$ and edges $uv, vz \in E(F)$
10. $\quad m_1^r(u,z) \leftarrow \min\{st^r(u,z) + un^r(u,v) + nz^r(v,z) + p_u^r + p_z^r,$
$\quad\quad st^r(u,z) + st^r(u,v) + dt^r(v,z) + p_u^r + p_v^r + p_z^r,$
$\quad\quad st^r(u,z) + dt^r(u,v) + st^r(v,z) + p_u^r + p_v^r + p_z^r,$
$\quad\quad dt^r(u,z) + st^r(u,v) + st^r(v,z) + p_u^r + p_v^r + p_z^r\}$
11. $\quad m_2^r(u,z) \leftarrow \min\{dt^r(u,z) + un^r(u,v) + nz^r(v,z) + p_u^r + p_z^r,$
$\quad\quad dt^r(u,z) + st^r(u,v) + dt^r(v,z) + p_u^r + p_v^r + p_z^r,$
$\quad\quad dt^r(u,z) + dt^r(u,v) + st^r(v,z) + p_u^r + p_v^r + p_z^r\}$

12. $m_3^r(u,z)\leftarrow\min\{un^r(u,z)+un^r(u,v)+p_u^r, un^r(u,z)+st^r(u,v)+vn^r(v,z)+p_u^r+p_v^r\}$
13. $m_4^r(u,z)\leftarrow\min\{nz^r(u,z)+nz^r(v,z)+p_z^r, nz^r(u,z)+nv^r(u,v)+st^r(v,z)+p_v^r+p_z^r\}$
14. $m_5^r(u,z)\leftarrow\min\{nn^r(u,z), nn^r(u,v), nn^r(v,z), nv^r(u,v)+vn^r(v,z)+p_v^r\}$
15. **for** $i=1:5$ **do begin**
16. Call PRC_UD with $m(r)=m_i^r(u,z)$, $r\in[r_l,r_u]$ to update $[r_l,r_u]$
17. **end-for**
18. $st^r(u,z)=st^r(z,u)\leftarrow m_1^r(u,z)$, $dt^r(u,z)=dt^r(z,u)\leftarrow m_2^r(u,z)$;
19. $un^r(u,z)=nu^r(z,u)\leftarrow m_3^r(u,z)$, $nz^r(u,z)=zn^r(z,u)\leftarrow m_4^r(u,z)$;
20. $nn^r(u,z)=nn^r(z,u)\leftarrow m_5^r(u,z)$.
21. $F\leftarrow F\backslash\{v,uv,vz\}$, $D\leftarrow$ {degree-2 vertices in F}
22. **end-while**
23. $\nu^r\leftarrow\min\{st^r(u,z), un^r(u,z), nz^r(u,z), nn^r(u,z)\}$, where $r\in[r_l,r_u]$, $E(F)=\{uz\}$
24. Find $r^*\in[r_l,r_u]$ s.t. $\nu^{r^*}=\mathsf{B}$
25. Output r^*

Procedure Update (PRC_UD) called by ALG_MMR

Input interval $[r_l,r_u]$, function $m(r)$ with $r\in[r_l,r_u]$, and input of ALG_MMR
Output updated interval $[r_l,r_u]$
1. Find the non-differentiable points $r_1\leq r_2\leq r_j$ of $m(r)$, where $j\leq 3$
2. **for** $i=1:j$ **do begin**
3. Call ALG_PCST to find optimal value $\nu(T^{r_i},x^{r_i},y^{r_i})$
4. **if** $\nu(T^{r_i},x^{r_i},y^{r_i})>\mathsf{B}$ **then** $r_l\leftarrow r_i$ **else** $r_u\leftarrow r_i$ and go to Step 6
5. **end-for**
6. Return $[r_l,r_u]$

Theorem 2. *Algorithm* ALG_MMR *outputs the optimal value* ν^* *of the MMR_PCST problem on 2-trees of n vertices in* $O(n^2)$ *time.*

3.2 PCST Problem under Min-Sum Risk Model

In this subsection, we first formulate the PCST problem under min-sum risk model denoted by MSR_PCST, and give a nice property of the optimal solutions, then we present a polynomial-time algorithm for MSR_PCST on 2-trees.

The goal of the MSR_PCST problem is to find a tree T along with payment x and prize y such that the sum of risks at edges and vertices in T is minimized and the value $\nu(T,x,y)$ is less than or equal to the given budget B. This problem can be formulated as follows:

$$(\text{MSR\_PCST})\qquad \min_{T\in\mathscr{T},\nu(T,x,y)\leq\mathsf{B}}\left(\sum_{e\in E(T)}\frac{c_e^+-x_e}{c_e^+-c_e^-}+\sum_{v\in V(T)}\frac{y_v-p_v^-}{p_v^+-p_v^-}\right)$$

$$\text{s.t.}\ \begin{cases} x_e\in[c_e^-,c_e^+],\ \forall\ e\in E;\\ y_v\in[p_v^-,p_v^+],\ \forall\ v\in V.\end{cases}$$

The following lemma exhibits a structural property of optimal solutions to the MSR_PCST problem, which plays an important role in our algorithm design.

Lemma 4. *There exists an optimal solution (T^*, x^*, y^*) to the MSR_PCST problem that includes an edge $f \in E(T^*)$ and a vertex $u \in V(T^*)$ such that $x_f \in [c_f^-, c_f^+]$, $y_u \in [p_u^-, p_u^+]$ and $x_e \in \{c_e^-, c_e^+\}$ for every edge $e \in E(T^*) \setminus \{f\}$, $y_v \in \{p_v^-, p_v^+\}$ for every vertex $v \in V(T^*) \setminus \{u\}$.*

To find an optimal solution specified in the above lemma, we will transform in the following three steps the original given graph $G = (V, E)$ with $c^-, c^+ \in \mathbb{R}_+^E$ and $p^-, p^+ \in \mathbb{R}_+^V$ to a new graph $\widetilde{G} = (\widetilde{V}, \widetilde{E})$ with $c, w \in \mathbb{R}_+^{\widetilde{E}}$ and $p, q \in \mathbb{R}_+^{\widetilde{V}}$.

Algorithm for Graph Transformation (ALG_GT)

Step 1 Construct graph $\bar{G} = (\bar{V}, \bar{E})$ with $\bar{c}, \bar{w} \in \mathbb{R}_+^{\bar{E}}$ as follows: Set $\bar{V} \equiv V$ and $\bar{E} \equiv \{\underline{e}, \overline{e} : e \in E\}$ in way that every edge $e \in E$ corresponds to two edges $\underline{e}, \overline{e} \in \bar{E}$ both having the same ends as e. For every $e \in E$, set $\bar{c}_{\underline{e}} \equiv c_e^-, \bar{c}_{\overline{e}} \equiv c_e^+$; $\bar{w}_{\underline{e}} \equiv 1, \bar{w}_{\overline{e}} \equiv 0$ if $c_e^- \neq c_e^+$ and set $\bar{w}_{\underline{e}} = \bar{w}_{\overline{e}} \equiv 0$ otherwise.

Step 2 Construct graph $\widehat{G} \equiv (\widehat{V}, \widehat{E})$ with $\widehat{c}, \widehat{w} \in \mathbb{R}_+^{\widehat{E}}$ and $\widehat{p}^-, \widehat{p}^+ \in \mathbb{R}_+^{\widehat{V}}$ as follows: Set $\widehat{V} \equiv \bar{V} \cup \{v_{\overline{e}} : e \in E\}$ and $\widehat{E} \equiv \{\underline{e} \in \bar{E} : e \in E\} \cup \{\overline{e}_1 \equiv v_{\overline{e}}u, \overline{e}_2 \equiv v_{\overline{e}}v : u, v \in \bar{V}, uv = \overline{e} \in \bar{E}\}$. For every $e \in E$, set $\widehat{c_{\overline{e}_1}} = \widehat{c_{\overline{e}_2}} \equiv \frac{1}{2}\bar{c}_{\overline{e}}$, $\widehat{w_{\overline{e}_1}} = \widehat{w_{\overline{e}_2}} \equiv \frac{1}{2}\bar{w}_{\overline{e}}$, $\widehat{p^+_{v_{\overline{e}}}} = \widehat{p^-_{v_{\overline{e}}}} \equiv 0$. For every $v \in V = \bar{V}$, set $\widehat{p_v^-} \equiv p_v^-$, $\widehat{p_v^+} \equiv p_v^+$.

Step 3 Construct graph $\widetilde{G} = (\widetilde{V}, \widetilde{E})$ with $c, w \in \mathbb{R}_+^{\widetilde{E}}$ and $p, q \in \mathbb{R}_+^{\widetilde{V}}$ as follows: Set $\widetilde{V} \equiv \{v_1, v_2 : v \in \widehat{V}\}$ and $\widetilde{E} \equiv \{u_1v_1, u_1v_2, u_2v_1, u_2v_2 : uv \in \widehat{E}\}$. For every $uv \in \widehat{E}$ and $i, j \in \{1, 2\}$, set $c_{u_iv_j} \equiv \bar{c}_{uv}$, $w_{u_iv_j} \equiv \bar{w}_{uv}$. For every $v \in \widehat{V}$, set $p_{v_1} \equiv \widehat{p_v^+}, p_{v_2} \equiv \widehat{p_v^-}$; set $q_{v_1} \equiv 1, q_{v_2} \equiv 0$ if $\widehat{p_v^-} \neq \widehat{p_v^+}$ and $q_{v_1} = q_{v_2} \equiv 0$ otherwise.

Note that there is a 1-1 correspondence between the set of pairs $(\widehat{T}, \widehat{y})$, where $\widehat{T}$ is a tree in $\widehat{G}$ and $\widehat{y} \in \mathbb{R}_+^{V(\widehat{T})}$ with $\widehat{y}v \in \{p_v^+, p_v^-\}$ for every $v \in V(\widehat{T})$, and set $\widetilde{\mathscr{T}} \equiv \{\widetilde{T}: \widetilde{T}$ is a tree in $\widetilde{G}$ and $|E(\widetilde{T}) \bigcap \{u_1v_1, u_1v_2, u_2v_1, u_2v_2\}| \leq 1, \forall\ uv \in \bar{E}\}$. The bijection is as follows: $pair_1(\widetilde{T}) \equiv (\widehat{T}, \widehat{y})$ if and only if $tree(\widehat{T}, \widehat{y}) \equiv \widetilde{T}$, which satisfies the following for every $uv \in \bar{E}$:

(1) $u_1v_1 \in E(\widetilde{T})$ if and only if $uv \in E(\widehat{T})$ and $\widehat{y}_u = \widehat{p_u^+}$, $\widehat{y}_v = \widehat{p_v^+}$;
(2) $u_1v_2 \in E(\widetilde{T})$ if and only if $uv \in E(\widehat{T})$ and $\widehat{y}_u = \widehat{p_u^+}$, $\widehat{y}_v = \widehat{p_v^-}$;
(3) $u_2v_1 \in E(\widetilde{T})$ if and only if $uv \in E(\widehat{T})$ and $\widehat{y}_u = \widehat{p_u^-}$, $\widehat{y}_v = \widehat{p_v^+}$;
(4) $u_2v_2 \in E(\widetilde{T})$ if and only if $uv \in E(\widehat{T})$ and $\widehat{y}_u = \widehat{p_u^-}$, $\widehat{y}_v = \widehat{p_v^-}$.

Lemma 5. *If $tree(\widehat{T}, \widehat{y}) = \widetilde{T}$, then $\nu(\widetilde{T}, c, p) = \nu(\widehat{T}, \widehat{c}, \widehat{y})$.*

In addition, there is a 1-1 correspondence between the pairs $(\bar{T}, y)$, where $\bar{T}$ is a tree in $\overline{G}$ and $y \in \mathbb{R}_+^{V(\bar{T})}$ with $y_v \in \{p_v^+, p_v^-\}$ for every $v \in V(\bar{T})$, and the pairs $(\widehat{T}, \widehat{y})$, where $\widehat{T} \in \widehat{\mathscr{T}} \equiv \{\widehat{T}' : \widehat{T}'$ is a tree in $\widehat{G}, E(\widehat{T}') \cap \{\overline{e}_1, \overline{e}_2, \underline{e}\} = \{\underline{e}\}$ or $\{\overline{e}_1, \overline{e}_2\}$ or $\emptyset, \forall\, e \in E\}$ and $\widehat{y} \in \mathbb{R}_+^{V(\widehat{T})}$ with $\widehat{y}_v \in \{\widehat{p_v^+}, \widehat{p_v^-}\}$ for every $v \in V(\widehat{T})$. The bijection is as follows: $add(\bar{T}, y) \equiv (\widehat{T}, \widehat{y})$ if and only if $con(\widehat{T}, \widehat{y}) \equiv (\bar{T}, y)$, which satisfies the following for every $e = uv \in E$:

(1) $\underline{e} \in E(\bar{T})$ if and only if $\underline{e} \in E(\widehat{T})$ and $\widehat{y}_u = y_u$, $\widehat{y}_v = y_v$;
(2) $\overline{e} \in E(\bar{T})$ if and only if $\{\overline{e}_1, \overline{e}_2\} \subseteq E(\widehat{T})$ and $\widehat{y}_u = y_u$, $\widehat{y}_v = y_v$, $\widehat{y}_{v_{\overline{e}}} = 0$.

Lemma 6. *If* $add(\bar{T}, y) = (\widehat{T}, \widehat{y})$, *then* $\nu(\widehat{T}, \widehat{c}, \widehat{y}) = \nu(\bar{T}, \bar{c}, y)$.

Moreover, there is a 1-1 correspondence between the pairs $(\bar{T}, y)$, where $\bar{T}$ is a tree in $\bar{G}$ and $y \in \mathbb{R}+^{V(\bar{T})}$ with $y_v \in \{p_v^+, p_v^-\}$ for every $v \in V(\bar{T})$, and the triples (T, x, y), where T is a tree in G and $x \in \mathbb{R}_+^{E(T)}$ with $x_e \in \{c_e^+, c_e^-\}$ for every $e \in E(T)$. The bijection is as follows: $triple(\bar{T}, y) \equiv (T, x, y)$ if and only if $pair_2(T, x, y) \equiv (\bar{T}, y)$, which satisfies $V(\bar{T}) = V(T)$ and the following for every $e \in E$:

(1) $\underline{e} \in E(\bar{T})$ if and only if $e \in E(T)$ and $x_e = c_e^-$;
(2) $\overline{e} \in E(\bar{T})$ if and only if $e \in E(T)$ and $x_e = c_e^+$.

Lemma 7. *If* $triple(\bar{T}, y) = (T, x, y)$, *then* $\nu(T, x, y) = \nu(\bar{T}, \bar{c}, y)$.

From Lemmas 5-7, we can establish a 1-1 correspondence between trees $\widetilde{T} \in \widetilde{\mathscr{T}}$ and triples (T, x, y) such that T is a tree in G, $x \in \mathbb{R}_+^{E(T)}$, $x_e \in \{c_e^+, c_e^-\}$ for every edge $e \in E(T)$ and $y \in \mathbb{R}_+^{V(T)}$, $yv \in \{p_v^+, p_v^-\}$ for every vertex $v \in V(T)$. The following theorem summarizes the correspondence.

Theorem 3. *Let* TREE $(T, x, y) \equiv tree(add(pair_2(T, x, y))) = \widetilde{T}$, *or equivalently* TRIPLE $(\widetilde{T}) \equiv triple(con(pair_1(\widetilde{T}))) = (T, x, y)$, *then* $\nu(T, x, y) = \nu(\widetilde{T}, c, p)$ *and* $\sum_{e \in E(T)} \frac{c_e^+ - x_e}{c_e^+ - c_e^-} + \sum_{v \in V(T)} \frac{y_v - p_v^-}{p_v^+ - p_v^-} = \sum_{e \in E(\widetilde{T})} w_e + \sum_{v \in V(\widetilde{T})} q_v$.

To present our algorithm for solving the MSR_PCST problem on the 2-*tree* $G = (V, E)$, we need to consider how to deal with the Weight Constrained PCST problem, denoted by WC_PCST. This problem can be formulated as follows: Given $(\widetilde{G}, c, p, w, q, \zeta)$ with $c, w \in \mathbb{R}_+^{E(\widetilde{G})}$ and $p, q \in \mathbb{R}_+^{V(\widetilde{G})}$ being vectors defined in the construction of $\widetilde{G}$ by ALG_GT, and ζ being an upper bound, let $W(T, w, q) \equiv \sum_{e \in E(T)} w_e + \sum_{v \in V(T)} q_v$, where $T \in \widetilde{\mathscr{T}}$. The goal is to find an optimal solution (a tree in $\widetilde{\mathscr{T}}$) to the following problem

$$\text{(WC\_PCST)} \qquad \min_{T \in \widetilde{\mathscr{T}}, W(T,w,q) \leq \zeta} \Big(\sum_{e \in E(T)} c_e - \sum_{v \in V(T)} p_v \Big).$$

we now describe how to extend the dynamic programming algorithm for the constrained SMT problem on 2-*tree* [3] to algorithm ALG_WC for the WC_PCST problem on 2-*tree*. First, we introduce more notations which work in a similar way to those introduced in Section 2. With each ordered pair (u, v) corresponding to an edge uv of $\widehat{G}$, we associate $(13\zeta + 13)$ measures which summarize the value incurred so far in the subgraph S which has been reduced onto the edge uv. For each $\xi = 0, 1, 2, \ldots, \zeta$, we define:

1. $st_{ij}(u, v, \xi)$ is the min value of trees T in S with $W(T, w, q) \leq \xi$ and u_i, $v_j \in V(T)$, for $i, j \in \{1, 2\}$.
2. $dt_{ij}(u, v, \xi)$ is the min value of two disjoint trees T_1 and T_2 in S with $W(T_1, w, q) + W(T_2, w, q) \leq \xi$ and $u_i \in V(T_1)$ while $v_j \in V(T_2)$, for $i, j \in \{1, 2\}$.

3. $un_i(u, v, \xi)$ is the min value of trees T in S with $W(T, w, q) \leq \xi$ and $u_i \in V(T)$ while v_1, $v_2 \notin V(T)$, for $i = 1, 2$.
4. $nv_i(u, v, \xi)$ is the min value of trees T in S with $W(T, w, q) \leq \xi$ and $v_i \in V(T)$ while u_1, $u_2 \notin V(T)$, for $i = 1, 2$.
5. $nn(u, v, \xi)$ is the min value of trees T in S with $W(T, w, q) \leq \xi$ and $u_i, v_j \notin V(T)$, for $i, j \in \{1, 2\}$.

Algorithm for WC_PCST Problem on 2-trees (ALG_WC)

Initially, set $\mathsf{L}= 2n\big(\sum_{e\in E} c_e^+ + \sum_{v\in V} p_v^+\big)$ and for (u, v, ξ) and for (u, v, ξ) with $uv \in E(G)$, set

$st_{ij}(u, v, \xi) \leftarrow c_{uv} - p_{u_i} - p_{v_j}$, if $w_{uv} + q_{u_i} + q_{v_j} \leq \xi$, L otherwise, $i, j = 1, 2$;
$dt_{ij}(u, v, \xi) \leftarrow -p_{u_i} - p_{v_j}$, if $q_{u_i} + q_{v_j} \leq \xi$; L otherwise, $i, j = 1, 2$;
$un_i(u, v, \xi) \leftarrow -p_{u_i}$, if $q_{u_i} \leq \xi$; L otherwise, $i = 1, 2$;
$nv_i(u, v, \xi) \leftarrow -p_{v_i}$, if $q_{v_i} \leq \xi$; L otherwise, $i = 1, 2$; $nn(u, v, \xi) \leftarrow 0$.

Then, update the measures when a degree-2 vertex in $\widehat{G}$ (corresponding to two degree-4 vertices in $\widetilde{G}$) is deleted. Suppose that at some stage there is a triangle of three vertices u, v, z with v of degree 2 in the current graph S. We use current measures for (u, v) and (v, z) (which have been computed) to compute the measures associated with (u, z, ξ) in the graph S for all $i, j \in \{1, 2\}$ using the following recurrences:

$$\begin{aligned}
st_{ij}(u, z, \xi) \leftarrow \min \big\{ & st_{ij}(u, z, \xi_1) + un_i(u, v, \xi_2) + nz_j(v, z, \xi_3) + p_{u_i} + p_{z_j},\\
& st_{ij}(u, z, \xi_{3h+1}) + st_{ih}(u, v, \xi_{3h+2}) + dt_{hj}(v, z, \xi_{3h+3}) + p_{u_i} + p_{v_h} + p_{z_j},\\
& st_{ij}(u, z, \xi_{3h+7}) + dt_{ih}(u, v, \xi_{3h+8}) + st_{hj}(v, z, \xi_{3h+9}) + p_{u_i} + p_{v_h} + p_{z_j},\\
& dt_{ij}(u, z, \xi_{3h+13}) + st_{ih}(u, v, \xi_{3h+14}) + st_{hj}(v, z, \xi_{3h+15}) + p_{u_i} + p_{v_h} + p_{z_j},\\
& h = 1, 2 \mid \textstyle\sum_{h=3k-2}^{3k} \xi_h = \xi, 1 \leq k \leq 7; \xi_1, \xi_2, \ldots, \xi_{21} \geq 0 \big\}
\end{aligned}$$

$$\begin{aligned}
dt_{ij}(u, z, \xi) \leftarrow \min \big\{ & dt_{ij}(u, z, \xi_1) + un_i(u, v, \xi_2) + nz_j(v, z, \xi_3) + p_{u_i} + p_{z_j},\\
& dt_{ij}(u, z, \xi_{3h+1}) + st_{ih}(u, v, \xi_{3h+2}) + dt_{hj}(v, z, \xi_{3h+3}) + p_{u_i} + p_{v_h} + p_{z_j},\\
& dt_{ij}(u, z, \xi_{3h+7}) + dt_{ih}(u, v, \xi_{3h+8}) + st_{1j}(v, z, \xi_{3h+9}) + p_{u_i} + p_{v_h} + p_{z_j},\\
& h = 1, 2 \mid \textstyle\sum_{h=3k-2}^{3k} \xi_h = \xi, 1 \leq k \leq 5; \xi_1, \xi_2, \ldots, \xi_{15} \geq 0 \big\}
\end{aligned}$$

$$\begin{aligned}
un_i(u, z, \xi) \leftarrow \min \big\{ & un_i(u, z, \xi_1) + un_i(u, v, \xi_2) + p_{u_i},\\
& un_i(u, z, \xi_{3h}) + st_{ih}(u, v, \xi_{3h+1}) + vn_h(v, z, \xi_{3h+2}) + p_{u_i} + p_{v_h}, h = 1, 2\\
& \mid \xi_1 + \xi_2 = \xi_3 + \xi_4 + \xi_5 = \xi_6 + \xi_7 + \xi_8 = \xi, \xi_1, \xi_2, \ldots, \xi_8 \geq 0 \big\}.
\end{aligned}$$

$$\begin{aligned}
nz_i(u, z, \xi) \leftarrow \min \big\{ & nz_i(u, z, \xi_1) + nz_i(u, v, \xi_2) + p_{z_i},\\
& nz_i(u, z, \xi_{3h}) + nv_h(u, v, \xi_{3h+1}) + st_{hi}(v, z, \xi_{3h+2}) + p_{v_h} + p_{z_i}, h = 1, 2\\
& \mid \xi_1 + \xi_2 = \xi_3 + \xi_4 + \xi_5 = \xi_6 + \xi_7 + \xi_8 = \xi, \xi_1, \xi_2, \ldots, \xi_8 \geq 0 \big\}.
\end{aligned}$$

$$\begin{aligned}
nn(u, z, \xi) \leftarrow \min\{ & \min\{nv_h(u, v, \xi_{2h-1}) + vn_h(v, z, \xi_{2h}) + p_{vh}, h = 1, 2 \mid \xi_{2h-1}, \xi_{2h} > 0,\\
& \xi_{2h-1} + \xi_{2h} = \xi, h = 1, 2\}, nn(u, z, \xi), nn(u, v, \xi), nn(v, z, \xi)\}.
\end{aligned}$$

In the end, $\widehat{G}$ is reduced to a single edge uv (corresponding to four edges u_1v_1, u_1v_2, u_2v_1, u_2v_2 in $\widetilde{G}$). Take the minimum, denoted as ν_ζ, among nine measures $st_{ij}(u, v, \zeta)$, $un_i(u, v, \zeta)$, $nv_i(u, v, \zeta)$, $nn(u, v, \zeta)$, $i, j \in \{1, 2\}$. Output ν_ζ and an optimal solution $\widetilde{T}_\zeta$ (which is a tree in $\widetilde{G}$ constructed by back tracing).

Theorem 4. *Algorithm* ALG_WC *outputs the an optimal solution $\widetilde{T}_\zeta$ and optimum value ν_ζ of the WC_PCST problem on 2-trees of n vertices in $O(n^3)$ time.*

Now we are ready to present our algorithm for the MSR_PCST problem on 2-trees. In the following pseudo-code, (T, x, y) denotes a solution to the MSR_PCST problem for which there exist an edge $f \in E(T)$ and a vertex $u \in V(T)$ such that $x_f \in [c_f^-, c_f^+]$, $y_u \in [p_u^-, p_u^+]$ and $x_e \in \{c_e^-, c_e^+\}$ for every edge $e \in E(T) \setminus \{f\}$, $y_v \in \{P_v^-, p_v^+\}$ for every vertex $v \in V(T) \setminus \{u\}$. In order to find such a solution, we first find for every $v \in V$ and every $e \in E$ a solution (T, x, y) with minimum risk sum such that $v \in V(T)$, $e \in E(T)$ and $x_g \in \{c_g^-, c_g^+\}$ for every edge $g \in E(T) \setminus \{e\}$, $y_z \in \{p_z^-, p_z^+\}$ for every vertex $z \in V(T) \setminus \{v\}$.

Algorithm for MSR_PCST on 2-tree (ALG_MSR)

Input 2-tree $G = (V, E)$ with $c^- \in \mathbb{R}_+^E$, $c^+ \in \mathbb{R}_+^E$, $p^- \in \mathbb{R}_+^V$, $p^+ \in \mathbb{R}_+^V$, $\mathsf{B} \in \mathbb{R}_+$ and $\widetilde{M} = 2\sum_{e \in E} c_e^+ + 2\sum_{v \in V} p_v^+ + 100$
Output an optimal solution (T^*, x^*, y^*) satisfying Lemma 4

1. Call ALG_GT to construct $\widetilde{G} = (\widetilde{V}, \widetilde{E})$, $w \in \{0,1\}^{\widetilde{E}}$, $q \in \{0,1\}^{\widetilde{V}}$, $c \in \mathbb{R}_+^{\widetilde{E}}$, $p \in \mathbb{R}_+^{\widetilde{V}}$
2. $\mathscr{T}^* \leftarrow \emptyset$, $\alpha \leftarrow 0$, $\beta \leftarrow 2n - 1$
3. Call ALG_WC to find opt value ν_{2n-1} for WC_PCST on $(\widetilde{G}, c, p, w, q, 2n-1)$
4. **if** $\nu_{2n-1} > \mathsf{B}$ **then** stop (No feasible solution!)
5. Call ALG_WC to find the optimum value ν_0 for WC_PCST on $(\widetilde{G}, c, p, w, q, 0)$
6. **if** $\nu_0 \le \mathsf{B}$ **then** output TRIPLE $(\widetilde{T_0})$, where $\widetilde{T_0} \in \widetilde{\mathscr{T}}$, $\nu(\widetilde{T_0}, c, p) = \nu_0$, and stop
7. **while** $\beta - \alpha > 1$ **do begin**
8. $\quad \gamma \leftarrow \lfloor (\beta + \alpha)/2 \rfloor$
9. $\quad$ Call ALG_WC to find opt value ν_γ for WC_PCST on $(\widetilde{G}, c, p, w, q, \gamma)$
10. $\quad$ **if** $\nu_\gamma \le \mathsf{B}$ **then** $\beta \leftarrow \gamma$ **else** $\alpha \leftarrow \gamma$
11. **end-while**
12. **for** every $v \in V$ **do begin**
13. $\quad k^- \leftarrow p_v^-$, $k^+ \leftarrow p_v^+$, $p_v^- \leftarrow \widetilde{M}$, $p_v^+ \leftarrow \widetilde{M}$
14. $\quad$ **for** every $e = ab \in E$ **do begin**
15. $\quad\quad p_a^+ \leftarrow p_a^+ + \widetilde{M}$, $p_a^- \leftarrow p_a^- + \widetilde{M}$, $p_b^+ \leftarrow p_b^+ + \widetilde{M}$, $p_b^- \leftarrow p_b^- + \widetilde{M}$, $t^- \leftarrow c_e^-$, $t^+ \leftarrow c_e^+$, $c_e^- \leftarrow 0$, $c_e^+ \leftarrow 0$
16. $\quad\quad$ Call ALG_GT to construct $\widetilde{G} = (\widetilde{V}, \widetilde{E})$, $c, w \in \mathbb{R}_+^{\widetilde{E}}$, $p, q \in \mathbb{R}_+^{\widetilde{V}}$
17. $\quad\quad$ **for** $i = 1 : 2$ **do begin**
18. $\quad\quad\quad$ Call ALG_WC to find an optimal solution (tree) $\widetilde{T}_{\beta-i}$ for WC_PCST on $(\widetilde{G}, c, p, w, q, \beta - i)$, where the tree in TRIPLE $(\widetilde{T}_{\beta-i})$ contains v, e
19. $\quad\quad$ **end-for**
20. $\quad\quad p_a^+ \leftarrow p_a^+ - \widetilde{M}$, $p_a^- \leftarrow p_a^- - \widetilde{M}$, $p_b^+ \leftarrow p_b^+ - \widetilde{M}$, $p_b^- \leftarrow p_b^- - \widetilde{M}$, $c_e^- \leftarrow t^-$, $c_e^+ \leftarrow t^+$
21. $\quad\quad$ **for** $i = 1 : 2$ **do begin**
22. $\quad\quad\quad$ **if** $\nu(\widetilde{T}_{\beta-i}, c, p) - c_e + p_v + t^- - k^+ \le \mathsf{B}$
23. $\quad\quad\quad$ **then** find the optimal solution (x_e^0, y_v^0) to the following LP:

$$\min \frac{t^+ - x_e}{t^+ - t^-} + \frac{y_v - k^-}{k^+ - k^-}$$

$$s.t. \begin{cases} x_e - y_v \le \mathsf{B} - \nu(\widetilde{T}_{\beta-i}, c, p) + c_e - p_v \\ t^- \le x_e \le t^+ \\ k^- \le y_v \le k^+ \end{cases}$$

24. $\quad\quad\quad (T, x, y) \leftarrow$ TRIPLE$(\widetilde{T}_{\beta-i})$; $x_e \leftarrow x_e^0$, $y_v \leftarrow y_v^0$

25. $\mathscr{T}^* \leftarrow \mathscr{T}^* \cup \{(T, x, y)\}$
26. **end-for**
27. **end-for**
28. $p_v^- \leftarrow k^-$, $p_v^+ \leftarrow k^+$
29. **end-for**
30. Take $(T^*, x^*, y^*) \in \mathscr{T}^*$ with minimum $\nu(T^*, x^*, y^*)$
31. Output (T^*, x^*, y^*)

Note that in Step 22, we have

$$\nu(\widetilde{T}_{\beta-i}, c, p) - c_e + p_v = \sum_{g \in E(\widetilde{T}_{\beta-i})} c_g - \sum_{z \in V(\widetilde{T}_{\beta-i})} p_z - c_e + p_v.$$

Furthermore, in Step 23, we do not need to solve the LP, because at least one of (c_e^+, y_v^*), (c_e^-, y_v^*), (x_e^*, p_v^+), (x_e^*, p_v^-) is an optimal solution to the LP. Suppose that (x_e^0, y_v^0) is an optimal solution to LP with $x_e^0 \in (c_e^-, c_e^+), y_v^0 \in (p_v^-, p_v^+)$. Then $x_e^0 - y_v^0 = \mathsf{B} - \nu(\widetilde{T}_{\beta-i}, c, p) + c_e - p_v$. In case of $c_e^+ - c_e^- \geq p_v^+ - p_v^-$, set $\delta \equiv \min\{x_e^0 - c_e^-, y_v^0 - p_v^-\}$; then $(x_e^0 - \delta, y_v^0 - \delta) = (c_e^-, y_v^0 - \delta)$ or $(x_e^0 - \delta, p_v^-)$ is an optimal solution to the LP. Similarly, if $c_e^+ - c_e^- \leq p_v^+ - p_v^-$, with $\lambda \equiv \min\{c_e^+ - x_e^0, p_v^+ - y_v^0\}$ we have $(c_e^+, y_v^0 + \lambda)$ or $(x_e^0 + \lambda, p_v^+)$ optimal. So we can solve the LP in $O(1)$ time by checking their feasibility and choosing the optimal one.

Theorem 5. *Algorithm* ALG_MSR *outputs an optimal solution to the MSR_PCST problem on 2-trees of n vertices in $O(n^5)$ time.*

4 Discussion

In this paper, we have established two models for the prize collecting Steiner tree problem in networks with interval data and proposed two polynomial-time algorithms for the corresponding problems on 2-trees, respectively. In fact, the obtained results could be extended to any *partial 2-tree* (also called *series parallel graph*), which is a spanning subgraph of a 2-tree. The polynomial-time solvability of our risk models exhibits the essential difference from the existing robust model [1,13], which yields NP-hard problems even if the original problems (with deterministic data) are polynomial-time solvable. Moreover, in real world, network designers may have their own preferences of money to risk depending on varying trade-offs between them. Our models and algorithms are very flexible in the sense that with different budget levels, they are usually able to produce a couple of candidates for selection by network designers, who are willing to take some risk to save some amount of budget (say, for future use) at their most preferred trade-off between money and risk.

It is not hard to come up with such an example that these two models could yield distinct solutions far away from each other. So further study is desirable to investigate the difference between these two models in real applications through numerical experiments.

References

1. Aron, I.D., Hentenryck, P.V.: On the complexity of the robust spanning tree problem with interval data. Operations Research Letters 32, 36–40 (2004)
2. Balas, E.: The prize cllecting travelling salesman problem. Network 19, 621–636 (1989)
3. Chen, G., Xue, G.: A PTAS for weight constrained Steiner trees in series parallel graphs. Theoretical Computer Science 304, 237–247 (2003)
4. Chen, X.J., Hu, J., Hu, X.D.: The polynomial solvable minimum risk spanning tree problem with interval data. European Journal Operational Research 198, 43–46 (2009)
5. Feofiloff, P., Fernandes, C.G., Ferreira, C.E., Pina, J.C.: Primal-dual approximation algorithms for the prize collecting Steiner tree problem. Information Processing Letters 103(5), 195–202 (2007)
6. Hu, J.: Minimizing maximum risk for fair network connection with interval data. Acta Mathematicae Applicatae Sinica 26(1), 33–40 (2010)
7. Hwang, F.K., Richards, D.S., Winter, P.: The Steiner Tree Problem, Amsterdam (1992)
8. Karp, R.M.: Reducibility among combinatorial problems. In: Miller, R.E., Tatcher, J.W. (eds.) Complexity of Computer Computations, pp. 85–103. Plenum Press, New York (1972)
9. Klau, G., Ljubic, I., Mutzel, P., Pferschy, U., Weiskircher, R.: The fractional prize collecting Steiner tree problem on trees. In: Di Battista, G., Zwick, U. (eds.) ESA 2003. LNCS, vol. 2832, pp. 691–702. Springer, Heidelberg (2003)
10. Lucena, A., Resende, M.G.: Strong lower bounds for the prize collecting Steiner tree problem in graphs. Discrete Applied Mathematics 141, 277–294 (1979)
11. Megiddo, N.: Combinatorial optimizaion with rational objective functions. Mathematics of Operations Research 4, 414–424 (1979)
12. Wald, J.A., Colbourn, C.J.: Steiner trees, partial 2-trees, and minimum IFI networks. Networks 13, 159–167 (1983)
13. Zielinski, P.: The computational complexity of the relative robust shortest path problem with interval data. European Journal Operational Research 158, 570–576 (2004)

The (K, k)-Capacitated Spanning Tree Problem

Esther M. Arkin[1,*], Nili Guttmann-Beck[2], and Refael Hassin[3]

[1] Department of Applied Mathematics and Statistics, State University of New York, Stony Brook, NY 11794-3600, USA
estie@ams.sunysb.edu

[2] Department of Computer Science, The Academic College of Tel-Aviv Yaffo, Yaffo, Israel
becknili@mta.ac.il

[3] School of Mathematical Sciences, Tel-Aviv University, Tel-Aviv 69978, Israel
hassin@post.tau.ac.il

Abstract. This paper considers a generalization of the capacitated spanning tree, in which some of the nodes have capacity K, and the others have capacity $k < K$. We prove that the problem can be approximated within a constant factor, and present better approximations when k is 1 or 2.

1 Introduction

Let $G = (V, E)$ be an undirected graph with nonnegative edge weights $l(e)$ $e \in E$ satisfying the triangle inequality. Let $1 \leq k \leq K$ be given integer *capacities*. Assume that $V = \{r\} \cup V_K \cup V_k$, where r is a *root node*, and V_K and V_k are the sets of nodes having capacity K and k, respectively. In THE (K, k) CAPACITATED SPANNING TREE PROBLEM we want to compute a minimum weight tree rooted at r such that for each $v \in V \backslash \{r\}$ the number of nodes in the subtree rooted at v is no bigger than its capacity.

We are motivated by the following: Nodes of the graph correspond to sensors collecting data that must be transported to a given base-station, the root of the tree. Each sensor forwards all of its data to another (single) node, thus forming a tree representing established data paths. Each node v is also responsible to keep an archive (backup, or data repository) for all of the data at all nodes in the subtree rooted at it (in case the link goes down to a child). The node's capacity represents a storage capacity, saying, e.g., how many nodes' worth of data can be stored at node v. So, we must build trees that obey this capacity constraint. Given costs of the edges, the goal is to build short ("cheap") trees.

The (K, k) CAPACITATED SPANNING TREE PROBLEM is NP-hard as it is a generalization of the CAPACITATED MINIMUM SPANNING TREE PROBLEM where $K = k$ (see [12]).

Our results are as follows:

- For $k = 1$:
  - For $K = 2$ we present a way to find the optimal solution.

* Partially supported by NSF CCF-0729019.

B. Chen (Ed.): AAIM 2010, LNCS 6124, pp. 25–34, 2010.

  - We present a $K - 1$ simple approximation algorithm, this algorithm is suitable for small values of K.
  - We also present a 6-approximation algorithm which is suitable for all values of K.
- For $k = 2$ we present a 10-approximation algorithm, suitable for all values of K.
- We present a 21-approximation algorithm suitable for all values of (K, k).
- We consider a generalization of the problem where each node $v \in V$ has its capacity k_v, we present an $(2 + \alpha)$-approximation algorithm for α which bounds the ratio between the maximal and minimal node capacities.

The Capacitated Minimum Spanning Tree Problem has been studied extensively in the Operations Research literature. It arises in practice in the design of local area telecommunication networks. See [5] for a survey. Various generalizations have also been considered, such as [3] who consider different types of edges, with costs depending on the edge type chosen.

Papadimitriou [12] proved that the capacitated spanning tree problem is NP-hard even with $k = 3$. In [1] Altinkemer and Gavish proposed a 3-approximation algorithm. Gavish, Li and Simchi-Levi gave in [6] worst case examples for the 3-approximation algorithm showing that the bound is tight. Gavish in [4] presented the directed version of the problem and gave a new linear integer programming formulation of the problem. This formulation led to a new Lagrangean relaxation procedure. This relaxation was used for deriving tight lower bounds on the optimal solution and heuristics for obtaining approximate solutions.

The most closely related model to ours seems to be the one considered by Gouveia and Lopes [7]. In their model, the children of the root are called *first-level nodes* and they are assigned capacities of, say K, while all the other *second level nodes* have smaller capacities, say $k < K$. The main difference between their model and our (K, k) model is that in our case the capacities are attached to the nodes as part of the input, whereas in their model the capacity of a node depends on its position in the solution. Gouveia and Lopes present heuristics and valid inequalities for their model supported by computational results.

Jothi and Raghavachari,[8], study the capacitated minimum spanning network problem, which asks for a minimum cost spanning network such that the removal of r and its incident edges breaks the network into 2-edge-connected components, each with bounded capacity. They show that this problem is NP-hard, and present a 4-approximation algorithm for graphs satisfying triangle inequality.

Jothi and Raghavachari in [9] study the capacitated minimum Steiner tree problem, looking for a minimum Steiner tree rooted at a specific node, in which the sum of the vertex weights in every subtree is bounded.

Könemann and Ravi present in [10] bicriteria approximation algorithms for the degree-bounded minimum cost spanning tree, a problem relevant to the one studied here, since bounding the out-degree of a node may imply bounds on the subtree descending from this node.

Morsy and Nagamochi study in [11] the Capacitated multicast tree routing problem. In this problem we search for a partition $\{Z_1, Z_2, \ldots, Z_l\}$ of a given terminal set and a set of subtrees $T_1, T_2, \ldots, T_l$ such that Z_i consists of at most k terminals and each T_i spans $Z_i \cup \{s\}$ (where s is the given source). The objective is to minimize the sum of lengths of the trees $T_1, T_2, \ldots, T_l$. They also propose a $(\frac{3}{2} + \frac{4}{3}\rho)$ approximation, where ρ is the best achievable approximation ratio for the Steiner tree problem.

Deo and Kumar in [2] suggest an iterative refinement technique to compute good suboptimal solutions in a reasonable time even for large instance of problems. They discuss how this technique may be effectively used for the capacitated minimum spanning tree problem.

2 The $(K, 1)$ Problem

In this case the nodes of V_k must be leaves of the tree.

2.1 The $(2, 1)$ Problem

An optimal solution can be obtained through a matching algorithm. We match pairs of nodes such that the root can be matched many times but any other node can be matched only once. Matching node v to the root costs $l(v, r)$. Matching non-root nodes u and v costs $l(u, v) + min\{l(r, u), l(r, v)\}$, for $u, v \in V_2$ and $l(u, v) + l(r, u)$ if $u \in V_2$ and $v \in V_1$.

2.2 The $(K, 1)$ Problem with Small K

When $K \leq 6$ the following simple idea gives a better approximation bound than the general one we present in the next subsection.

Remark 1. It follows easily from the triangle inequality that a star (where all the nodes are directly connected to the root) is a K-approximation.

Lemma 1. *The matching solution described for the case $K = 2$ is a $(K - 1)$-approximation.*

2.3 The $(K, 1)$ Problem with General K

We present now an approximation algorithm for the general $(K, 1)$ problem.

Algorithm $(K, 1)$_Tree

1. Compute a minimum weight matching M from the nodes of V_k to $V_K \cup \{r\}$ such that each node in V_K may be assigned at most $K - 1$ nodes, and all the remaining nodes are assigned to r. The matching cost is the weight of the connecting edges in G. M defines a set of stars in the graph, each star is rooted at one of the nodes in $V_K \cup \{r\}$, and the leaves of this star are the nodes from V_k matched to the root of the star (see Figure 1 top left).

2. For every star rooted at a node from V_K with at least $\frac{K}{2}$ nodes, [By Step 1 the number of nodes in this star is at most K.] connect this star to r using the shortest possible edge (see Figure 1 top right). [Later (in Step 5) we will change this connection to be a feasible connection, as the nodes from V_k must be leaves of the tree.]
3. Compute an MST, T_s, on r and the nodes from V_K that were not connected to r in Step 2 (see Figure 1 middle left). [The optimal solution contains a tree T on $V_K \cup \{r\}$. T is a steiner tree on $V(T_s)$, hence $l(T_s) \leq 2l(T)$.]

   Figure 1 middle right shows $T_s \cup M$, which includes all the connections made so far.

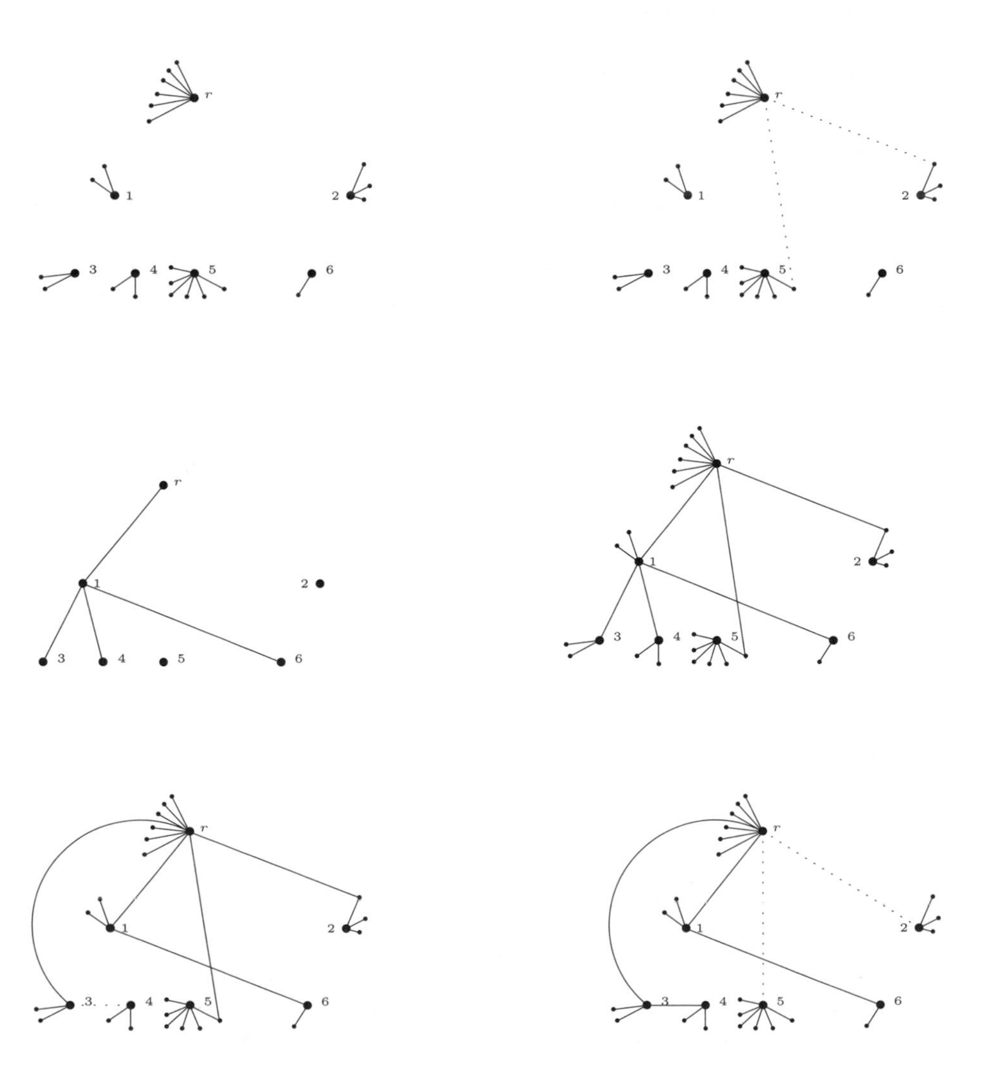

Fig. 1. The different steps in Algorithm $(K, 1)$_Tree with $K = 7, k = 1$

4. Scan T_s from bottom to top and for every node $v \in V_K$ we make the following changes (to guarantee that the subtree rooted at v has at most K nodes):
   - Denote by $y_1, \ldots, y_m \in V_K$ and $u_1, \ldots, u_l \in V_k$ the sons of v, and denote the subtree rooted at y_i by T_i. [Since the tree is scanned from bottom to top $|V(T_i)| \leq K$.]
   - While $\sum_{i=1}^{m} |V(T_i)| \geq \frac{K}{2}$ let p satisfy $\frac{K}{2} \leq \sum_{i=1}^{p} |V(T_i)| \leq K$, disconnect $T_i, (i \in \{1, \ldots, p\})$ from T_s, add the edges $\{(y_q, y_{q+1}) | 1 \leq q \leq p-1\}$, and connect this new tree to r using the shortest possible edge. Renumber the nodes $y_{p+1}, \ldots, y_m$ to $y_1, \ldots, y_{m-p}$ and set $m = m - p$.
   [After the change the number of descendants of v going through V_K nodes is smaller than $\frac{K}{2}$, and the number of sons of v from V_k is smaller than $\frac{K}{2}$, giving that overall v has less than K descendants.]
   - If the subtree rooted at v (including v) contains at least $\frac{K}{2}$ nodes, disconnect this subtree from T_s, and connect to the root using the shortest possible edge.

   (See Figure 1 bottom left)
5. In all cases of connecting a subtree to r by the end edge (r, u) where $u \in V_k$, change this connection to connect the subtree to r using the parent of u. Note that the parent of u is always included in the subtree and is always a node in V_K. (See Figure 1 bottom right.)

Theorem 2. *Denote by* apx *the solution returned by Algorithm* $(K,1)$*_Tree, and let* opt *be the optimal value, then:* $l(\text{apx}) \leq 6\text{opt}$.

3 The $(K, 2)$ Problem

Theorem 3. *Assume* $k = 2$ *and denote by* apx *the solution returned by Algorithm* $(K,1)$*_Tree, and let* opt *be the value of an optimal solution. Then* $\text{apx} \leq 10\text{opt}$.

4 The (K, k) Problem

We now turn to the (K,k) CAPACITATED SPANNING TREE PROBLEM, and consider first a naïve algorithm for the problem: Solve (optimally or approximately) two separate problems. One on $\{r\} \cup V_K$ and the second on $\{r\} \cup V_k$. Then hang the two separate trees on r. This clearly yields a feasible solution.

The following simple example shows that the value of this solution can be as much as $\frac{K-1}{k} + 1$ times the optimal value (even if both separate problems are solved optimally). In this example we assume that $\frac{K-1}{k}$ is integer. The graph has a single node of capacity K, and $K-1$ nodes of capacity k, all at distance 0 from each other, and distance 1 from the root. The first tree is a single edge of length 1, and the second tree includes $\frac{K-1}{k}$ unit length edges from the root.Thus yielding a solution of cost $\frac{K-1}{k} + 1$ while an optimal solution has all nodes in V_k hanging off the single node in V_K, and thus is of cost 1.

In this section we show how to obtain a constant factor approximation algorithm for the (K, k) problem. We first show that any feasible solution F can be transformed into another feasible solution F' with restricted structure, without increasing the weight "too much".

4.1 Ordered Tree

Definition 4. *In an* ordered tree *the capacities of nodes in every path starting at the root are nonincreasing.*

Lemma 5. *Consider a (K, k) capacitated spanning tree problem, Let* opt *be the length of an optimal solution. There is a feasible* ordered tree *with length no greater than* 3opt.

Remark 2. There is an instance of the (K, k) capacitated spanning tree problem such that $l(T_o) = 3l(T)$ where T_o is a minimal length feasible ordered tree and T is an optimal solution.

4.2 The Approximation Algorithm

In our algorithm we use the algorithm for the minimum capacitated tree problem described in [1]. This algorithm computes a 3 approximation solution where each subtree is a path.

We offer the following algorithm:

Algorithm (K, k)_Tree

1. Compute a minimum spanning tree in the graph induced by $r \cup V_K$, call it T_1. An example of T_1 is shown in Figure 2 top-left.
2. Contract the nodes $r \cup V_K$ into a single node R, and find an approximate capacitated spanning tree on $R \cup V_k$ with capacities k, using the method of [1]. Call this tree T_2. Note that in T_2, each subtree of nodes of V_k hanging on R is a path of length exactly k, except for possibly one shorter path. An example of T_2 is shown in Figure 2 top-right.
3. "Uncontract" the node R in T_2, obtaining a forest in which each connected component is a *rooted-spider*, a node in $r \cup V_K$ with paths of nodes from V_k, all of length k except possibly for one shorter path. Let F_2 denote the forest created from T_2 edges after the 'uncontraction' . Consider Figure 2 middle-left for an example of F_2, where the bold nodes denote $r \cup V_K$.
4. Define a matching problem on a complete bipartite graph $B = (S_1, S_2, S_1 \times S_2)$: In the first side, S_1, of our bipartite graph B, we have a node for each "leg" (path) of a spider. Each node in S_1 should be matched exactly once. In the second side of B we have nodes $S_2 = r \cup V_K$, nodes of V_K have capacity $\lfloor K/k \rfloor - 1$ (meaning that each can be matched at most that many times) and r has unbounded capacity. The cost of matching a node in S_1 to node in S_2 is the length of the edge from a node in the spider leg closest to the destination node.

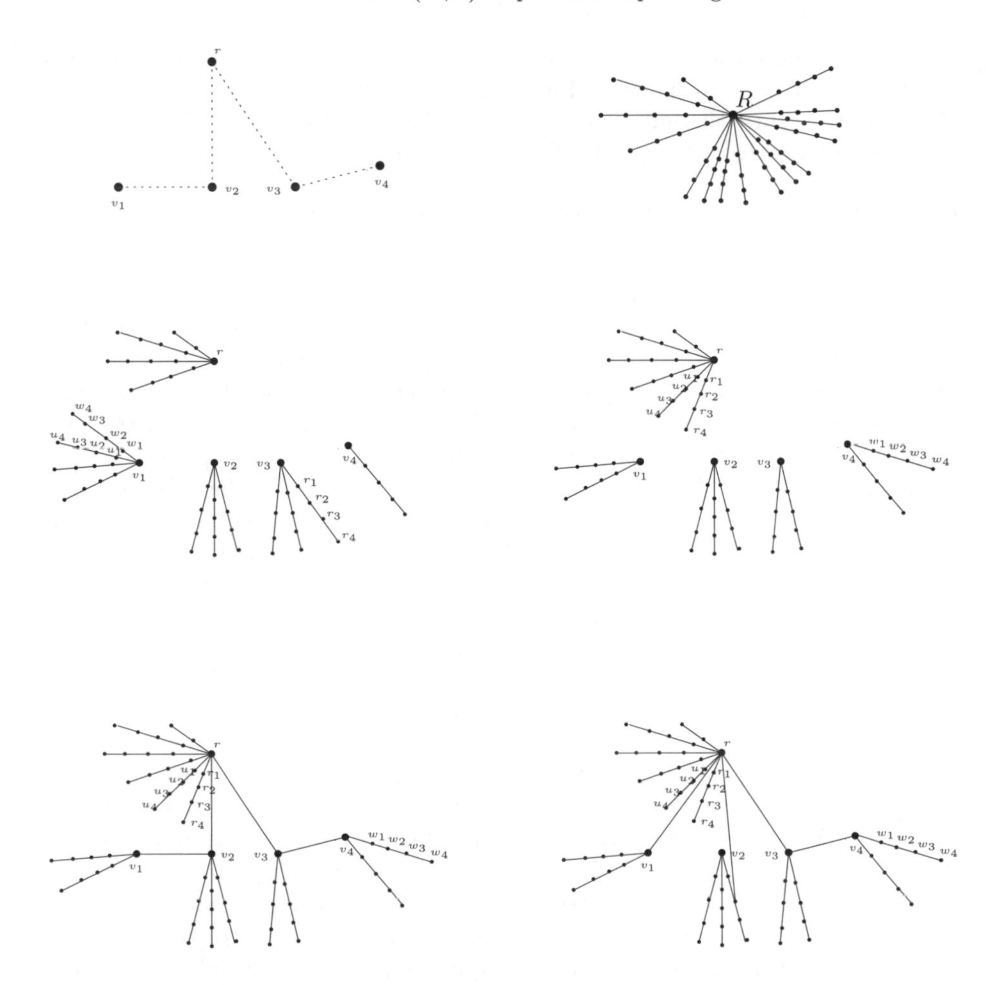

Fig. 2. The different steps in Algorithm (K, k)_Tree with $K = 18$ and $k = 4$

5. Solve the matching problem and change F_2 in the following way: Each spider leg will be attached to the node in $r \cup V_K$ it is assigned to it in the matching problem. The attachment is done by connecting the node in the path closest to the node (i.e., the edge which defines the cost used in the matching). Denote the new forest as F_2'. The forest F_2' is illustrated in Figure 2 middle-right.
6. Consider $T_1 \cup F_2'$ (this graph can be shown in Figure 2 bottom-left). For every $v \in V_K$ with legs $P_1, \ldots, P_l \in V_k$ and $\sum_{i=1}^{l} |V(P_i)| \geq \frac{K}{2} - 1$, disconnect $v \cup \{P_1, \ldots, P_l\}$ from $T_1 \cup F_2'$. [By the way the algorithm works $\sum_{i=1}^{l} |V(P_i)| \leq K - 1$.] Connect this subtree to r using the shortest possible edge. This step is applied to the subtree rooted at v_2 in the bottom figures of Figure 2.
7. The tree T_1 was disconnected in the previous step, reconnect it using only edges between nodes in $V_K \cup \{r\} \setminus \{$nodes that were disconnected in previous

step }. Denote the new tree induced on V_K as T_3. The graph after applying this change to v_1 is shown in Figure 2 bottom-right.
8. Finally, to turn this into a feasible solution for all nodes of V_K, we follow Steps 4,5 of Algorithm $(K,1)$_*Tree* in Section 2.3.

Theorem 6. *Denote by* opt *the value of an optimal solution, and* apx *the solution returned by* Algorithm (K,k)_Tree, *then* $l(\text{apx}) \leq 21\text{opt}$.

Proof: By construction, $l(T_1) \leq 2\text{opt}$ and $l(T_2) \leq 3\text{opt}$, $l(T_3) \leq 2l(T_1) \leq 4\text{opt}$. Next, we bound the length of the edges in the matching. Consider another bipartite graph $B' = (S_1, S_2', E)$, with the same nodes on the first side as B, namely S_1, and nodes on the second side S_2' each corresponding to maximal subtrees induced by V_k in opt$'$ where opt$'$ is the best feasible ordered tree. By Lemma 5 $opt' \leq 3opt$. There is an edge in this bipartite graph between a node in S_1 and a node in S_2' if the two sets of nodes (the leg and the subtree) have at least one node in common (B' is their intersection graph). We now show that B' has a matching in which all nodes of S_1 are matched, using Hall's Theorem. Recall that all legs of apx are of length exactly k, except possibly for one shorter leg, whereas all the subtrees from opt$'$ have length at most k.

In our graph $B' = (S_1, S_2', E)$, we want to show that Hall's condition holds, and therefore there is a matching saturating all nodes of S_1. Let $X \subseteq S_1$. X represents $|X|$ disjoint paths each of length k except possibly one shorter, therefore it represents more than $k(|X|-1)$ nodes of V_k. Call this set $V_k(X)$, and so we have $|V_k(X)| > k(|X|-1)$. Similarly, $N(X)$ also represents disjoint subtrees of nodes in V_k, each subtree contains at most k nodes. Call this set $V_k(N(X))$. Therefore $|V_k(N(X))| \leq k|N(X)|$. By construction $V_k(X) \subseteq V_k(N(X))$ and therefore $k(|X|-1) < |V_k(X)| \leq |V_k(N(X))| \leq k|N(X)|$, resulting in $|X|-1 < |N(X)|$, or equivalently $|X| \leq |N(X)|$ as required by Hall's Theorem.

Observe that our graph G can be thought of as a subgraph of the graph for which the algorithm finds a matching, simply merge subtrees (nodes of S_2') that are attached to the same node in $r \cup V_K$ to obtain S_2. Thus, the matching in graph G is a feasible solution to the matching found by our approximation algorithm, and our algorithm picked the best such matching. Thus the connections in the last step of our algorithm have total length $l(conn)$ which is at most opt$'$.

When disconnecting subtrees from the tree and connecting them directly to r we add three kinds of edges:

- Connecting brothers (adding edges (y_i, y_{i+1}) to the tree). The sum of lengths of these edges can be bounded by the length of T_3.
- We add edges connecting trees with at least $\frac{K}{2}$ nodes to r. As in the proof of Theorem 2 we can bound the length of the edges with 2opt.
- In the last step we change some of the connecting edges from $(r, u), u \in V_k$ to (r, A_u), $u \in V_k$, where A_u is the closest ancestor of u which is in V_K. By the triangle inequality $l(r, A_u) \leq l(r, u) + l(u, A_u)$, where (u, A_u) is a part of leg added to the tree in the matching. Thus, this step adds at most the length of all the edges from nodes in V_k to their ancestors in V_K, with total length at most $l(conn)$.

Summing all this, $l(\text{apx})$ consists of:

- $l(T_1) \leq 2\text{opt}$, $l(T_2) \leq 3\text{opt}$, $l(T_3) \leq 4\text{opt}$.
- $l(conn) \leq opt' \leq 3\text{opt}$.
- The edges added connecting brothers with length $\leq l(T_3) \leq 4\text{opt}$.
- Edges connecting subtrees to r with length $\leq 2\text{opt}$.
- Changing the connecting edges to ancestors from V_K with maximal length $l(conn) \leq 3\text{opt}$.

Altogether, $l(\text{apx}) \leq 21\text{opt}$. ∎

5 Concluding Remarks: General Capacities

A natural extension of our model allows more than two capacity types. In the extreme case, each node v may have a different capacity, k_v. We leave this generalized problem for future research, and observe that a straightforward extension of the naïve algorithm of Section 4 is possible, as follows: Let k_M be the maximal capacity bound and k_m the minimal capacity bound, and let $\alpha = \frac{k_M}{k_m}$. W.l.o.g., assume that $\frac{|V|}{k_m}$ is an integer, otherwise add an appropriate number of nodes with zero distance from r without affecting the solution.

The algorithm: Compute an MST, T. Double its edges to create an Eulerian cycle. By shortcutting the cycle form a Hamiltonian cycle in the standard way. Partition the cycle into subpaths, each containing k_m nodes. Connect each subpath to r using the shortest possible edge. (This is actually the approximation algorithm suggested in [1] for $k_M = k_m$.)

Theorem 7. *Denote by* opt *the value of the optimal solution and by* apx *the approximation solution, then* $l(\text{apx}) \leq (2+\alpha)\text{opt}$.

References

1. Altinkemer, K., Gavish, B.: Heuristics with constant error guarantees for the design of tree networks. Management Sci. 34, 331–341 (1988)
2. Deo, N., Kumar, N.: Computation of constrained spanning trees: a unified approach. Lecture Notes in Econ. and Math. Systems, vol. 450, pp. 194–220. Springer, Berlin (1997)
3. Gamvros, I., Raghavan, S., Golden, B.: An evolutionary approach to the multi-level Capacitated Minimum Spanning Tree problem. Technical report (2002)
4. Gavish, B.: Formulations and algorithms for the capacitated minimal directed tree problem. J. Assoc. Comput. Mach. 30, 118–132 (1983)
5. Gavish, B.: Topological design of telecommunication networks - local access design methods. Annals of Operations Research 33, 17–71 (1991)
6. Gavish, B., Li, C.L., Simchi-Levi, D.: Analysis of heuristics for the design of tree networks. Annals of Operations Research 36, 77–86 (1992)
7. Gouveia, L., Lopes, M.J.: Using generalized capacitated trees for designing the topology of local access networks. Telecommunication Systems 7, 315–337 (1997)

8. Jothi, R., Raghavachari, B.: Survivable network design: the capacitated minimum spanning network problem. Inform. Process. Let. 91, 183–190 (2004)
9. Jothi, R., Raghavachari, B.: Approximation algorithms for the capacitated minimum spanning tree problem and its variants in network design. ACM Trans. Algorithms 1, 265–282 (2005)
10. Könemann, J., Ravi, R.: Primal-dual meets local search: approximating MSTs with nonuniform degree bounds. SIAM J. Comput. 34, 763–773 (2005)
11. Morsy, E., Nagamochi, H.: An improved approximation algorithm for capacitated multicast routings in networks. Theoretical Comput. Sci. 390, 81–91 (2008)
12. Papadimitriou, C.H.: The complexity of the capacitated tree problem. Networks 8, 217–230 (1978)

Optimal Algorithms for the Economic Lot-Sizing Problem with Multi-supplier*

Qing-Guo Bai[1] and Jian-Teng Xu[2,**]

[1] School of Operations Research and Management Sciences, Qufu Normal University, Rizhao, Shandong, 276826, China
qfnubaiqg@163.com
[2] School of Management, Harbin Institute of Technology, Harbin, Heilongjiang, 150001, China
qingniaojt@163.com

Abstract. This paper considers the economic lot-sizing problem with multi-supplier in which the retailer may replenish his inventory from several suppliers. Each supplier is characterized by one of two types of order cost structures: incremental quantity discount cost structure and multiple set-ups cost structure. The problem is challenging due to the mix of different cost structures. By analyzing the optimal properties, we reduce the searching range of the optimal solutions and develop several optimal algorithms to solve all cases of this multi-supplier problem.

Keywords: Economic lot-sizing; optimal algorithm; dynamic programming.

1 Introduction

The classical Economic Lot-Sizing (ELS) problem was first introduced in [1] and it has been widely extended during recent years. The extended ELS problem becomes the focus of extensive studies and continues to receive considerable attention. Many versions of the ELS problem build on different cost structures. For example, the ELS problems with both all-unit quantity discount and incremental quantity discount cost structures were proposed and solved by dynamic programming(DP) algorithms in [2] with complexity $O(T^3)$ and $O(T^2)$, respectively. T is the length of the planning horizon. See also [3] for additional observations. Zhang et al.[4] presented the general model with multi-break point all-unit quantity discount cost structure, and designed a polynomial time DP algorithm. Indeed there are some other theoretical results for the ELS problem with other cost structures and assumptions (see[5],[6],[7],[8],[9] for example).

However, most of the literature discussed above assumes that products can be ordered from only one supplier. Sometimes this is not a valid assumption in real

* This work was supported by the National Science Foundation Grant of China (No. 70971076, 10926077 and 70901023).
** Corresponding author.

B. Chen (Ed.): AAIM 2010, LNCS 6124, pp. 35–45, 2010.

life. In order to replenish inventory with economical cost, the retailer usually faces not a single supplier but many suppliers. He should determine from which supplier and how many units to order in each period. Thus the ELS problem with multiple suppliers has more wider domain of applications.

In this paper, the ELS problem with multiple suppliers is called multi-supplier ELS problem. To the best of our knowledge, there are few results about the multi-supplier ELS problem. The first result about the multi-supplier problem we are aware of is the one proposed in [10]. Their ELS problem with multi-mode replenishment is equivalent to a multi-supplier ELS problem. They analyzed two structural properties for the $N(N > 2)$ suppliers problem, and presented a DP algorithm without a detail description for its calculation. Instead, they focused on a two-supplier problem. For the two-supplier problem, they discussed three special cases according to two types of cost structures (fixed set-up cost structure and multiple set-ups cost structure) and designed a polynomial time algorithm for each case. The ELS problem with multiple products and multiple suppliers was considered in [11]. In the model, the purchase cost and holding cost are stationary in each period, and they are simple linear functions about purchase and holding quantities. These linear cost functions help to prove two properties: there is no period where an order is made and inventory is carried into the period for each product, and no product is ordered from two (or more) suppliers in the same period. Based on the properties, enumerative and heuristic algorithms were given to solve the problem. However, the two properties are not true for piecewise linear cost functions, such as multiple set-ups cost function.

In this paper, we propose a multi-supplier ELS problem in which each supplier is characterized by different order cost structures including the incremental quantity discount cost structure and the multiple set-ups cost structure. In this problem, the purchase cost and holding cost vary from period to period, and are more general cost structures. This multi-supplier ELS problem can be divided into three cases according to the different combinations of the order cost structures. They are: (1) each supplier has an incremental quantity discount cost structure; (2) each supplier has a multiple set-ups cost structure; (3) some suppliers have incremental quantity discount cost structures, others have multiple set-ups cost structures. Since only two optimality properties are given for the case (2) in [10], this paper will continue discussing this case and give an optimal algorithm to solve it. We develop two polynomial time algorithms for the case (1) and a special case of case (3), two optimal algorithms for case (2) and case (3). Some previous literature results are the special cases of this multi-supplier ELS problem, such as [1], [10], [12], and so on.

2 Notations and Formulations

The multi-supplier ELS problem proposed in this paper consists of N suppliers and one retailer, $N \geq 2$. Each supplier is characterized by a different order cost structure. The retailer is the decision maker. He must determine: 1)from which supplier to order; 2)how many units to order and 3)when to order so as to minimize his total cost within a finite time planning horizon. Let T be the length of the planning horizon, and N be the total number of the suppliers. For each $t = 1, \dots, T$ and $n = 1, \dots, N$, we define the following notations.

- d_t = the demand in period t.
- x_{nt} = the order quantity from supplier-n in period t.
- x_t = the total order quantity in period t. It is easy to know that $x_t = \sum_{n=1}^{N} x_{nt}$.
- c_{nt} = the unit order cost from supplier-n in period t.
- I_t = the inventory level of the retailer at the end of period t. Without loss of generality, assume the initial and the final inventory during the planning horizon are zero. That is, $I_0 = 0$, $I_T = 0$.
- h_t = the unit inventory holding cost in period t.
- S_1 = the set of the suppliers who offer an incremental quantity discount cost structure.
- S_2 = the set of the suppliers who offer a multiple set-ups cost structure.
- K_{nt} = the fixed set-up cost when order from suppler-n in period t.
- A_{nt} = the fixed set-up cost per standard container when order from supplier-n in period t, $n \in S_2$. $A_{nt} = 0$ if $n \in S_1$.
- W_n = the standard container capacity when order from supplier-n, $n \in S_2$. $W_n = 0$ if $n \in S_1$.
- r_{nt} = the discount rate given by supplier-n in period t, $n \in S_1$. $r_{nt} = 0$ if $n \in S_2$. If the order quantity in period t is greater than the critical value Q_n (Q_n is a positive integer and is determined by supplier-n), the discounted unit order cost is implemented to the excess quantity $x_{nt} - Q_n$.
- $\lceil a \rceil$ = the smallest integer that is greater than or equal to a.
- $\lfloor a \rfloor$ = the largest integer that is less than or equal to a.
- $\delta(a) = 1$ if and only if $a > 0$; otherwise, $\delta(a) = 0$.
- $C_{nt}(x_{nt})$ = the cost of ordering x_{nt} units from supplier-n in period t. When $n \in S_1$, $C_{nt}(x_{nt})$ belongs to an incremental quantity discount cost structure,

$$C_{nt}(x_{nt}) = \begin{cases} K_{nt}\delta(x_{nt}) + c_{nt}x_{nt}, & x_{nt} \le Q_n \\ K_{nt} + c_{nt}Q_n + c_{nt}(1 - r_{nt})(x_{nt} - Q_n), & x_{nt} > Q_n \end{cases}. \tag{1}$$

When $n \in S_2$, $C_{nt}(x_{nt})$ belongs to a multiple set-ups cost structure,

$$C_{nt}(x_{nt}) = K_{nt}\delta(x_{nt}) + A_{nt}\lceil \frac{x_{nt}}{W_n} \rceil + c_{nt}x_{nt}. \tag{2}$$

With above notations, this multi-supplier ELS problem can be formulated as

$$\begin{aligned} \min \quad & \sum_{t=1}^{T} \Big[\sum_{n=1}^{N} C_{nt}(x_{nt}) + h_t I_t \Big] \\ s.t. \quad & I_{t-1} + x_t - d_t = I_t, \quad t = 1, \dots, T \\ & x_t = \textstyle\sum_{n=1}^{N} x_{nt}, \quad t = 1, \dots, T \\ & I_0 = 0, \quad I_T = 0 \\ & I_t \ge 0, \quad x_{nt} \ge 0 \end{aligned}$$

This paper will discuss three cases of the multi-supplier ELS problem:

(1) each supplier offers an incremental quantity discount cost structure, P_1 problem for short;

(2) each supplier offers a multiple set-ups cost structure, P_2 problem for short;

(3) some supplier offer incremental quantity discount cost structures, others offer multiple set-ups cost structures, P_3 problem for short.

For convenience, we use the traditional definitions of most ELS problems. Period t is called an **order period** if $x_t > 0$. If $I_t = 0$, period t is a **regeneration point**. x_{nt} is called a **Full-Truck Load(FTL) shipment** if $x_{nt} = lW_n$ for some positive integer l, $n \in S_2$, otherwise, it is a **Less-than-Truck Load(LTL) shipment**. $x_t > 0$ is called a **full order** if $\sum_{n \in S_1} x_{nt} = 0$, and x_{nt} is zero or an FTL shipment for all suppliers $n \in S_2$; otherwise, it is called a **partial order**.

For $1 \le i \le j \le T$, let

$$h(i,j) = \sum_{l=i}^{j} h_l, \quad d(i,j) = \sum_{l=i}^{j} d_l, \quad H(i,j) = \sum_{k=i+1}^{j} h(i,k-1)d_k.$$

If $i > j$, we define $d(i,j) = 0$, $h(i,j) = 0$ and $H(i,j) = 0$. By this definition, we can calculate $d(i,j)$, $h(i,j)$ and $H(i,j)$ in $O(T^2)$ time for all i and j with $1 \le i \le j \le T$.

In practical situation, the more frequently an item is ordered or dispatched, the more favorable its relevant cost is. Considering this situation, we assume that for each $n \in S_1 \cup S_2$, K_{nt}, A_{nt}, c_{nt} and h_t are non-increasing functions on t, and r_{nt} is a non-decreasing function on t. In other words, for $1 \le t \le T-1$, we have $K_{nt} \ge K_{n,t+1}$, $A_{nt} \ge A_{n,t+1}$, $c_{nt} \ge c_{n,t+1}$, $h_t \ge h_{t+1}$ and $r_{nt} \le r_{n,t+1}$.

Let $F(j)$ denote the minimum total cost of satisfying the demand from period 1 to period j and $C(i,j)$ denote the minimum cost of satisfying the demand from period i to period j, where $i-1$ and j are two consecutive regeneration points with $1 \le i \le j \le T$. Set $F(0) = 0$, the multi-supplier ELS problem can be solved by the following DP algorithm

$$F(j) = \min_{1 \le i \le j} \{F(i-1) + C(i,j)\}, \quad 1 \le i \le j \le T \tag{3}$$

Obviously, the objective function of P_1 problem is $F(T)$. If the value of $C(i,j)$ for all $1 \le i \le j \le T$ is known, the value of $F(T)$ can be computed in no more than $O(T^2)$ time via the formula (3). Hence the remaining task is how to compute the value of $C(i,j)$ in an efficient time.

3 Optimality Properties and Algorithm for P_1 Problem

In this section, the first case of the multi-supplier ELS problem in which each supplier offers an incremental quantity discount cost structure is discussed, that is, $S_1 = \{1, \ldots, N\}$ and $S_2 = \emptyset$ hold in this section. To simplify the proof, we first analyze the optimality properties of P_1 problem with two-supplier. The optimality property of P_1 problem with N ($N > 2$) suppliers can be proved via the induction on the number of suppliers.

Lemma 1. *There exists an optimal solution for the two-supplier P_1 problem such that if period t is an order period then the retailer orders products only from one supplier.*

Proof. Suppose that there is an optimal solution for P_1 problem such that $0 < x_{1t} < x_t$ and $0 < x_{2t} < x_t$. Order x_{2t} units in period t from supplier-1 instead of from supplier-2 if the unit order cost from supplier-1 is less than the one from supplier-2. Otherwise, order x_{1t} units from supplier-2 instead of from supplier-1. After this perturbation, we can obtain a solution with non-increasing total cost.

Using above Lemma, Theorem 1 can be proven by induction on the number of suppliers.

Theorem 1. *There exists an optimal solution for P_1 problem such that if period t is an order period then the retailer orders products only from one supplier.*

Theorem 2. *There exists an optimal solution for P_1 problem such that $I_{t-1}x_t = 0$, $t = 1, \dots, T$.*

Proof. Suppose that there is an optimal solution for P_1 problem such that $x_t > 0$ and $I_{t-1} > 0$. Since $I_{t-1} > 0$, there exists an order before period t. Let s be latest order before period t. Then we have $I_k \geq I_{t-1} > 0$ for all $s \leq k \leq t-1$. Using Theorem 1, we assume that $x_s = x_{ms}$, $x_t = x_{nt}$, $m, n \in S_1$.

The proof can be completed via discussing four subcases: (1) $x_{nt} > Q_n$ and $x_{ms} > Q_m$; (2) $0 < x_{nt} \leq Q_n$ and $x_{ms} > Q_m$; (3) $x_{nt} > Q_n$ and $0 < x_{ms} \leq Q_m$; (4) $0 < x_{nt} \leq Q_n$ and $0 < x_{ms} \leq Q_m$. The discussion of these cases is similar, so we only discuss the first case in detail.

In case (1), if $c_{ms}(1 - r_{ms}) - c_{nt}(1 - r_{nt}) + h(s, t-1) \geq 0$, we decrease the value x_{ms} by Δ with $\Delta = \min\{I_{t-1}, x_{ms}\}$ and increase x_{nt} by the same amount. After the perturbation, we obtain a new solution with either $x_{ms} = 0$ or $I_{t-1} = 0$. The total cost of this new solution is either reduced by at least $[c_{ms}(1 - r_{ms}) - c_{nt}(1 - r_{nt}) + h(s, t-1)]\Delta \geq 0$, or not changed. Otherwise, we cancel the order from supplier n in period t and increase the value x_{ms} by x_{nt} units. The total cost of the new solution is reduced by $K_{nt} + c_{nt}r_{nt}Q_n + [c_{nt}(1 - r_{nt}) - c_{ms}(1 - r_{ms}) - h(s, t-1)]x_{nt} > 0$ after this perturbation.

For the other three cases, we can decrease the value x_{ms} by $\min\{I_{t-1}, x_{ms}\}$ and increase x_{nt} by the same amount if $c_{ms}(1-r_{ms}) - c_{nt} + h(s, t-1) \geq 0$ in case (2), or $c_{ms} - c_{nt}(1-r_{nt}) + h(s, t-1) \geq 0$ in case (3), or $c_{ms} - c_{nt} + h(s, t-1) \geq 0$ in case (4). Otherwise, we let $x_{nt} = 0$ and increase the value x_{ms} by x_{nt} units. The total cost of the new solution is not increased after the perturbations and we finish the proof.

Basing on the above theorems, we develop a polynomial time algorithm to calculate all $C(i, j)$. From Theorem 1, we know that the order cost in period t is $\min_{n \in S_1} C_{nt}(x_t)$. By Theorem 2, there exists only one order period between the two consecutive regeneration points $i - 1$ and j. This means that the order quantity in order period i is exactly $d(i, j)$. Thus for each pair of i and j with $1 \leq i \leq j \leq T$, the value of $C(i, j)$ can be computed by

$$C(i,j) = \min_{n \in S_1} C_{ni}\big(d(i,j)\big) + H(i,j), \quad 1 \le i \le j \le T \tag{4}$$

Obviously, the computational complexity of the formula (4) is no more than $O(NT^2)$ for all $1 \le i \le j \le T$. So the total computational complexity of the DP algorithm for the P_1 problem is $O(NT^2)$.

4 Optimality Properties and Algorithm for P_2 Problem

In this subsection, each supplier offers a multiple set-ups cost structure, that is, $S_2 = \{1, \ldots, N\}$, $S_1 = \emptyset$. We propose two optimality properties and design an optimal algorithm to solve it.

Theorem 3. *[10] There exists an optimal solution for P_2 problem such that there is at most one partial order during two consecutive regeneration points $i-1$ through j.*

Theorem 4. *There exists an optimal solution for P_2 problem such that if x_t is a partial order, then only one shipment is an LTL shipment in period t.*

Proof. Suppose the statement of this theorem is not true. That is, there exists an optimal solution of P_2 problem such that x_{mt} and x_{nt} are two LTL shipments in period t, $m, n \in S_2$. Without loss of generality, we suppose $c_{mt} \ge c_{nt}$. We extract $\min\{x_{mt}, \lceil \frac{x_{nt}}{W_n} \rceil W_n - x_{nt}\}$ units from x_{mt}, and add them to x_{nt}, then we can obtain a new solution with equal or more lower total cost.

Corollary 1. *There exists an optimal solution for P_2 problem such that if x_{nt} is an LTL shipment then $c_{nt} = \max_{k \in S_2} c_{kt}$.*

Suppose $c_{n_0 t} = \max_{n \in S_2} c_{nt}$, that is, $x_{n_0 t}$ is a potential LTL shipment in order period t. Let $x_{nt} = m_n W_n$, $m_n = 0, 1, \ldots, \lceil \frac{x_t}{W_n} \rceil$ for all $n \in S_2 \setminus \{n_0\}$. Then, for $i \le t \le j$ we have

$$x_t = \sum_{n \in S_2 \setminus \{n_0\}} m_n W_n + x_{n_0 t}.$$

Recall that $i-1$ and j are two consecutive regeneration points, we have

$$d(i,j) = x_i + \cdots + x_j.$$

The value of $C(i,j)$ is the minimal total cost to satisfying demand $d(i,j)$ among all combinations of x_{nt}, $n \in S_2$, $t = i, \ldots, j$. That is, we can calculate the value of $C(i,j)$ in $O(T(\lceil \frac{d(1,T)}{W^*} \rceil + 1)^{NT-1})$ time for each pair of i and j, where $W^* = \min_{n \in S_2} W_n$. After computing all value of $C(i,j)$, we can solve P_2 problem by the formula (3) in $O(T^2)$. So the total computational complexity of solving P_2 problem is $O\big(T^3(\lceil \frac{d(1,T)}{W^*} \rceil + 1)^{NT-1} + T^2\big)$. Obviously, it is not a polynomial time algorithm but an optimal one.

5 Optimality Properties and Algorithms for P_3 Problem

In this section we propose some optimality properties for P_3 problem, in which some suppliers offer incremental quantity discount cost structures, others offer multiple set-ups cost structures. That is, $S_1 \neq \emptyset$, $S_2 \neq \emptyset$ and $S_1 \cup S_2 = \{1, \ldots, N\}$. After that we give an optimal algorithm for P_3 problem whose running time is non-polynomial. Then we show that there exists an polynomial time algorithm for a special case of the P_3 problem, we denoted it by SP_3 problem in this section.

Theorem 5. *There exists an optimal solution for P_3 problem such that if t is an order period then only one of the following two situations will happen:*
(1) There is at most onc n_1 with $x_{n_1t} > 0$ in set S_1, and x_{nt} is zero or an FTL shipment for all $n \in S_2$.
(2) $x_{nt} = 0$ for all $n \in S_1$, and there is at most one n_2 such that x_{n_2t} is an LTL shipment in set S_2.

Proof. Suppose that the statement is not true, that is, there exists an optimal solution for P_3 problem with $x_{n_1t} > 0$ and $x_{n_2t} \neq lW_{n_2}$. According to Theorem 1, we know $x_{n_1t} = x_t$. According to Theorem 4, we know that x_{nt} is zero or an FTL shipment for all $n \in S_2 \setminus \{n_2\}$. Let c'_{n_1t} be the unit purchase cost, $c'_{n_1t} = c_{n_1t}$ when $x_{n_1t} \leq Q_{n_1}$, $c'_{n_1t} = c_{n_1t}(1 - r_{n_1t})$ when $x_{n_1t} > Q_{n_1}$. If $c_{n_2t} \geq c'_{n_1t}$, we increase the value x_{n_1t} and decrease x_{n_2t} by x_{n_2t} units. After that, we can obtain a new solution with $x_{n_1t} > 0$, $x_{n_2t} = 0$ and non-increasing total cost. If $c_{n_2t} < c'_{n_1t}$, we increase x_{n_2t} and decrease x_{n_1t} by $\min\{x_{n_1t}, \lceil \frac{x_{n_2t}}{W_{n_2}} \rceil W_{n_2} - x_{n_2t}\}$ units. After that, we can obtain a new solution in which $x_{n_1t} = 0$ or x_{n_2t} is an FTL shipment. The total cost of the new solution is much lower.

Theorem 6. *There exists an optimal solution for P_3 problem such that there is at most one $n_1 \in S_1$ with $x_{n_1s} > 0$ or at most one $n_2 \in S_2$ with an LTL shipment x_{n_2t} between two consecutive regeneration points $i-1$ and j, $i \leq s, t \leq j$.*

Proof. When $s = t$, the Theorem reduces to Theorem 5. So we only prove the case where $s \neq t$. Without loss of generality, let $s < t$. Suppose that this Theorem is not true, that is, there exists an optimal solution for P_3 problem such that $x_{n_1s} > 0$ and x_{n_2t} is an LTL shipment, $n_1 \in S_1$, $n_2 \in S_2$, $i \leq s < t \leq j$. If $c_{n_2t} - c'_{n_1s} - h(s, t-1) \leq 0$, decrease the value x_{n_1s} and incrcasc x_{n_2t} by $\min\{x_{n_1s}, \lceil \frac{x_{n_2t}}{W_{n_2}} \rceil W_{n_2} - x_{n_2t}\}$, we can obtain a new solution with a non-increasing total cost, in which $x_{n_1s} = 0$ or x_{n_2t} is an FTL shipment. Otherwise, decrease the value x_{n_2t} and increase x_{n_1s} by x_{n_2t}, we can obtain a new solution with a lower total cost, in which $x_{n_1s} > 0$ and $x_{n_2t} = 0$.

It is easy to verify that Theorem 3 and Theorem 4 are true for P_3 problem. Recall that periods $i-1$ and j are two consecutive regeneration points, we have

$$d(i,j) = x_{1i} + \cdots + x_{Ni} + \cdots + x_{1j} + \cdots + x_{Nj} \tag{5}$$

According to Theorem 6, at most one period could be the potential partial order between two consecutive regeneration points $i-1$ and j. Let period t is the potential partial order, $t = i, \ldots, j$. For each such period t between two consecutive regeneration points $i-1$ and j, either there is at most one $n_1 \in S_1$ with $x_{n_1 t} > 0$, $x_{nu} = 0$ for all $n \in S_1 \setminus \{n_1\}$, and x_{nu} is zero or an FTL shipment for all $n \in S_2$, $u = i, \ldots, j$, or there is at most one $n_2 \in S_2$ with an LTL shipment $x_{n_2 t}$, $x_{nu} = 0$ for all $n \in S_1$, and x_{nu} is zero or an FTL shipment for all $n \in S_2 \setminus \{n_2\}$, $u = i, \ldots, j$. So the number of purchase policies between two consecutive regeneration points $i-1$ and j is not more than $O\big(T|S_1|(\lceil \frac{d(i,j)}{W^*} \rceil + 1)^{|S_2|T} + T|S_2|(\lceil \frac{d(i,j)}{W^*} \rceil + 1)^{|S_2|T-1}\big)$, where $W^* = \min_{n \in S_2} W_n$. In other words, it takes at most $O\big(T|S_1|(\lceil \frac{d(i,j)}{W^*} \rceil + 1)^{|S_2|T} + T|S_2|(\lceil \frac{d(i,j)}{W^*} \rceil + 1)^{|S_2|T-1}\big)$ time to calculate $C(i,j)$ for each pair of i and j. So the total complexity to calculate $C(i,j)$ for all i and j is at most $O\big(T^3|S_1|\lceil \frac{d(1,T)}{W^*} \rceil^{|S_2|T} + T^3|S_2|\lceil \frac{d(1,T)}{W^*} \rceil^{|S_2|T-1}\big)$. After that, we can use formula (3) to find the optimal value of P_3 problem in $O(T^2)$ time. Obviously, it is a non-polynomial time algorithm.

Fortunately, the P_3 problem have more optimality properties when $|S_2| = 1$. These optimality properties help to explore a polynomial time algorithm. Let SP_3 problem denote this special case of P_3 problem. In other words, in SP_3 problem only one supplier has a multiple set-ups cost structure, the rest of $N-1$ suppliers have incremental quantity discount cost structures. Without loss of generality, assume that supplier 1 has a multiple set-ups cost structure in the SP_3 problem. For this SP_3 problem, we obtain several optimality properties in addition.

Theorem 7. *There exists an optimal solution for SP_3 problem such that $x_{1t} \in \{0, x_t, \lfloor \frac{x_t}{W_1} \rfloor W_1\}$ for any $t = 1, \ldots, T$.*

Proof. Suppose that there exists an optimal solution for SP_3 problem such that $0 < x_{1t} < x_t$ and $x_{1t} \neq \lfloor \frac{x_t}{W_1} \rfloor W_1$. According to Theorem 6, if x_{1t} is an FTL shipment, there exists at most one $n_0 \in S_1$ with $x_{n_0 t} > 0$, and $x_{nt} = 0$ for all $n \in S_1 \setminus \{n_0\}$. If x_{1t} is an FTL shipment, and there is no $n \in S_1$ with $x_{nt} > 0$, then we have $x_{nt} = 0$ for all $n \in S_1$. If x_{1t} is an LTL shipment, we have $x_{nt} = 0$ for all $n \in S_1$. The above two cases mean that $x_{1t} = x_t$, which contradicts the fact $0 < x_{1t} < x_t$.

Now we assume that x_{1t} is an FTL shipment but $x_{1t} \neq \lfloor \frac{x_t}{W_1} \rfloor W_1$, and there is only one $n_0 \in S_1$ with $x_{n_0 t} > 0$. Without loss of generality, we let $n_0 = 2 \in S_1$, that is, $x_{2t} > 0$. Since x_{1t} is an FTL shipment and $x_{1t} \neq \lfloor \frac{x_t}{W_1} \rfloor W_1$, Let $x_{1t} = lW_1$, then we have $l \neq \lfloor \frac{x_t}{W_1} \rfloor$, and $x_{2t} \geq W_1$. If $\frac{A_{1t}}{W_1} + c_{1t} \geq c'_{2t}$ ($c'_{2t} = c_{2t}$ when $x_{2t} \leq Q_2$, $c'_{2t} = c_{2t}(1 - r_{2t})$ when $x_{2t} > Q_2$), we cancel the order from supplier 1 in period t, and increase x_{2t} by x_{1t} units, then we get a new solution with a non-increasing cost. Otherwise, let $\Delta = \lfloor \frac{x_{2t}}{W_1} \rfloor W_1$, increase x_{1t} by Δ units and decrease x_{2t} by the same amount. After this perturbation, the total cost is reduced.

Theorem 8. *There exists an optimal solution for SP_3 problem such that for any $t = 1, \ldots, T$,*

(1) $x_{1t} > 0$ only if $I_{t-1} < \min\{d_t, W_1\}$;

(2) For some supplier n with $n \in S_1$, $x_{nt} > 0$ only if $I_{t-1} < \min\{d_t, W_1\}$ or $d_t - W_1 < I_{t-1} < d_t$.

Proof. Since [10] provided the same property for their model, and the proof of result (1) in this theorem can be completed using the similar approach, so we only prove the proposition (2).

Following Theorem 1, the retailer orders products only from one supplier in set S_2 in period t. Here we suppose that there exists an optimal solution such that $x_{nt} > 0$ and $\min\{d_t, W_1\} \leq I_{t-1} \leq d_t - W_1$, $n \in S_1$, which means that $d_t \geq W_1$. Furthermore, we have $W_1 \leq I_{t-1} \leq d_t - W_1$. This means that there exists an order period before period t. Let s be the latest order period before period t. Then we have $x_s = x_{1s} \geq I_s \geq \cdots \geq I_{t-1}$. By proposition (1) of this theorem, $x_{1t} = 0$ holds if $I_{t-1} > W_1$, that is, $x_{nt} = x_t \geq W_1$ since $I_{t-1} \leq d_t - W_1$ and $I_t \geq 0$. If $c'_{nt} W_1 \geq A_{1s} + W_1 c_{1s} + h(s, t-1) W_1$, we increase x_{1s} by W_1 units and decrease x_{nt} by the same amount. $c'_{nt} = c_{nt}$ when $x_{nt} \leq Q_n$, $c'_{nt} = c_{nt}(1 - r_{nt})$ when $x_{nt} > Q_n$. The total cost will not increase after this perturbation. Otherwise, we increase x_{nt} by $\lfloor \frac{I_{t-1}}{W_1} \rfloor W_1$ units and decrease x_{1s} by the same amount. After this perturbation, the new solution has a lower cost which is a contradiction.

The above properties for SP_3 problem also hold for the two-supplier problem in which supplier 1 has a multiple set-ups and supplier 2 has a fixed set-up cost structure studied in [10], since the fixed set-up cost is a special case of incremental quantity discount cost with $Q_n = +\infty$. In other words, [10] becomes a special case of our problem. Using their optimality properties, they develop a polynomial time algorithm with complexity $O(T^4)$ based on the dynamic programming-based shortest-path-network approach to solve their problem. With the Theorems 3, 7 and 8, it is easy to show that SP_3 problem can be solved by their algorithm, except the calculation formula of order cost. Here we use the following formula to calculate the order cost.

$$\begin{aligned} C_t(x_t) = \min \Big\{ & C_{1t}(x_t), \min_{n \in S_1} C_{nt}(x_t), \\ & C_{1t}(\lfloor \frac{x_t}{W_1} \rfloor W_1) + \min_{n \in S_1} C_{nt}(x_t - \lfloor \frac{x_t}{W_1} \rfloor W_1) \Big\}. \end{aligned} \tag{6}$$

So there exists a polynomial algorithm with running time $O(T^4 + NT)$ for SP_3 problem.

6 Numerical Example

In this section, we illustrate the optimal algorithm for P_3 problem with an example. The algorithm is written in the runtime environment MATLAB 7.0, and is achieved and executed on an Lenovo personal computer with a 2.16 GHz Intel Core 2 processor and 1 GB RAM. The running time of the algorithm is 39.15 seconds. The planning horizon of the considered example contains 4 periods, that is, $T = 4$. There are 3 suppliers in this example, that is, $N = 3$. Supplier 1

Table 1. The value of parameters

Parameter	Value	Parameter	Value
T	4	$\boldsymbol{K}_1$	(21,17,10,8)
N	3	$\boldsymbol{K}_2$	(20,16,12,7)
S_1	{1}	$\boldsymbol{K}_3$	(19,19,9,9)
S_2	{2,3}	$\boldsymbol{A}_1$	(0,0,0,0)
$\boldsymbol{Q}$	(20,0,0)	$\boldsymbol{A}_2$	(45,45,45,45)
$\boldsymbol{d}$	(14,9,28,13)	$\boldsymbol{A}_3$	(30,30,30,30)
$\boldsymbol{h}$	(3,3,3,2)	$\boldsymbol{c}_1$	(3,3,3,3)
$\boldsymbol{W}$	(0,20,15)	$\boldsymbol{c}_2$	(2,2,2,2)
$\boldsymbol{r}_1$	(0.2,0.3,0.3,0.4)	$\boldsymbol{c}_3$	(2,2,2,2)

Table 2. The results of $C(i, j)$

$i \backslash j$	1	2	3	4
1	63.0000	115.2000	276.6000	344.6000
2	0	44.0000	171.2000	237.5000
3	0	0	86.8000	153.1000
4	0	0	0	47.0000

Table 3. The results of $F(j)$

j	1	2	3	4
$F(j)$	63.0000	107.000	193.8000	240.8000

Table 4. The value of i correspond with $F(j)$

$F(j)$	$F(1)$	$F(2)$	$F(3)$	$F(4)$
i	1	2	3	4

is in set S_1, suppliers 2 and 3 are in set S_2, that is, $S_1 = \{1\}$, $S_2 = \{2, 3\}$. The other parameters are expressed by vectors (see Table 1).

The computation results of $C(i, j)$ and $F(j)$ are in Table 2, Table 3 and Table 4. From the computation results, we know that the optimal value of P_3 problem is $F(4) = F(3) + C(4, 4) = 240.8$, $F(3) = F(2) + C(3, 3) = 193.8$, $F(2) = F(1) + C(2, 2) = 107$, $F(1) = C(1, 1) = 63$. The optimal solution of P_3 problem is $x_{11} = 14$, $x_{12} = 9$, $x_{13} = 28$, $x_{14} = 13$, $x_{nt} = 0$ for all $n = 2, 3$, $t = 1, 2, 3, 4$.

7 Conclusion

This paper extends the classical economic lot-sizing problem to the multi-supplier ELS problem. Each supplier has one of the two types of order cost structures: incremental quantity discount cost structure and multiple set-ups cost structure. We analyzed all possible cases for the multi-supplier ELS problem. After proposing corresponding optimal properties for each case, we find that there exists a polynomial time algorithm for P_1 problem, in which each supplier offers an incremental quantity discount cost structure. It is difficult to find a polynomial time optimal algorithm for P_2 problem and P_3 problem. The optimal algorithm for P_2 and P_3 problems given in this paper can find an optimal solution in a short time for a small size of the problem. However, there exists a polynomial time algorithm for SP_3 problem, which is a special case of P_3 problem with $|S_2| = 1$. More future research includes multi-echelon economic lot-sizing problem with multi-delivery modes problem and economic lot-sizing problem with multi-supplier multi-item and multiple cost structures.

References

1. Wagner, H.M., Whitin, T.: Dynamic Version of the Economic Lot Size Model. Manage. Sci. 14, 429–450 (1958)
2. Federgruen, A., Lee, C.Y.: The Dynamic Lot Size Model with Quantity Discount. Nav. Res. Logist. 37, 707–713 (1990)
3. Xu, J., Lu, L.L.: The Dynamic Lot Size Model with Quantity Discount: Counterexample and Correction. Nav. Res. Logist. 45, 419–422 (1998)
4. Zhang, Y.Z., Xu, J.T., Bai, Q.G.: Dynamic Lot-Sizing Model with a Class of Multi-Breakpoint Discount Cost Structures. In: The Sixth International Symposium on Operations Research and Its Applications, pp. 264–269. World Publishing Corporation, China (2006)
5. Chu, L.Y., Hsu, V.H., Shen, Z.J.: An Economic Lot Sizing Problem with Perishable Inventory and Economic of Scale Costs: Approximation Solution and Worst Case Analysis. Nav. Res. Logist. 52, 536–548 (2005)
6. Lee, C.Y.: Inventory Replenishment Model: Lot Sizing Versus Just-In-Time. Oper. Res. Lett. 32, 581–590 (2004)
7. Li, C.L., Hsu, V.N., Xiao, W.Q.: Dynamic Lot Sizing with Batch Ordering and Truckload Discounts. Oper. Res. 52, 639–654 (2004)
8. Solyal, O., Süral, H.: A Single SupplierCSingle Retailer System with an Order-Up-To Level Inventory Policy. Oper. Res. Lett. 36, 543–546 (2008)
9. Atamtürk, A., Küçükyavuz, S.: An $O(n^2)$ Algorithm for Lot Sizing with Inventory Bounds and Fixed Costs. Oper. Res. Lett. 36, 297–299 (2008)
10. Jaruphongsa, W., Çetinkaya, S., Lee, C.Y.: A Dynamic Lot-Sizing Model with Multi-Mode Replenishments: Polynomial Algorithms for Special Cases with Dual and Multiple Modes. IIE Trans. 37, 453–467 (2005)
11. Basnet, C., Leung, J.M.Y.: Inventory lot-sizing with supplier selection. Comput. Oper. Res. 32, 1–14 (2005)
12. Ekşioğlu, S.D.: A Primal-Dual Algorithm for the Economic Lot-Sizing Problem with Multi-Mode Replenishment. Eur. J. Oper. Res. 197, 93–101 (2009)

Synthetic Road Networks*

Reinhard Bauer, Marcus Krug, Sascha Meinert, and Dorothea Wagner

Karlsruhe Institute of Technology (KIT), Germany
{Reinhard.Bauer,Marcus.Krug,Sascha.Meinert,Dorothea.Wagner}@kit.edu

Abstract. The availability of large graphs that represent huge road networks has led to a vast amount of experimental research that has been custom-tailored for road networks. There are two primary reasons to investigate graph-generators that construct synthetic graphs similar to real-world road-networks: The wish to theoretically explain noticeable experimental results on these networks and to overcome the commercial nature of most datasets that limits scientific use. This is the first work that experimentally evaluates the practical applicability of such generators. To this end we propose a new generator and review the only existing one (which until now has not been tested experimentally). Both generators are examined concerning structural properties and algorithmic behavior. Although both generators prove to be reasonably good models, our new generator outperforms the existing one with respect to both structural properties and algorithmic behavior.

1 Introduction

During the last two decades, advances in information processing led to the availability of large graphs, that accurately represent the road networks of whole continents in full detail. Today, these networks are omnipresent in applications like route planning software, geographical information systems or logistics planning. While there is a vast amount of research on algorithms that work on (and often are custom-tailored for) road networks, the natural structure of these networks is still not fully understood.

Aims. In this work we aim to synthetically generate graphs that replicate real-world road networks. The motivation of doing so is manifold: Firstly, the existing data is often commercial and availability for research is only restricted. In those situations, graph generators are a good and established way to obtain test-data for research purposes. Additionally, it seems likely that datasets that represent the road-network of the entire world will be available in a few years. It will be shown later that the size of the road-network has a crucial (non trivial, non-linear) influence on the performance of algorithms on it. Using graph generators that are able to generate graphs of appropriate size and structure one can then do algorithmic research on such networks.

Secondly, we want to improve the understanding of the structure within road-networks. This may support a theoretical analysis of algorithms that have been

* Partially supported by the DFG (project WAG54/16-1).

B. Chen (Ed.): AAIM 2010, LNCS 6124, pp. 46–57, 2010.

custom-tailored for road-networks. A well known example is the development of route-planning techniques during the last decade that yield impressive runtimes by exploiting a special 'hierarchy' in road networks [5]. Many of these algorithms have recently been analyzed in [2]. There, the intuition of hierarchy has been formalized using the notion of 'highway dimension' and a generator for road networks. In [2] no evidence is given that this generator is a good model for road networks. We check that in this work.

Further, there is not only one 'road network'. Hence, we want to compare graphs originating from different sources with each other. This helps practitioners to assess the value of a given experimental work for their specific data and problem. Finally, as generators usually involve some tuning parameters, experimentalists can use them to generate interesting instances and steer the properties these instances have. This can help to understand custom-tailored algorithms better.

Related Work. There is a huge amount of work on point-to-point route-planning techniques. These are mostly custom-tailored for road networks. An overview can be found in [5]. In [2] a graph generator for road-networks is proposed.

The work [7] studies properties of real-world road-networks. Firstly, road networks are characterized as a special class of geometric graphs. This characterization is evaluated on the Tiger/Line road networks. It could be a starting point for a possible graph generator, but too many degrees of freedom are left open to use it directly. Furthermore, road networks are analyzed concerning planarity: The typical number of crossings (of the embedding given by the GPS-coordinates) of an n-vertex road-network is reported to be $\Theta(\sqrt{n})$.

Contribution. This is the first work that experimentally generates synthetic road networks and tests their practical applicability. In Chapter 2 we survey existing networks that are (partly) available to the scientific community. Chapter 3 introduces a new generator for road networks and describes the generator given in [2]. In order to assess the quality of these custom-tailored generators, we also describe two standard graph generators, namely those for Unit-Disk and Grid-graphs.

Chapter 4 evaluates the generators and compares their output with real-world networks. There, structural properties and algorithmic behavior are taken into account. As structural properties we consider connectedness, directedness, degree distribution, density and distance distribution. We found out that the custom-tailored generators are a good model for the real-world data. The generator of Abraham et al. [2] performs also well, with the exception that it produces graphs that are too dense and incorporate nodes of too high degree. For testing the algorithmic behavior we focus on point-to-point shortest path computation as this area uses techniques that have been highly custom-tailored for road-networks. One unexpected outcome is that the real-world graphs significantly differ in their algorithmic behavior. Further, the standard-generators approximate road-networks only to a limited extend (which is not surprising). Finally, the custom-tailored generators seem to be reasonably good approximations for the real-world instances.

2 Real-World Road-Networks

Graphs that represent road networks are typically constructed from digital maps. Digital mapping is expensive and mostly done by companies like Google, Teleatlas and NAVTEQ. Hence, this data is hard or expensive to obtain. We are aware of only three sources for real-world road-networks that are (almost) free for scientific use.

The dataset PTV is commercial and property of the company PTV-AG (`http://www.ptv.de/`). It is based on data of the Company NavTeq and not fully free, but has been provided to participants of the 9th Dimacs Implementation Challenge [6]. We use slightly updated data from the year 2006. The U.S. Census Bureau publishes the Tiger/Line datasets TIGER (`http://www.census.gov/geo/www/tiger/`). We use the version available at the 9th Dimacs Implementation Challenge (`http://www.dis.uniroma1.it/~challenge9/`). OpenStreetMap (`http://www.openstreetmap.org/`) is a collaborative project that uses GPS-data gathered by volunteers to assemble a digital map of the whole world. We use Europe-data of December 2009 and remove all items that do not correspond to roads (more details can be found in the appendix). See Table 1, page 52 for basic information on these three datasets, nomination is as follows: *Density* denotes the number of edges divided by the number of nodes, *directedness* denotes the number of edges (u, v) for which there either is no edge (v, u) or for which (v, u) has a different length than (u, v) divided by the total number of edges. In the experimental evaluation we always use Euclidean distances as edge weights. This results in better comparability and is well justified as changing to other metrics like travel time has only low impact on the considered algorithms [3]. Edges are always counted as directed, i.e. edges (u, v) and (v, u) both contribute to the overall number of edges.

3 Graph Generators

Voronoi-Based Generator. Our generator is based on the following assumptions about road networks which are well motivated from real-world road networks: (1) Road networks are typically built so as to interlink a given set of resources, such as, for instance, natural resources or industrial agglomerations of some sort. (2) Road networks exhibit a hierarchical, nested structure: The top level of this hierarchy is defined by the highways, which form a grid-like pattern. Each cells of this grid-like network is subdivided by a network of smaller roads, which again exhibits a grid-like structure. (3) Shortest paths in road networks do not typically exhibit a large dilation with respect to the Euclidean distance: Although dilation may be very large in the presence of long and thin obstacles, such as rivers and lakes, it is rather low for the larger part of the network. We further assume that in the presence of two or more sites of resources it is best to build roads along the bisectors of these sites, since any two sites incident to a road can access this road at the same cost.

Our generator is a generic framework which works in two phases: At first, we generate a random graph based on recursively defined Voronoi diagrams in the Euclidean plane. In the second phase we then compute a sparser subgraph since the graph computed in the first phase is rather dense as compared to real-world networks.

A *Voronoi diagram of a set of points* P is a subdivision of the plane into convex Voronoi regions $\text{vreg}(p)$ for all $p \in P$. The Voronoi region $\text{vreg}(p)$ contains all points whose distance to p is smaller than their distance to any other point $q \in P \setminus \{p\}$. The union of all points which are equally far from at least two points form a plane graph, which we call the *Voronoi graph* $\mathcal{G}(P)$. The vertices of this graph are exactly the set of points which are equally far from at least three points in P. Each face f of this graph corresponds to the Voronoi region of some point p. By $P(f)$ we denote the simple polygon which forms the boundary of f. The Voronoi diagram for a set of n points can be computed in $\mathcal{O}(n \log n)$ points by a simple sweep line algorithm [9]. In our implementation, we used the CGAL library in order to compute the Voronoi diagram of a set of points [1].

The first phase of our generator is a recursive procedure whose core is a routine called SUBDIVIDE-POLYGON$(P, n, \mathcal{D}, \mathcal{R})$. The pseudo-code for this routine is listed in Algorithm 1. Invoked on a simple polygon P this routine computes a Voronoi diagram inside P from a set of points which are chosen as follows: First, we choose a set of n uniformly distributed points in P, which we call center sites. For each of the center sites x we choose a density parameter α according to the distribution $\mathcal{D}$ as well as a radius r according to the distribution $\mathcal{R}$. As distributions we used the uniform distribution with density function $\text{Unif}_{[a,b]}(x) = \frac{1}{b-a}$ for all $x \in [a, b]$ with $a < b \in \mathbb{R}$ as well as the exponential distribution with density function $\text{Exp}_\lambda(x) = \lambda e^{-\lambda x}$ for all $x \in \mathbb{R}_0^+$. Then we choose $\lceil r^\alpha \rceil$ points in the disc centered at x with radius r by choosing radial coordinates uniformly at random. Thus, we create a set of points as agglomerations around uniformly distributed centers.

Algorithm 1. SUBDIVIDE-POLYGON

```
   Input: P, n, D, R
1  C ← ∅;
2  for i = 1 to n do
3      x ← choose uniform point inside P;
4      α ← random value chosen according to D;
5      r ← random value chosen according to R;
6      m ← ⌈r^α⌉;
7      C ← C ∪ {x};
8      for j = 1 to m do
9          p ← choose random point in R(x, r);
10         C ← C ∪ {p};
11     compute Voronoi diagram of C in P;
```

In our implementation we choose points inside the bounding box of P uniformly at random by rejecting points not in P until we have found the desired number of points.

The pseudo-code of the first phase is listed in Algorithm 2. The input consists of an initial polygon P, a number ℓ of levels for the recursion and for each recursion level $1 \leq i \leq \ell$ a fraction γ_i of cells to be subdivided along with distributions $\mathcal{C}_i$, $\mathcal{R}_i$ and $\mathcal{D}_i$. We then proceed as follows: First, we choose a set of Voronoi regions to subdivide among the regions which were produced in the previous iteration of the algorithm. Let S be the set of Voronoi regions which were produced in the previous iteration, then we subdivide the $\gamma_i|S|$ smallest regions in the current iteration of the algorithm. Hence, the distribution of points will be concentrated in areas with many points. Therefore, we simulate the fact that sites with many resources will attract even more resources.

For each Voronoi region R which has been chosen to be subdivided, we first choose an associated number of centers n according to the distribution $\mathcal{C}_i$. Then we call Subdivide-Polygon on input R, n and the distributions corresponding to the current level in the recursion.

Algorithm 2. Voronoi-Roadnetwork

Input: $\ell, \gamma_i, \mathcal{C}_i, \mathcal{R}_i, \mathcal{D}_i$, $1 \leq i \leq \ell$

```
1  ℓ_0 ← 1;
2  S ← P;
3  for i = 1 to ℓ do
4  |  m ← γ_{i−1}|S|;
5  |  S ← smallest m faces in S;
6  |  S′ ← ∅;
7  |  for f ∈ S do
8  |  |  n ← choose according to distribution C_i;
9  |  |  S′ ← S′ ∪ Subdivide-Polygon(P(f), n, D_i, R_i);
10 |  S ← S′
```

In the second phase we greedily compute a sparse graph spanner of the graph computed in phase one. Given a graph G a *graph spanner* H of G with stretch t is a subgraph of G such that for each pair of vertices u, v in G we have $\text{dist}_H(u,v) \leq t \cdot \text{dist}_G(u,v)$. We would like H to contain as few edges as possible. Determining if a graph G contains a t spanner with at most m edges is NP-hard [8].

In order to compute a sparse graph spanner greedily, we iterate over the edges sorted by non-increasing length and add only those edges to the graph whose absence would imply a large dilation in the graph constructed so far. Note that the order in which we consider the edges differs from the greedy algorithm for graph spanners discussed, e.g., in [4]. Let H be the graph we have obtained after considering m edges. Then we insert the $(m+1)$-st edge $\{u, v\}$ if and only if $\text{dist}_H(u,v)$ is larger than $t \cdot \text{len}(u,v)$. We assume $\text{len}(u,v)$ to be infinity if u and v are not in the same component. Hence, at each step H is a t-spanner for all pairs of vertices which are connected in H. At the end we will obtain a connected graph, and therefore, a t-spanner for G.

In order to determine $\text{dist}_H(u,v)$, we use Dijkstra's algorithm for computing shortest paths, which can be pruned whenever we have searched the complete graph-theoretic ball of radius $t \cdot \text{len}(u,v)$. Since we consider the edges in sorted order with non-increasing length, we will heuristically consider only few edges, as long edges are considered at the beginning, when the graph is still very sparse. Since the larger part of the edges in the graph is short compared to the diameter, the running time for this algorithm is not too large. In order to speed up computation for the cases in which u and v are not connected, we use a union-find data structure to keep track of the components of H.

The Generator of Abraham et al. This approach is due to [2] and is based on an arbitrary underlying metric space (M, d) with diameter D. In this work M will always be a rectangle in the Euclidean plane with $d(u, v)$ being the Euclidean distance between points u and v. Further, the generator requires a random generator rand() for points in M. No distribution is given in [2], we will report the distributions used in this work later in this section.

Algorithm 3. Generator of Abraham et al. [2]

input : number of vertices n
output: Graph (V, E)

1 initialize $V = C_0, \ldots, C_{\log D}, E$ to be $\emptyset$;
2 **for** $t = 1$ *to* n **do**
3 $\quad v_t \leftarrow$ rand() ;
4 $\quad V \leftarrow V \cup \{v_t\}$;
5 $\quad$ **for** $i = \log D$ *to* 1 **do**
6 $\quad\quad$ **if** $d(v_t) > 2^i$ *for each* $w \in C_i$ **then**
7 $\quad\quad\quad C_i \leftarrow C_i \cup \{v_t\}$;
8 $\quad\quad\quad$ **for** $w \in C_i$ **do**
9 $\quad\quad\quad\quad$ **if** $d(v_t, w) \leq k \cdot 2^i$ **then** $E \leftarrow E \cup \{v_t, w\}$
10 $\quad\quad w \leftarrow$ closest point of v_t in C_{i+1} ;
11 $\quad\quad$ **if** $v_t \neq w$ **then** $E \leftarrow E \cup \{v_t, w\}$
12 set edge weights such that $\text{len}(u, v) = d(u, v)^{1-\delta}$ for $\delta = 1/8$

The generator starts with the empty graph $(V, E) = (\emptyset, \emptyset)$ and iteratively adds new vertices to V. Their location in M is distributed according to rand(). A 2^i-cover is a set C_i of vertices such that for $u, v \in C_i$, $d(u, v) \geq 2^i$ and such that for each $u \in V$ there is a $v \in C_i$ with $d(u, v) \leq 2^i$. During the process of adding vertices, the generator maintains for each i with $1 \leq i \leq \log D$, a 2^i-*cover* C_i: After a vertex v_t has been added, the lowest index i is computed such that there is a $w \in C_i$ with $d(v_t, w) \leq 2^i$. Then v_t is added to all C_j with $0 \leq j < i$. If no such i exists, v_t is added to all sets C_j. Then, given a tuning-parameter k, for each $C_i \ni v_t$ and each $w \in C_i$ an edge (w, v_t) is added if $d(w, v_t) \leq k \cdot 2^i$. Further, for each $C_i \ni v_t$ with $i < \log D$ such that $v_t \notin C_{i+1}$, an edge from v_t to its nearest neighbor in C_{i+1} is added. Finally, edge lengths are fixed such that $\text{len}(u, v) = d(u, v)^{1-\delta}$ for $\delta = 1/8$. Pseudocode of the generator can be found in Algorithm 3. Throughout the rest of the paper, we fill the remaining tuning parameters as follows. We choose the number of levels to be 25 and therefore set $D := 2^{25}$. The aspect ratio of the rectangle representing M is 0.75. The parameter $k = \sqrt{2}$ (deviating from the original description where $k = 6$). We tried two point sources rand(). Firstly, we sampled points uniformly at random (which will not be used later on). Secondly, we used an improved point source that mimics city-like structures: We use a 2-phase approach. In the preparation phase we iteratively compute special points within M called *city centers*. This is done as follows. A new city center c is chosen uniformly at random within M. We assign a value r_c to c which is chosen from an exponential distribution with parameter $\lambda = 1/(0.05 \cdot s)$ where s is the length of the longer border of M.

We then assign a population p_c to c which is $r_c^{1.1}$. The preparation-phase stops, when the overall population exceeds the number of requested nodes n.

Afterwards, a new point x is generated on request as follows: Firstly, a center c with positive population p_c is chosen uniformly at random. We then set $p_c := p_c - 1$. The location of x is determined in polar-coordinates with center c by choosing an angle uniformly at random and by choosing the distance from a normal distribution with mean 0 and standard deviation r_c. Whenever we sampled points not lying in M these get rejected.

Unit Disk. Given a number of nodes n, a unit disk graph is generated by randomly assigning each of the n nodes to a point in the unit square of the Euclidean plain. There is an edge (u, v) in case the Euclidean distance between u and v is below a given radius r. We adjusted r such that the resulting graph has approximately $7n$ edges. We use the Euclidean distances as edge weights.

Grid. These graphs are based on two-dimensional square grids. The nodes of the graph correspond to the crossings in the grid. There is an edge between two nodes if these are neighbors on the grid. Edge weights are randomly chosen integer values between 1 and 1000.

4 Experimental Evaluation

In this section we experimentally assess the proposed generators. To that end we generated the datasets VOR (originating from the generator our Voronoi-based generator) and ABR (originating from the Abraham et al. generator). More information on the generation-process can be found in the appendix.

Graph Properties. We now have a look at some basic structural properties of the networks. We first observe that all real-world graphs are undirected or almost undirected and have an almost equal density of 2.05 to 2.4 (Table 1). By construction the synthetic graphs are undirected, the density of VOR is similar to the real-world graphs, the density of ABR is slightly too high. Further, all three real-world networks consist of one huge and many tiny strongly connected components. The OSM-data deviates in the size of the biggest strongly connected component. An explanation for this is the unfinished state of the underlying

Table 1. Overview of origin, size, density and directedness of the available real-world data and the generated datasets

dataset	origin	represents	#nodes	#edges	density	dir'ness
TIGER	U.S administration	USA	24,412,259	58,596,814	2.40	.0 %
PTV	commercial data	Europe	31,127,235	69,200,809	2.22	4.9 %
OSM	collaborative project	Europe	48,767,450	99,755,206	2.05	3.5 %
ABR	Abraham et al	synthetic	15,000,000	43,573,536	2.90	.0 %
VOR	Voronoi generator	synthetic	42,183,476	93,242,474	2.21	.0 %

map and we expect the deviation to decrease with increasing level-of detail in the OSM-data in the future (Table 2). By construction the synthetic graphs are strongly connected.

The distribution of the node-degrees is very similar for PTV and TIGER but again deviates significantly for OSM. All three graphs have in common that almost no nodes of degree of at least 5 exists. The difference in the number of degree-2 nodes can be explained by a higher level of detail of the OSM-data (Table 3). The distribution of VOR is quite similar to OSM while ABR significantly deviates from all other distributions: 7% of the nodes have degree 6 or higher, the maximum degree is 58.

Table 2. Relative sizes (measured in number of nodes) of the k biggest SCCs

dataset	k = 1	2	5	20	100	total # of SCCs
OSM	.80	.91	.94	.96	.97	541264
PTV	.97	.97	.97	.97	.97	924561
TIGER	.98	.98	.99	.99	.99	89796

Table 3. Degree distribution of the datasets: Relative frequency according to degree

dataset	degree 0	1	2	3	4	5	≥6
OSM	.001	.097	.773	.114	.015	0	0
PTV	.030	.247	.245	.429	.049	.001	0
TIGER	0	.205	.301	.386	.106	.001	0
ABR	0	.324	.254	.160	.093	.056	.077
VOR	0	.100	.601	.292	.005	.002	0

Figure 1 shows the distribution of the distances between pairs of nodes in the networks. The distributions are hard to interpret but can be used as a fuzzy fingerprint for the structure within the graphs (for instance to separate them from different graph classes like small-world graphs). We observe that OSM and PTV have a quite similar distribution, differing from the TIGER-data. A possible explanation for that could be the geographical origin of the data. Both OSM and PTV map the European road-network while TIGER maps the U.S. road-network. The VOR-dataset is a good approximation for the TIGER-data, the ABR-dataset is a good approximation for OSM and PTV.

For the sake of completeness, we also report the distribution of the according edge weights (Figure 1). Note that the edge-weight distribution has only small impact on the considered algorithms [3]. For better comparability, we always applied Euclidean distances instead of the original weights.

Algorithmic Behavior. In this section we compare the algorithmic behavior of both real-world and synthetically generated graphs. For this purpose we analyze the speedups which can be achieved using speedup techniques for point-to-point shortest path queries, such as Bidirectional Dijkstra's algorithm, ALT (16 landmarks computed by max-cover strategy), Arc-Flags (128 Voronoi-cells), Reach-Based Routing ($\epsilon = 0.5$, $\delta = 16$) as well as Contraction Hierarchies (CH, original code), as compared to standard Dijkstra's algorithm. See [5] for more information. These techniques have specifically been designed for the task of heuristically speeding up Dijkstra's algorithm on real-world road networks and,

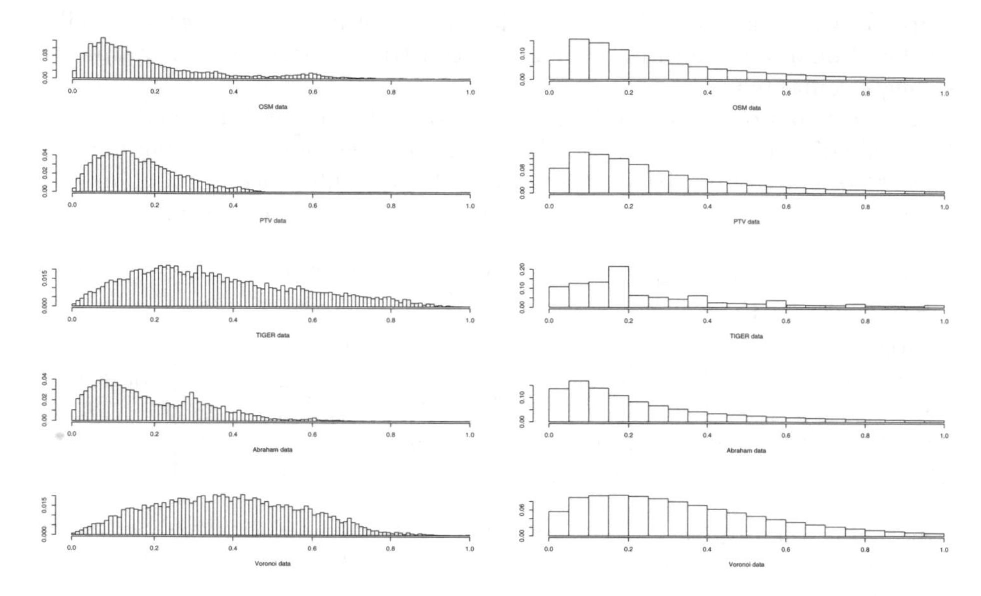

Fig. 1. Left: Distance-distribution sampled by considering 1000 source nodes and, for each source node, 1000 target nodes. Unconnected pairs have been removed, values have been normalized. Right: Edge-weight distributions. Outliers have been removed, values have been normalized. Graphs are from top: OSM, PTV, TIGER, ABR, VOR.

thus, we will use their performance as an indicator for the similarity of the networks.

As a benchmark set we analyze the data from the real-world instances PTV, TIGER and OSM as well as the synthetically generated graphs VOR and ABR. Since the speedup of the various techniques is non-trivially correlated with the size of the graph, we sampled random snapshots with sizes ranging from 1,000 up to 512,000 from our benchmark set in order to be able to capture the underlying correlation. Sampling has been performed exploiting the given geometrical layout and extracting square-sized subgraphs around random points. Thereby, the diameter has been adapted, such that the resulting graph has the desired size. In order to assess the quality of the synthetic data, we also included data from generators that are not custom-tailored for road-networks: Grid-graphs and unit-disk-graphs (a description of the generators is given at the end of the previous section). The respective data-sets are named GRID and UD. The measured speedups are summarized in Table 4.

The speedups we measured using a bidirectional Dijkstra search range between 1.1 and 1.8 and are concentrated around 1.5 for all graph sizes. There does not seem to be any significant trend in this data and, hence, we omit a detailed analysis of this speedup technique. Our results can be summarized as follows: Real-world graphs significantly differ in their algorithmic behavior: Although OSM and TIGER behave similarly with respect to the speedup techniques ALT, Arc-Flags and CH, the two speedup techniques differ by almost a factor

Table 4. Speedups of the tested Point-To-Point Shortest Path Algorithms

technique	#nodes	OSM	TIGER	PTV	VOR	ABR	UD	GRID
ALT	2000	1.9	3.9	5.1	4.9	9.3	5.9	8.1
ALT	32000	5.5	9.6	10.7	11.3	16.4	12.3	12.6
ALT	128000	10.8	10.8	18.2	13.8	31.4	21.7	22.2
ALT	256000	15.3	13.4	26.9	16	35.8	26.1	25.6
ALT	512000	16	17.7	30.6	21.6	37.3	28.8	24.2
Arc-Flags	2000	4.4	14.5	20.7	20.5	48.6	29.1	35.7
Arc-Flags	32000	21.8	55.9	68.5	89.7	131.5	44.1	32.7
Arc-Flags	128000	48.7	65.7	142.2	111.7	219.9	64.4	52.3
Arc-Flags	256000	69.5	76.2	219.9	143.2	245.6	74	72
CH	2000	24.3	16.5	20.7	13.9	13.7	9	7.5
CH	32000	100.4	80.5	102.5	114.9	42.2	38.3	29.1
CH	128000	229	209.2	390.2	346.4	184.6	134.2	74.5
CH	256000	435.2	357.3	626.1	684.7	327.8	195.7	128.2
CH	512000	459.5	497.9	831.4	1565.5	474.2	271.8	175.6
Reach	2000	3.9	3.3	3.8	3.3	4.2	1.8	1.2
Reach	32000	10.1	7.5	9.1	9.9	5.9	2.9	2.1
Reach	128000	28.2	15.7	24.2	22.3	10.1	4.2	2.9
Reach	256000	37.8	22.8	31.8	32.9	10	4.4	4
Reach	512000	34.3	16.3	27.1	49.2	10.7	5	4.7

of two with respect to Reach. Even worse, the PTV-dataset achieves speedups almost twice as large as OSM and TIGER with ALT and CH, and it shows a totally different trend concerning its behavior with Arc-Flags.

GRID and UD approximate streetgraphs only to a limited extent: Both GRID and UD behave similarly for all speedup techniques we considered. Although both graphs seem to be rather good approximations for PTV with respect to ALT and for both OSM and TIGER with respect to Arc-Flags, they seem to be rather bad estimates for CH and Reach. This may be explained by the lack of hierarchy inherent to both generators and the fact that both techniques are based on hierarchical decompositions of the underlying graphs.

ABR and VOR are reasonably good approximations: Contrary to GRID and UD, ABR and VOR seem to capture both the magnitude and the trend exhibited by real-world instances quite well. The speedups measured for ABR are slightly too large for ALT and Arc-Flags and they are slightly too small for Reach. For CH, on the other hand, they are very close to the speedups measured for OSM and TIGER. Except for CH, the speedups measured for Voronoi are well in between the speedups measured for the real-world instances. For CH, the speedups are very close to the speedups for PTV which are larger than those for OSM and TIGER by roughly a factor of two.

Note that parameters of the synthetic data have not been finetuned for approximating the given results. Since there is considerable amount of deviation in the behavior of the real-world data we consider doing so to be over-fitting.

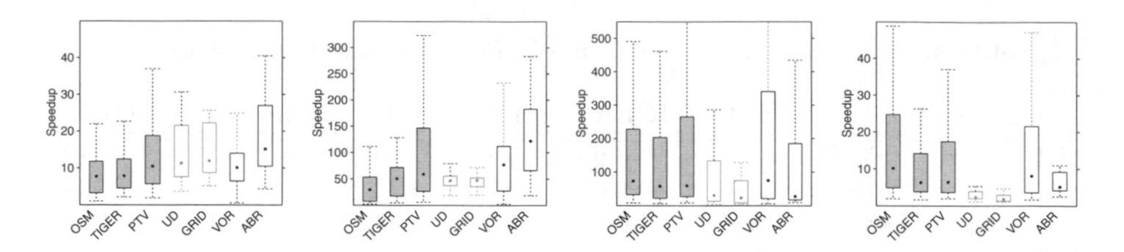

Fig. 2. Speedups observed for different techniques: ALT, Arc-Flags, Contraction Hierarchies and Reach (from left to right)

Summary. Figure 2 shows a summary of the speedups measured for the different techniques. GRID and UD perform poorly for Arc-Flags, CH and Reach, since either the median is far too small as in the case of CH and Reach or the interquartile range is far too small as in the cases of Arc-Flags. Although both may be considered good estimates for the PTV-data with respect to ALT, VOR seems to be the better choice. A similar, albeit slightly better behavior can be observed for ABR: Although it captures the interquartile range rather well for all techniques, except Reach, the medians do not fit too well. VOR, on the other hand, seems to be the best fit with respect to both median and interquartile range. Although the median is slightly too large for Arc-Flags and the interquartile range is too large for CH it seems to capture the behavior of the real-world data with respect to the speedup techniques best.

References

1. Cgal, Computational Geometry Algorithms Library, `http://www.cgal.org`
2. Abraham, I., Fiat, A., Goldberg, A.V., Werneck, R.F.: Highway Dimension, Shortest Paths, and Provably Efficient Algorithms. In: Proceedings of the 21st Annual ACM–SIAM Symposium on Discrete Algorithms, SODA 2010 (2010)
3. Bauer, R., Delling, D., Wagner, D.: Experimental Study on Speed-Up Techniques for Timetable Information Systems. In: Liebchen, C., Ahuja, R.K., Mesa, J.A. (eds.) Proceedings of the 7th Workshop on Algorithmic Approaches for Transportation Modeling, Optimization, and Systems (ATMOS 2007), Internationales Begegnungs- und Forschungszentrum für Informatik (IBFI), Schloss Dagstuhl, Germany, pp. 209–225 (2007)
4. Bose, P., Carmi, P., Farshi, M., Maheshwari, A., Smid, M.: Computing the Greedy Spanner in Near-Quadratic Time. In: Gudmundsson, J. (ed.) SWAT 2008. LNCS, vol. 5124, pp. 390–401. Springer, Heidelberg (2008)
5. Delling, D., Sanders, P., Schultes, D., Wagner, D.: Engineering Route Planning Algorithms. In: Lerner, J., Wagner, D., Zweig, K.A. (eds.) Algorithmics of Large and Complex Networks. LNCS, vol. 5515, pp. 117–139. Springer, Heidelberg (2009)
6. Demetrescu, C., Goldberg, A.V., Johnson, D.S. (eds.): The Shortest Path Problem: Ninth DIMACS Implementation Challenge. DIMACS Book, vol. 74. American Mathematical Society, Providence (2009)

7. Eppstein, D., Goodrich, M.T.: Studying (non-planar) road networks through an algorithmic lens. In: Proceedings of the 16th ACM SIGSPATIAL International Conference on Advances in Geographic Information Systems. ACM Press, New York (2008)
8. Peleg, D., Schäffer, A.A.: Graph spanners. Journal of Graph Theory 13(1), 99–116 (1989)
9. Preparata, F.P., Shamos, M.I.: Computational geometry: an introduction. Springer, New York (1985)

A Additional Technical Information

Extraction of the OSM-Dataset. To distill the OSM-dataset, we used the Europe-data available at `http://download.geofabrik.de/osm/` of December 2009. The graph includes all elements for which the highway-tag equals one of the following values: residential, motorway_link, trunk, trunk_link primary, primary_link, secondary, secondary_link, tertiary, unclassified, road, residential, living_street, service, services.

Generation of the Synthetic Data. The graph VOR is a 4-level graph computed using the following parameters: Level 1 has 1700 centers, density of 0.2 The Voronoi streetgraph was computed using the parameters listed in Table 5. In the second phase we greedily computed a 4-spanner subgraph.

The first phase of the algorithm took 122 minutes and consumed up to 19 GB of space; the second phase of the algorithm took 28 minutes consumed up to 21 GB of space. The ABR streetgraph has 15 mio. nodes and more than 43 mio. edges. It was generated within 63 minutes and consumed up to 8.39 GB of disk space.

Table 5. Parameters of the Voronoi streetgraph

Level i	# of centers $\mathcal{C}_i$	density $\mathcal{D}_i$	radius $\mathcal{R}_i$	fraction γ_i
1	1700	.2	$\text{Exp}_{.01}$	.95
2	$\text{Unif}_{[2,40]}$	.5	$\text{Exp}_{.1}$	.9
3	$\text{Unif}_{[2,70]}$	.9	Exp_{2}	.7
4	$\text{Unif}_{[4,40]}$	.0	0	0

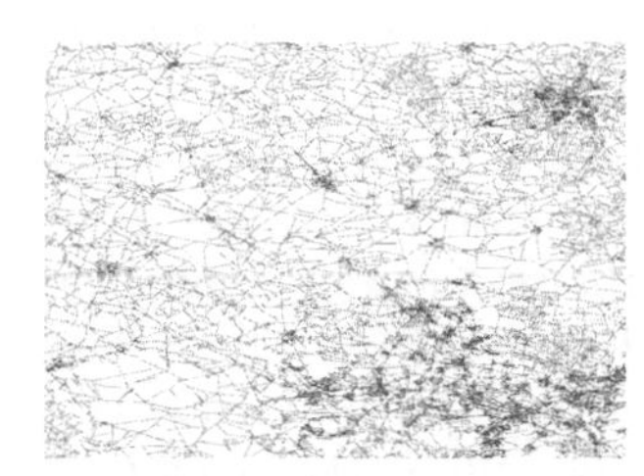

Fig. 3. PTV graph

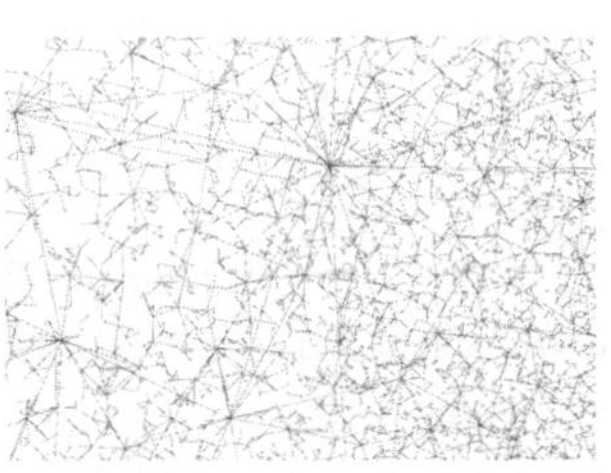

Fig. 4. ABR graph

Fig. 5. VOR graph

Computing Exact and Approximate Nash Equilibria in 2-Player Games*

Vittorio Bilò[1] and Angelo Fanelli[2]

[1] Department of Mathematics, University of Salento
Provinciale Lecce-Arnesano, P.O. Box 193, 73100 Lecce - Italy
vittorio.bilo@unisalento.it
[2] School of Physical and Mathematical Sciences,
Nanyang Technological University, 637371 Singapore
angelo.fanelli@ntu.edu.sg

Abstract. The problem of computing a Nash equilibrium in a normal form 2-player game (or bimatrix games) is PPAD-complete in general, while it can be efficiently solved in a special subclass which we call regular bimatrix games. The current best approximation algorithm, proposed in [19], achieves a guarantee of 0.3393. In this paper we design a polynomial time algorithm for computing exact and approximate Nash equilibria for bimatrix games. The novelty of this contribution is twofold. For regular bimatrix games, it allows to compute equilibria whose payoffs optimize any objective function and meet any set of constraints which can be expressed through linear programming, while, in the general case, it computes α-approximate Nash equilibria, where α is the maximum difference between any two payoffs in the same strategy of any player. Hence, our algorithm improves the best know approximation guarantee for the bimatrices in which $\alpha < 0.3393$.

1 Introduction

The complexity of computing Nash equilibria in normal form games has been for ages one of the most interesting and challenging problems lying at the boundary of P and NP. The Lemke-Howson algorithm [14] for the 2-player case, introduced in 1964, was shown to require an exponential running time in the worst case only 40 years later by Savani and von Stengel [18]. After that, a fast escalation of results appeared showing the PPAD-completeness of the problem of computing Nash equilibria in any n-player game. In particular, Daskalakis, Goldberg and Papadimitriou [8] proved it for any $n \geq 4$, Daskalakis and Papadimitriou [11] settled the 3-player case and finally Chen and Deng [4] solved the remaining 2-player case.

Under the widely believed conjecture that $P \subset PPAD$, next challenge is to understand the hardness of computing approximate Nash equilibria in normal form

* This research was partially supported by the grant NRF-RF2009-08 "Algorithmic aspects of coalitional games".

B. Chen (Ed.): AAIM 2010, LNCS 6124, pp. 58–69, 2010.

games and to detect subcases for which exact Nash equilibria can be computed in polynomial time. To the best of our knowledge, all the research up to date has only focused on the 2-player case, the so called *bimatrix games*. In such a setting, a result achieved by Abbot, Kane and Valiant [1], implies that computing Nash equilibria remains PPAD-complete even when restricting to win-lose bimatrix games, that is, the case in which all the entries in the bimatrix are either zero or one.

It is well-know that computing Nash equilibria can be done in polynomial time by exploiting linear programming in the zero-sum case, that is, when the sum of any pair of entries in the bimatrix is equal to zero. By exploiting reduction among games which preserves the set of Nash equilibria, it is possible to use such an algorithm to compute Nash equilibria also in the more general case of regular bimatrix games (see Sections 2 and 3 for more details). We stress that in this case, however, there is no way to ask such an algorithm for the computation of a particular Nash equilibrium when the game possesses several ones (see Example 1 in the appendix). Recently, two tractable cases were discovered for the class of win-lose bimatrix games. In particular, Codenotti, Leoncini and Resta [7] gave an algorithm for computing a Nash equilibrium when the number of winning positions per strategy of each of the players is at most two, while Addario-Berry, Oliver and Vetta [2] solved the case in which the graph representation of the game is planar.

As to approximation, we say that a pair of mixed strategies is an ϵ-*approximate Nash equilibrium*, where $\epsilon > 0$, if for each player all strategies have expected payoff that is at most ϵ more than the expected payoff of the given strategy. The first approximation algorithms for bimatrix games has been given by Lipton, Markakis and Mehta [15] who showed that an ϵ-approximation can be found in time $O(n^{\frac{\log n}{\epsilon^2}})$ by examining all supports of size $\frac{\log n}{\epsilon^2}$. It is quite easy to obtain a 0.75-approximate Nash equilibrium and a slightly improved result has been achieved by Kontogiannis, Panagopoulou and Spirakis [13]. Almost contemporaneously, Daskalakis, Mehta and Papadimitriou [9] gave a very simple algorithm computing a 0.5-approximate Nash equilibrium with support size at most two. On the negative side, Chen, Deng and Teng [5] showed that, for any $\epsilon > 0$, computing a $\frac{1}{n^{\epsilon+1}}$-approximate Nash equilibrium is PPAD-complete. Hence no FPTAS exists for this problem unless P=PPAD. Hardness of approximation carries over also to win-lose games as Chen, Teng and Valiant [6] showed that correctly computing a logarithmic number of bits of the equilibrium strategies is as hard as the general problem. Furthermore, Feder, Nazerzadeh and Saberi [12] claimed no algorithm examining only strategies with support of size smaller than about $\log n$ can achieve an approximation guarantee better than 0.5 even for zero-sum games. Because of this result, in order to break the 0.5 barrier, approximate equilibria with large support size need to be computed.

The first breakthrough was given by Daskalakis, Mehta and Papadimitriou [10] who contributed an algorithm based on linear programming achieving an approximation guarantee of $\frac{3-\sqrt{5}}{2} + \epsilon \approx .38 + \epsilon$ in time $n^{O(1/\epsilon^2)}$. Despite its theoretical importance, we note that, in order to slightly improve on the 0.5

approximation, we need to run such an algorithm with $\epsilon \leq 0.1$ thus obtaining a prohibitive, although polynomial, running time. Recently, again two different algorithms have been proposed contemporaneously. Bosse, Byrka and Markakis [3] derived a 0.36392-approximation at the cost of solving a linear program, while the best known approximation guarantee of 0.3393 has been achieved by Tsaknakis and Spirakis [19] by exploiting a more involved use of LP formulations. Different tests performed by Tsaknakis, Spirakis and Kanoulas [20] seem to show that the "real" approximation guarantee of this last algorithm is very close to zero.

Our Contribution. We propose an LP formulation for the problem of computing Nash equilibria in bimatrix games whose number of variables and constraints is linear in the dimensions of the bimatrix. We demonstrate that any feasible solution to this formulation is always a Nash equilibrium in the case of regular bimatrix games. The novelty of this approach is in the fact that our technique allows to compute equilibria whose payoffs optimize any objective function and meet any set of constraints which can be expressed through linear programming (i.e., maximizing the minimum payoff per player, minimizing the maximum payoff per player, minimizing or maximizing the sum of the players' payoffs, minimizing or maximizing the payoff of one player when the one of the other player is constrained to be greater or smaller than a given threshold, deciding whether there exists a Nash equilibrium whose payoffs fall in certain ranges, etc.), while the previous known algorithm for this case can only output a generic equilibrium. Our formulation can also be used to compute approximate Nash equilibria in general bimatrix games. We show that the approximation guarantee of this algorithm is upper bounded by the maximum difference between any two payoffs in the same strategy for any player. This means that for all bimatrices in which such a value is smaller than 0.3393, our algorithm achieves the best provable worst-case guarantee.

2 Definitions and Notation

A 2-player game in normal form, also called a bimatrix game, is modeled as a pair of $n \times n$ payoff matrices (R, C), one for the first player (the row player) and the other for the second one (the column player). Clearly, the assumption that both players have n available strategies is without loss of generality since it is always possible to add dummy strategies (i.e., always giving to both players the lowest possible payoff) to the ones possessed by the player having the lowest number of strategies.

The pure strategies available for the row player are rows, while those for the column player are columns. If the row player plays pure strategy i and the column player plays pure strategy j, the row player receives a payoff equal to R_{ij}, while the column player gets C_{ij}. A mixed strategy for player i is a probability distribution over the set of her strategies. When dealing with mixed strategies the payoff of each player is computed in expectation, that is, if the row and the column player play respectively mixed strategies x and y, then the row player receives a payoff equal to $x^T Ry$, while the column player gets $x^T Cy$. The support of a mixed strategy x is the set of pure strategies played with positive probability in x. Clearly, pure strategies are mixed strategies whose support has cardinality one.

Let us define $\Delta_n = \{x \in \mathbb{R}^n : x_i > 0\ \forall i \in [n] \text{ and } \sum_{i=1}^n x_i = 1\}$ as the set of all mixed strategies which can be defined over a set of n pure strategies.

Definition 1 (Nash Equilibrium). *A pair of mixed strategies* $(x^*, y^*) \in \Delta_n \times \Delta_n$ *is a* Nash equilibrium *if no player can improve her payoff by deviating unilaterally to another mixed strategy, that is,* $x^{*T} Ry^* \geq x^T Ry^*$, $\forall x \in \Delta_n$ *and* $x^{*T} Cy^* \geq x^{*T} Cy$, $\forall y \in \Delta_n$.

Definition 2 (Approximate Nash Equilibrium). *For any* $\epsilon > 0$, *a pair of mixed strategies* $(x^*, y^*) \in \Delta_n \times \Delta_n$ *is an* ϵ-approximate Nash equilibrium *if no player can improve her payoff of more than an additive factor of* ϵ *by deviating unilaterally to another mixed strategy, that is,* $x^{*T} Ry^* + \epsilon \geq x^T Ry^*$, $\forall x \in \Delta_n$ *and* $x^{*T} Cy^* + \epsilon \geq x^{*T} Cy$, $\forall y \in \Delta_n$.

We consider the following special classes of bimatrix games.

Definition 3 (Zero-Sum Games). *A bimatrix game* (R, C) *is a zero-sum game if* $R_{ij} + C_{ij} = 0$ *for any* $i, j \in [n]$, *that is,* (R, C) *can be rewritten as* $(R, -R)$.

Definition 4 (Regular Bimatrix Games). *A bimatrix game* (R, C) *is regular if one of the following two conditions holds:*

- $R_{ij} + C_{ij} = \overline{b}_i$ *for any* $i, j \in [n]$ *(row regularity);*
- $R_{ij} + C_{ij} = \overline{a}_j$ *for any* $i, j \in [n]$ *(column regularity).*

Definition 5 (Smoothed Bimatrix Games). *A bimatrix game* (R, C) *is* α-*smoothed if* $\alpha \geq \max_{i,j,j' \in [n]}\{R_{ij} - R_{ij'}\}$ *and* $\alpha \geq \max_{i,i',j \in [n]}\{C_{ij} - C_{i'j}\}$.

3 Preliminaries

Reductions among bimatrix games. For each $a, b \in \mathbb{R}^+$, consider the bimatrix games $\mathcal{G} = (R, C)$ and $\mathcal{G}' = (aR, bC)$. For each Nash equilibrium (x^*, y^*) for $\mathcal{G}$ we have $x^{*T}(aR)y^* = ax^{*T} Ry^* \geq ax^T Ry^* = x^T(aR)y^*$, $\forall x \in \Delta_n$ and $x^{*T}(bC)y^* = bx^{*T} Cy^* \geq bx^{*T} Cy = x^{*T}(bC)y$, $\forall y \in \Delta_n$. Similarly, we obtain that for each ϵ-approximate Nash equilibrium (x^*, y^*) for $\mathcal{G}$ it holds $x^{*T}(aR)y^* + a\epsilon \geq x^T(aR)y^*$, $\forall x \in \Delta_n$ and $x^{*T}(bC)y^* + b\epsilon \geq x^{*T}(bC)y$, $\forall y \in \Delta_n$. Hence, $\mathcal{G}$ and $\mathcal{G}'$ have the same set of Nash equilibria; moreover, any ϵ-approximate Nash equilibrium for $\mathcal{G}$ is an $\ell\epsilon$-approximate Nash equilibrium for $\mathcal{G}'$, with $\ell = \max\{a, b\}$, and vice versa.

Now let A and B be two $n \times n$ matrices such that $a_{ij} = a_j$ and $b_{ij} = b_i$ for all $i, j \in [n]$. Consider the game $\mathcal{G}'' = (A + R, B + C)$; for each Nash equilibrium

(x^*, y^*) for $\mathcal{G}$, by exploiting $x^T Ay = \sum_{j=1}^n a_j y_j$ and $x^T By = \sum_{i=1}^n b_i x_i$ for any $x, y \in \Delta_n$, we have $x^{*T}(A+R)y^* = \sum_{j=1}^n a_j y_j^* + x^{*T} Ry^* \geq x^T Ay^* + x^T Ry^* = x^T(A+R)y^*$, $\forall x \in \Delta_n$ and $x^{*T}(B+C)y^* = \sum_{i=1}^n b_i x_i^* + x^{*T} Cy^* \geq x^{*T} By + x^{*T} Cy = x^{*T}(B+C)y$, $\forall y \in \Delta_n$. Similarly, for each ϵ-approximate Nash equilibrium (x^*, y^*) for $\mathcal{G}$ it holds $x^{*T}(A+R)y^* + \epsilon \geq x^T(A+R)y^*$, $\forall x \in \Delta_n$ and $x^{*T}(B+C)y^* + \epsilon \geq x^{*T}(B+C)y$, $\forall y \in \Delta_n$. Hence $\mathcal{G}$ and $\mathcal{G}''$ have the same set of Nash equilibria and the same set of approximate Nash equilibria.

Because of these results, when studying the complexity of computing Nash equilibria and approximate Nash equilibria in bimatrix games we can restrict our attention to the case in which all the elements of R and C belong to $[0; 1]$.

A polytime algorithm for zero-sum games. Consider a zero-sum bimatrix game $(R, -R)$ and let us examine the problem under the column player's viewpoint. Since anything the row player wins is lost by the column player, the latter wants to play a strategy y such that the row player cannot get too much, no matter what she does. More formally, the column player wants a mixed strategy $y \in \Delta_n$ and a value ω such that $\sum_{j \in [n]} R_{ij} y_j \leq \omega$, $\forall i \in [n]$. This means that the column player should solve the linear program

$$\begin{aligned} &\min\ \omega \\ s.t. &\sum_{j \in [n]} R_{ij} y_j \leq \omega \qquad \forall i \in [n], \\ &\sum_{j \in [n]} y_j = 1, \\ &y_j \geq 0 \qquad \forall j \in [n]. \end{aligned}$$

Furthermore, from the row player's viewpoint, the goal is to solve the linear program

$$\begin{aligned} &\min\ \psi \\ s.t. -&\sum_{i \in [n]} R_{ij} x_i \leq \psi \qquad \forall j \in [n], \\ &\sum_{i \in [n]} x_i = 1, \\ &x_i \geq 0 \qquad \forall i \in [n]. \end{aligned}$$

By the Duality Theorem and the Min-max Theorem, any pair (x^*, ω^*) and (y^*, ψ^*) of optimal solutions to these two linear programs must obey $\omega^* = \psi^*$, hence (x^*, y^*) is a Nash equilibrium for the zero-sum game $(R, -R)$. This analysis gives us a polynomial time algorithm for computing a generic Nash equilibrium in zero-sum bimatrix games; moreover, it also implies that all Nash equilibria yield the same payoffs for both players.

Extension to regular bimatrix games. When given a row (resp. column) regular bimatrix game (R, C) let us define the matrix B (resp. A) in such a way that $b_{ij} = -\overline{b}_i$ (resp. $a_{ij} = -\overline{a}_j$) for all $i, j \in [n]$. By the definition of row (resp. column) regular bimatrix games, it is easy to see that $(R, B + C)$ (resp. $(A + R, C)$) is a zero-sum game. Because of this and the fact that the pair of games (R, C) and $(R, B + C)$ (resp. $(A + R, C)$) have the same set of Nash equilibria, it is possible to compute in polynomial time a generic Nash equilibrium for regular bimatrix games.

4 Our LP Formulation

We first formulate the problem of computing a Nash equilibrium in bimatrix games by using quadratic programming. Let us call 2-Nash the problem of computing a Nash equilibrium in bimatrix games.

A quadratic formulation for 2-Nash. An important combinatorial characterization of improving defections states that for any tuple of mixed strategies, if one of the players possesses an improving defection, then there always exists a best response which is also a pure strategy for the defecting player. This claim easily follows by linearity of expectation. Thus, if we denote with $e_i \in \Delta_n$ the vector with a 1 in its i-th coordinate and 0 elsewhere, we obtain that (x^*, y^*) is a Nash equilibrium if and only if $\forall i \in [n]$, $x^{*T} R y^* \geq e_i^T R y^*$ and $x^{*T} C y^* \geq x^{*T} C e_i$, (see, for instance, [9]).

Thanks to this characterization, it is possible to formulate 2-Nash as a quadratic program with $2n$ variables and $4n+2$ constraints as shown in Figure 1.

$$\begin{array}{llr}
NLP: & \min 1 & \\
s.t. & \sum_{i\in[n]}\sum_{j\in[n]} R_{ij}x_i y_j \geq \sum_{j\in[n]} R_{ij} y_j & \forall i \in [n], \\
 & \sum_{i\in[n]}\sum_{j\in[n]} C_{ij}x_i y_j \geq \sum_{i\in[n]} C_{ij} x_i & \forall j \in [n], \\
 & \sum_{i\in[n]} x_i = 1, & \\
 & \sum_{j\in[n]} y_j = 1, & \\
 & x_i \geq 0 & \forall i \in [n], \\
 & y_j \geq 0 & \forall j \in [n].
\end{array}$$

Fig. 1. A quadratic formulation for 2-Nash

It is quite easy to see that NLP is an exact, although non-linear, formulation of 2-Nash. In fact, the last four constraints assure that the pair of vectors (x, y) constitutes a pair of mixed strategies, while the first two constraints coincide with the characterization of Nash equilibria given above.

The linear relaxation. We now propose a linear relaxation of NLP having $n^2 + 2n$ variables and $n^2 + 6n + 2$ constraints as shown in Figure 2.

$$
\begin{array}{llll}
LR: & & \min 1 & \\
s.t. & \sum_{i\in[n]}\sum_{j\in[n]} R_{ij} z_{ij} \geq \sum_{j\in[n]} R_{ij} y_j, & & \forall i \in [n] \\
& \sum_{i\in[n]}\sum_{j\in[n]} C_{ij} z_{ij} \geq \sum_{i\in[n]} C_{ij} x_i, & & \forall j \in [n] \\
& \sum_{j\in[n]} z_{ij} = x_i & & \forall i \in [n], \\
& \sum_{i\in[n]} z_{ij} = y_j & & \forall j \in [n], \\
& \sum_{i\in[n]} x_i = 1, & & \\
& \sum_{j\in[n]} y_j = 1, & & \\
& x_i \geq 0 & & \forall i \in [n], \\
& y_j \geq 0 & & \forall j \in [n], \\
& z_{ij} \geq 0 & & \forall i, j \in [n].
\end{array}
$$

Fig. 2. The linear relaxation for 2-Nash

As it can be easily noted, in LR we use the variables z_{ij} to approximate the values $x_i y_j$ occurring in NLP. The third and fourth constrains, together with $z_{ij} \geq 0$, guarantee that $z_{ij} \leq \min\{x_i, y_j\}$. In fact $x_i, y_j \in [0; 1]$ implies $x_i y_j \leq \min\{x_i, y_j\}$. It is not difficult to see that any Nash equilibrium is a feasible solution of LR, hence LR is a relaxation of NLP. Since z_{ij} only provides an approximation of $x_i y_j$, we have that the payoffs yielded by each pair of mixed strategies (x, y) corresponding to feasible solutions of LR may be over or underestimated. Let us denote with $feas(LR)$ the set of feasible solutions of a certain linear relaxation LR, we show in the following that underestimation can never occur.

Lemma 1. *For any* $s = (x, y, z) \in feas(LR)$ *it holds* $\sum_{i\in[n]}\sum_{j\in[n]} R_{ij}(z_{ij} - x_i y_j) \geq 0$ *and* $\sum_{i\in[n]}\sum_{j\in[n]} C_{ij}(z_{ij} - x_i y_j) \geq 0$.

Proof. Let $k = \text{argmax}_{i\in[n]}\{\sum_{j\in[n]} R_{ij} y_j\}$ be a pure best response for the row player to the column player's mixed strategy y. Because of the first constraint of LR we have

$$\sum_{i\in[n]}\sum_{j\in[n]} R_{ij} z_{ij} \geq \sum_{j\in[n]} R_{kj} y_j.$$

Moreover, we also have

$$\sum_{i\in[n]}\sum_{j\in[n]} R_{ij} x_i y_j \leq \sum_{i\in[n]} x_i \sum_{j\in[n]} R_{kj} y_j = \sum_{j\in[n]} R_{kj} y_j.$$

Combining these two inequalities we obtain

$$\sum_{i\in[n]}\sum_{j\in[n]} R_{ij} z_{ij} \geq \sum_{i\in[n]}\sum_{j\in[n]} R_{ij} x_i y_j.$$

A completely symmetric argument shows that $\sum_{i\in[n]}\sum_{j\in[n]} C_{ij}z_{ij} \geq \sum_{i\in[n]}\sum_{j\in[n]} C_{ij}x_iy_j$. □

For any $s = (x, y, z) \in feas(LR)$ define

$$\epsilon(s) = \max\{\sum_{i\in[n]}\sum_{j\in[n]} R_{ij}(z_{ij} - x_iy_j), \sum_{i\in[n]}\sum_{j\in[n]} C_{ij}(z_{ij} - x_iy_j)\}.$$

Lemma 2. *For any $s = (x, y, z) \in feas(LR)$ we have that the pair of mixed strategies (x, y) is an $\epsilon(s)$-approximate Nash equilibrium.*

Proof. For any $k \in [n]$ we have that

$$\sum_{i\in[n]}\sum_{j\in[n]} R_{ij}x_iy_j + \epsilon(s) \geq \sum_{i\in[n]}\sum_{j\in[n]} R_{ij}z_{ij} \geq \sum_{j\in[n]} R_{kj}y_j.$$

Analogously, for any $k \in [n]$ we have that

$$\sum_{i\in[n]}\sum_{j\in[n]} C_{ij}x_iy_j + \epsilon(s) \geq \sum_{i\in[n]}\sum_{j\in[n]} C_{ij}z_{ij} \geq \sum_{i\in[n]} C_{ik}x_i.$$

Hence, (x, y) is an $\epsilon(s)$-approximate Nash equilibrium. □

Clearly, if $\epsilon(s) = 0$, the pair of mixed strategies (x, y) induced by s is a Nash equilibrium. We now show one of the two results of the paper, that is the fact that our formulation can be used to compute any Nash equilibria for regular bimatrix games satisfying certain desiderata.

Theorem 1. *Any feasible solution of LR is a Nash equilibrium when (R, C) is a regular bimatrix game.*

Proof. For any $s = (x, y, z) \in feas(LR)$, we show that $\epsilon(s) = 0$ when (R, C) is a row regular bimatrix game. The proof for column regular games can be obtained by switching the role of rows and columns in what follows. Assume that $\epsilon(s) > 0$, otherwise, by Lemma 2, (x, y) is a Nash equilibrium. Consider a row i such that $\sum_{j\in[n]} R_{ij}z_{ij} - \sum_{j\in[n]} R_{ij}x_iy_j = \delta_i(s) > 0$, that is, such that the payoff of the row player is overestimated in s. By the row regularity of (R, C) we obtain that $\sum_{j\in[n]} C_{ij}z_{ij} - \sum_{j\in[n]} C_{ij}x_iy_j = \sum_{j\in[n]}(\bar{b}_i - R_{ij})(z_{ij} - x_iy_j) = \bar{b}_i\sum_{j\in[n]}(z_{ij} - x_iy_j) - \sum_{j\in[n]} R_{ij}(z_{ij} - x_iy_j) = \bar{b}_i(\sum_{j\in[n]} z_{ij} - x_i) - \sum_{j\in[n]} R_{ij}(z_{ij} - x_iy_j) = -\delta_i(s)$. Hence, for each row on which the payoff of the row player is overestimated in s by a quantity δ, we have that the column player gets an underestimation equal to δ. Because of the fact that none of the two players can have an underestimated payoff in s, it follows that $\epsilon(s) = 0$, that is, (x, y) is a Nash equilibrium. □

As a consequence of the above theorem, we have that any Nash equilibrium satisfying any desiderata and optimizing any objective function involving the players' payoffs (modelled through the variables z_{ij}) which can be expressed by linear programming can be computed in polynomial time by using our relaxation LR. Notice that a similar result cannot be obtained by using the algorithm for computing a Nash equilibrium in zero-sum games since in this case all Nash equilibria are equivalent, in the sense that they all yield the same payoffs, that is, when transforming a regular bimatrix game into a zero-sum game we maintain the same set of equilibria but lose information on their payoffs. For the sake of clarity and completeness, we illustrate this discussion in the following example.

Example 1. Consider the column regular bimatrix game defined by the following bimatrix.

$$\begin{pmatrix} 2,1 & 1,-1 & 0,0 & 0,0 \\ 4,-1 & -1,1 & 0,0 & 0,0 \\ 3,0 & 0,0 & 1,-1 & -1,1 \\ 3,0 & 0,0 & -1,1 & 1,-1 \end{pmatrix}$$

It possesses four Nash equilibria:

1. $(\frac{1}{2},\ \frac{1}{2},\ 0,\ 0)$ and $(\frac{1}{2},\ \frac{1}{2},\ 0,\ 0)$ yielding the pair of payoffs $(\frac{3}{2},\ 0)$,
2. $(\frac{1}{2},\ \frac{1}{2},\ 0,\ 0)$ and $(0,\ 0,\ \frac{1}{2},\ \frac{1}{2})$ yielding the pair of payoffs $(0,\ 0)$,
3. $(0,\ 0,\ \frac{1}{2},\ \frac{1}{2})$ and $(\frac{1}{2},\ \frac{1}{2},\ 0,\ 0)$ yielding the pair of payoffs $(\frac{3}{2},\ 0)$,
4. $(0,\ 0,\ \frac{1}{2},\ \frac{1}{2})$ and $(0,\ 0,\ \frac{1}{2},\ \frac{1}{2})$ yielding the pair of payoffs $(0,\ 0)$.

If, for instance, we are interested in computing a Nash equilibrium maximizing the sum of the players' payoffs, by adding the suitable objective function to our formulation LR we will obtain equilibrium number 1 or equilibrium number 3. Analogously, if we want to minimize the players' payoffs we are able to compute one between equilibrium number 2 and equilibrium number 4.

By using the technique of reducing a regular game to a zero-sum game, by subtracting 3 from the first entry of the first column, we obtain the zero-sum game defined by the following bimatrix.

$$\begin{pmatrix} -1,1 & 1,-1 & 0,0 & 0,0 \\ 1,-1 & -1,1 & 0,0 & 0,0 \\ 0,0 & 0,0 & 1,-1 & -1,1 \\ 0,0 & 0,0 & -1,1 & 1,-1 \end{pmatrix}$$

Its set of Nash equilibria remains the same, but all of them now yield the pair of payoffs $(0,\ 0)$. Hence, by using the techniques known so far in the literature, we are unable to choose a particular equilibrium among the various ones (eventually) possessed by a regular bimatix games. This illustrates the novelty and usefulness of our contribution in the computation of exact equilibria in bimatrix games.

As a consequence of Lemma 2, we have that the approximation guarantee provided by LR in general bimatrix games can be equal to $\max_{s \in feas(LR)} \epsilon(s)$ in the worst-case. We now show the second of our results, stating that our algorithm computes an α-approximation Nash equilibrium for any α-smoothed bimatrix game.

Theorem 2. *If (R, C) is α-smoothed then any $s \in feas(LR)$ is an α-approximate Nash equilibrium.*

Proof. For any $i \in [n]$ define $\underline{R}_i = \min_{j\in[n]}\{R_{ij}\}$ and $\overline{R}_i = \max_{j\in[n]}\{R_{ij}\}$, while for any $j \in [n]$ define $\underline{C}_j = \min_{i\in[n]}\{C_{ij}\}$ and $\overline{C}_j = \max_{i\in[n]}\{C_{ij}\}$.

It is quite easy to see that

$$\sum_{i\in[n]}\sum_{j\in[n]} R_{ij}z_{ij} \geq \sum_{i\in[n]} x_i \cdot \underline{R}_i$$

and

$$\sum_{i\in[n]}\sum_{j\in[n]} R_{ij}x_iy_j \leq \sum_{i\in[n]} x_i \cdot \overline{R}_i.$$

Analogously, $\sum_{i\in[n]}\sum_{j\in[n]} C_{ij}z_{ij} \geq \sum_{j\in[n]} y_j \cdot \underline{C}_j$ and $\sum_{i\in[n]}\sum_{j\in[n]} C_{ij}x_iy_j \leq \sum_{j\in[n]} y_j \cdot \overline{C}_j$.

The claim follows from the definitions of $\epsilon(s)$ and α-smoothness. □

Thus, for α-smoothed bimatrix games with $\alpha < 0.3393$, our technique allows to compute approximate Nash equilibria with the currently best provable worst-case guarantee. For the sake of completeness, we show in the following theorem that good approximations cannot be achieved in the general case, by providing a win-lose game (which is clearly a 1-smoothed bimatrix game) for which our algorithm might return a 1-approximate Nash equilibrium.

Theorem 3. *There exists a bimatrix game such that* $\max_{s\in feas(LR)} \epsilon(s) = (1 - \frac{3}{n} + \frac{2}{n^2})$.

Proof. The instance giving the result is the following.

$$\begin{pmatrix}
1,0 & 0,1 & 1,0 & 1,0 & \ldots\ldots & 1,0 & 1,0 \\
 & 1,0 & 0,1 & & & & \\
 & & 1,0 & 0,1 & & & \\
 & & & 1,0 & & & \\
 & & & & \vdots & & \\
 & & & & & 0,1 & \\
 & & & & & 1,0 & 0,1 \\
0,1 & & & & & & 1,0
\end{pmatrix}$$

In every cell (i, j) of the bimatrix, the payoffs (R_{ij}, C_{ij}) are shown. We assume a payoff of $(0, 0)$ for the empty cells. Let us consider the solution $s = (x, y, z)$ of LR, defined as follows:

- $x_i = y_i = \frac{1}{n}$ for any $1 \leq i \leq n$,
- $z_{ij} = \begin{cases} \frac{1}{n} - \frac{1}{n^2} & \text{if } i = j, \\ \frac{1}{n^2} & \text{if } j = (i+1)\ mod\ n, \\ 0 & \text{otherwise.} \end{cases}$

We show that $s \in feas(LR)$ for any $n > 2$. In fact,

$$\sum_{i=1}^{n}\sum_{j=1}^{n} R_{ij}z_{ij} = n\left(\frac{1}{n} - \frac{1}{n^2}\right) = 1 - \frac{1}{n}$$

while

$$\sum_{j=1}^{n} R_{ij}y_j = \begin{cases} 1 - \frac{1}{n} & \text{if } i = 1, \\ \frac{1}{n} & \text{if } 2 \leq i \leq n. \end{cases}$$

Moreover

$$\sum_{i=1}^{n}\sum_{j=1}^{n} C_{ij}z_{ij} = n\frac{1}{n^2} = \frac{1}{n},$$

while

$$\sum_{i=1}^{n} C_{ij}x_i = \frac{1}{n} \text{ for any } 1 \leq j \leq n.$$

The expected payoff of the row player in (x, y) is $\sum_{i=1}^{n}\sum_{j=1}^{n} R_{ij}x_iy_j = 2(n-1)\frac{1}{n^2} = 2(\frac{1}{n} - \frac{1}{n^2})$. Since, for any $n > 2$, deviating to strategy $x_1 = 1$ gives the row player an expected payoff equal to $(n-1)\frac{1}{n} = 1 - \frac{1}{n}$, we have that (x, y) is a $(1 - \frac{3}{n} + \frac{2}{n^2})$-approximate Nash equilibrium. □

References

1. Abbott, T., Kane, D., Valiant, P.: On the Complexity of Two-player Win-lose Games. In: Proc. of the 46th Annual IEEE Symposium on Fondations of Computer Science (FOCS), pp. 113–122. IEEE Computer Society, Los Alamitos (2005)
2. Addario-Berry, L., Olver, N., Vetta, A.: A Polynomial Time Algorithm for Finding Nash Equilibria in Planar Win-Lose Games. Journal of Graph Algorithms and Applications 11(1), 309–319 (2007)
3. Bosse, H., Byrka, J., Markakis, E.: New Algorithms for Approximate Nash Equilibria in Bimatrix Games. In: Deng, X., Graham, F.C. (eds.) WINE 2007. LNCS, vol. 4858, pp. 17–29. Springer, Heidelberg (2007)
4. Chen, X., Deng, X.: Settling the Complexity of 2-Player Nash Equilibrium. In: Proc. of the 47th Annual IEEE Symposium on Foundations of Computer Science (FOCS), pp. 252–261. IEEE Computer Society, Los Alamitos (2006)
5. Chen, X., Deng, X., Teng, S.H.: Computing Nash Equilibria: Approximation and Smoothed Complexity. In: Proc. of the 47th Annual IEEE Symposium on Foundations of Computer Science (FOCS), pp. 603–612. IEEE Computer Society, Los Alamitos (2006)
6. Chen, X., Teng, S., Valiant, P.: The Approximation Complexity of Win-Lose Games. In: Proc. of the 18th Annual ACM-SIAM Symposium on Discrete Algorithms (SODA), pp. 159–168. SIAM, Philadelphia (2007)
7. Codenotti, B., Leoncini, M., Resta, G.: Efficient Computation of Nash Equilibria for Very Sparse Win-Lose Bimatrix Games. In: Azar, Y., Erlebach, T. (eds.) ESA 2006. LNCS, vol. 4168, pp. 232–243. Springer, Heidelberg (2006)

8. Daskalakis, K., Goldberg, P.W., Papadimitriou, C.H.: The Complexity of Computing a Nash Equilibrium. In: Proc. of the 38th Annual ACM Symposium on Theory of Computing (STOC), pp. 71–78. ACM Press, New York (2006)
9. Daskalakis, K., Mehta, A., Papadimitriou, C.H.: A Note on Approximate Nash Equilibria. In: Spirakis, P.G., Mavronicolas, M., Kontogiannis, S.C. (eds.) WINE 2006. LNCS, vol. 4286, pp. 297–306. Springer, Heidelberg (2006)
10. Daskalakis, K., Mehta, A., Papadimitriou, C.H.: Progress in Approximate Nash Equilibria. In: Proc. of the 8th ACM Conference on Electronic Commerce (EC), pp. 355–358. ACM Press, New York (2007)
11. Daskalakis, K., Papadimitriou, C.H.: Three-Player Games Are Hard. In: Electronic Colloquium on Computational Complexity (ECCC), vol. (139) (2005)
12. Feder, T., Nazerzadeh, H., Saberi, A.: Personal Communication
13. Kontogiannis, S.C., Panagopoulou, P.N., Spirakis, P.G.: Polynomial Algorithms for Approximating Nash Equilibria of Bimatrix Games. In: Spirakis, P.G., Mavronicolas, M., Kontogiannis, S.C. (eds.) WINE 2006. LNCS, vol. 4286, pp. 286–296. Springer, Heidelberg (2006)
14. Lemke, C., Howson, J.: Equilibrium Points of Bimatrix Games. Journal of the Society of Industrial and Applied Mathematics 12, 413–423 (1964)
15. Lipton, R., Markakis, E., Mehta, A.: Playing Large Games Using Simple Strategies. In: Proc. of the 4th ACM Conference on Electronic Commerce (EC), pp. 36–41. ACM Press, New York (2003)
16. Porter, R., Nudelman, E., Shoham, Y.: Simple Search Methods for Finding a Nash Equilibrium. In: Proc. of the 19th National Conference on Artificial Intelligence (AAAI), pp. 664–669. MIT Press, Cambridge (2004)
17. Sandholm, T., Gilpin, A., Conitzer, V.: Mixed-Integer Programming Methods for Finding Nash Equilibria. In: Proc. of the 20th National Conference on Artificial Intelligence (AAAI), pp. 495–501. MIT Press, Cambridge (2005)
18. Savani, R., von Stengel, B.: Exponentially Many Steps for Finding a Nash Equilibrium in a Bimatrix Game. In: Proc. of the 45th Annual IEEE Symposium on Foundations of Computer Science (FOCS), pp. 258–267. IEEE Computer Society, Los Alamitos (2004)
19. Tsaknakis, H., Spirakis, P.: An Optimization Approach for Approximate Nash Equilibria. In: Deng, X., Graham, F.C. (eds.) WINE 2007. LNCS, vol. 4858, pp. 42–56. Springer, Heidelberg (2007)
20. Tsaknakis, H., Spirakis, P., Kanoulas, D.: Performance Evaluation of a Descent Algorithm for Bimatrix Games. In: Papadimitriou, C., Zhang, S. (eds.) WINE 2008. LNCS, vol. 5385, pp. 222–230. Springer, Heidelberg (2008)

Where Would Refinancing Preferences Go?

Yajun Chai and Bo Liu*

School of Management and Economics,
University of Electronic Science and Technology of China
No.4, Section 2, North Jianshe Rd., Chengdu 610054, China
Liubor@gmail.com

Abstract. We study the relation between the non-tradable shares reform and the refinancing preferences. From the viewpoints of change in market and policy environments led by the reform, we find that right issues dominate before the reform, however, public offerings (including private placement) dominate after reform, which could be attributed to more money encirclement induced by the shift of the public offering mechanism from in discount to in premium after reform and no requirements for large shareholders' participation commitments in public offerings.

Keywords: non-tradable shares reform; refinancing preferences; refinancing.

1 Introduction

In China stock market, the listed companies could refinance, i.e., further raise the capital after IPO (Initial Public Offering), mainly via convertible bonds and Seasoned Equity Offerings (SEOs) which includes public offerings, private placement and right issues. As a fundamental institutional reform in China stock market, where refinancing preferences would go after the non-tradable shares reform has straightforwardly attracted much attention from the academics and policy makers.

The extant literatures about refinancing preferences mainly focus on following three dimensions: equity refinancing preference, operating performances of exiting refinancing ways, and market reaction to refinancing announcements.

In the first dimension, Zhang (2005) [1] point out that, different from other mature markets, the financing order of "internal financing priority, followed by debt financing, then equity financing" is significant in China stock market. Equity refinancing of listed companies in China share a great scale. Zhang (2005) also uses methods of theoretic analysis and statistical analysis to verify the equity refinancing prejudice of China's listed companies. Specifically, he constructs a model on the equity refinancing and the model shows that there exists wealth transfer of tradable shareholders to non-tradable shareholders during rights issue and seasoned equity offering in China, therefore the non-tradable shareholders can acquire riskless high

* Correspondence author.

B. Chen (Ed.): AAIM 2010, LNCS 6124, pp. 70–77, 2010.

abnormal returns quickly (the book value per share increases); in addition, the managers can obtain positive utility through rights issue and seasoned equity offering. Therefore, the controllers have prejudice on the equity refinancing since the firms are usually controlled by the managers or the non-tradable shareholders; and they issue convertible bonds when there have good investment opportunities; furthermore the statistical results on the samples in 2001 find that the firms have prejudice on the equity refinancing, and it is uncorrelated with the market condition. Wu and Ruan (2004) [2] empirically researches the preference of equity financing of China's listed companies and conclude that there is no distinctive evidence between industry location, competition structure and financing behavior, but have positive correlation between share concentration ratio, the structure of split shares and financing behavior. At the meantime, from the viewpoint of cost and return of financing, their test results show that under the special market environment in China, not only the cost of equity financing is much lower than debt financing, but also the equity financing makes hugu return for non-circulating shareholders. The huge return made by the structure if split share is the key factor of financing behavior of China's listed companies.

In the second dimension, Du and Wang (2006) [3] investigates the changes and causes of operating performance of China's firms conducting rights issues. They documented that, rights issuers experience a sharp, statically significant decrease in post-offering operating performance; from the pre-offering years to the post-offering year 2, the operating performance of issuing firms is higher than that of the non-issuer matched by industry, asset size, and operating performance, however, in the year 3 following the offering, the operating performance of issuing firms is lower than that of the non- issuer. In the years after the offering, the median issuer's operating performance rapidly deteriorates relative to the non - issuer. Consistent with the implication of the free cash flow hypothesis, the decline in post-offering operating performance is greater for issuing firms that have higher pre-offering free cash flow, and issuing firms with high ownership concentration that invest more in new assets perform better than issuing firms with low ownership concentration that do less. Among rights equity issuers, firms with more growth opportunities have larger post-offering performance declines. Finally, firm size appears to have negative impact on the post-offering performance for issuers. Li et al. (2008) [4] selects 495 refinancing firms from 1999 to 2004 to examine the performance evaluation of listed companies in China. From demonstration analysis, they find that performance of refinancing company decreased obviously in the first six years after refinancing, and there are huge differences on performance in different years; the methods of refinancing greatly affects performance: convertible bond is the best, public offerings is following and right issues is the last; after refinancing, capital debt ration increased and performance decreased; the total capital scale of listed company is not related to the performance. Deng (2007) [5] compares the characteristics of equity refinancing, reviewed the history of the development process in various refinancing ways, and analyzes the motivations and the refinancing of listed companies and its impact on the capital market. He (2007) [6] investigates the market performance to the public offerings and right issues, and finds that the market performance are very low both during and after the non-tradable shares reform, thus the ownership structure is not the root cause of low performance.

In the third dimension, the latest literatures are mostly concerned about the market reaction to the innovative ways to refinance. As one of the most important innovative ways to refinance, private placement is launched soon by the academic attention. Liu et al. (2003) [7] empirically study the market reaction to seasoned equity offerings (SEO) and the underlying reasons in Chinese share market. They find that SEOs caused significant negative abnormal returns on announcement days, but the abnormal returns cannot be explained by offering size, financial leverage or earning prospect, as was expected by classical financial theories. They put forward a duality ownership structure theory of Chinese listed companies and point out that under this structure, non-liquid equity owners can make use of their controlling power to "make money" by issuing new "liquid power" and thus cause negative stock price effect . At the same time, they analyze some deficiency in samples and methodology of previous works.

From above, we could easily see that few of the extant researches have put the refinancing preferences into the big picture of the institutional background of China stock market. Actually, it is well-known that the non-tradable-share is the unique characteristics of China stock market. Before the reform, non-tradable shares in China stock market occupy almost 2/3 of the total issued shares and mainly consist of state-owned and legal person shares. The original rationale for those non-tradable shares was to keep the property of the state-owned enterprises. It has been generally accepted among the Chinese investors that it is just this institutional defect that has contributed many anomalies which would not emerge in western mature markets and lowered down the entire development of China stock market. To improve the health of the China financial system, the government finally made the decision to clear this huge overhang. On April 29, 2005, with the guidelines of the Circular on Issues relating to Pilot Reform of Listed Companies' Non-Tradable Shares by China Securities Regulatory Commission (CSRC), the Pilot Reform of non-tradable shares initialized, which stipulates that non-tradable shareholders have to bargain with tradable shareholders for compensation for gaining liquidity. By the end of the year of 2009, almost all of the listed companies of China stock market have successfully accomplished the reform. But what has the reform really brought to the refinancing preferences has not yet been deeply addressed.

Hence, taking the realities of China stock market into account, from the viewpoints of change in environments of the market and policy and in behaviors of the firms and investors led by the reform, we study how the reform would affect refinancing preferences, thus contribute to the literatures both on non-tradable shares reform and refinancing preferences areas.

The remainder of the paper is organized as follows. Section 2 shows the dynamics of refinancing preferences during 2001-2009. Section 3 compares the refinancing preferences before and after the non-tradable reform, and Section 4 concludes.

2 Dynamics of Refinancing Preferences during 2001-2009

Table 1 and 2 summarize the refinancing announcements (count 1 for one firm with one announcement date) and projects (count 1 for one firm with one project) during year of 2001-2009 respectively.

Table 1. Number of refinancing announcements in three refinancing preferences categories

Year	CB	PO	RI	Total
2001	NA	22	96	118
2002	5	28	20	53
2003	15	15	23	53
2004	11	11	22	44
2005	NA	4	1	5
2006	7	54	28	63
2007	14	158	7	179
2008	15	122	8	145
2009	NA	30	NA	30

Note: NA denotes Not Available; CB denotes Convertible Bond; PO denotes public offerings; RI denotes Right Issues.

Table 2. Number of refinancing projects in three refinancing preferences categories

Year	CB	PO	RI	Total
2001	NA	183	652	835
2002	39	164	115	318
2003	60	109	103	272
2004	52	55	77	184
2005	NA	12	1	13
2006	38	166	8	212
2007	48	450	36	534
2008	52	372	28	452
2009	NA	80	NA	80

Fig.1 and 2 better exhibits the dynamics of refinancing announcements and projects in three refinancing preferences during year of 2001 to 2009 demonstrated in Table 1 and 2.

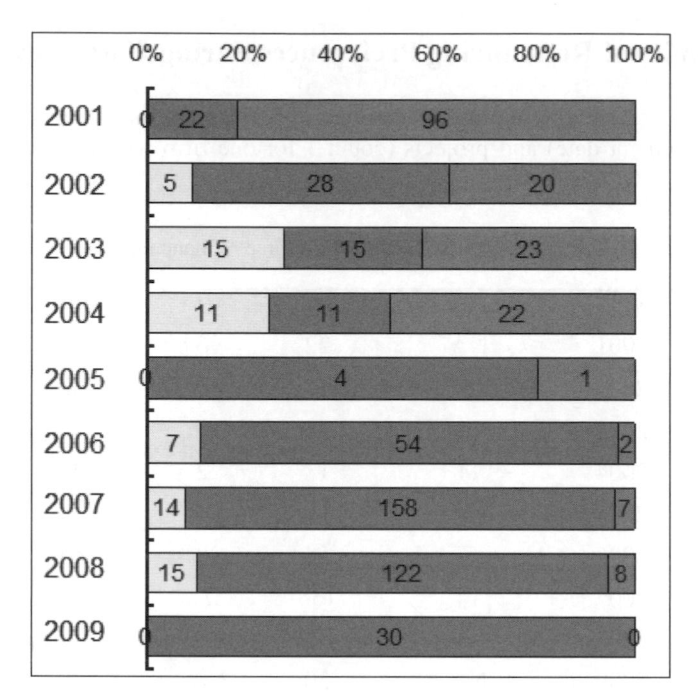

Fig. 1. Dynamics of refinancing announcements in three refinancing preferences. It denotes Convertible Bond, Public Offerings, and Right Issues from the left to the right.

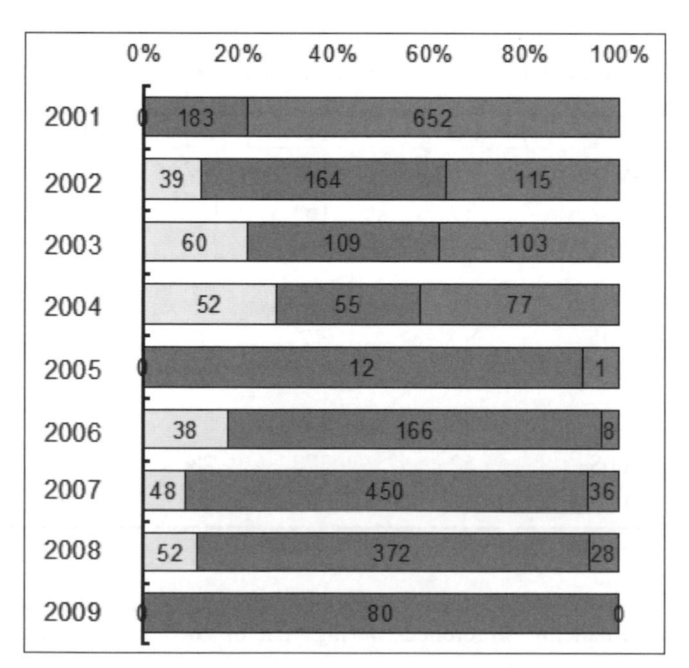

Fig. 2. Dynamics of refinancing projects in three refinancing preferences. It denotes Convertible Bond, Public Offerings, and Right Issues from the left to the right.

It can be seen from Table 1 and 2, Fig.1 and 2 that, before 2006 (almost right before the time that the reform was initialized), public offerings and right issues were dominant, and their refinancing announcements and projects were almost equal. However, after 2006 (almost right after the time that the reform was initialized), public offerings dominate others. In addition, except Not Available in 2001, 2005 and 2009, there were no significant differences in between the years for convertible bond.

We notice that the dynamics of the refinancing preferences nearly match the process of the non-tradable reform. It is straightforward to ask what role the reform plays in the dynamics? Thus, we divide the sample from Jan.2, 2001 to Jun. 4, 2009 into two sub-samples of before and after the reform for further analysis. Taking the time heterogeneity of each firm's reform into account, we define the criteria for determination of before and after reform as follows: "after reform" if the announcement dates are after resumption date of the reform; "before reform" otherwise.

3 Comparison of Refinancing Preferences before and after the Non-tradable Reform

From Table 3 and 4, we find that the right issues dominate before the reform while the public offerings dominate after the reform. There are no significant changes in convertible bond.

We therefore formally explain the pattern above from the viewpoints of change in market and policy environments led by the reform to explore the relation between the non-tradable shares reform and the refinancing preferences. Before the reform, right issues and public offerings are in very strict examination for approval by China Securities Regulatory Commission (CSRC). However, the May 8, 2006 version of "The Issuance of Securities by Listed Companies" was a latest specification of important

Table 3. Number of refinancing announcements in three refinancing preferences categories before and after the reform

Categories	CB	PO	RI	Total
Whole sample	67	444	179	690
Before reform	31	98	162	291
After reform	36	346	17	399

Table 4. Number of refinancing projects in three refinancing preferences categories before and after the reform

Categories	CB	PO	RI	Total
Whole sample	289	1591	1020	2900
Before reform	151	545	948	1644
After reform	138	1046	72	1256

documents for refinancing, intentionally according to the fundamental change on the market mechanism brought by the reform. The general provisions for listed companies' securities issuance include: sound organization, good running, persistent profit capacities (such as be profitable in the last three fiscal years), good financial condition, and no false financial and accounting records within last 30 months, raising funds not exceeding the amount of project requirements, establishing special funds storage system. In particular, for preparation by three traditional refinancing ways of public offerings, right issues, convertible bonds, some additional and specific conditions are also required.

From the May 8, 2006 version of "The Issuance of Securities by Listed Companies", we can see that the financial situation requirements for right issues of listed companies were reduced: firstly, abolish the "the weighted average annual rate of 6% return on net assets within last three years" restrictions and only require "be profitable in last three fiscal years"; secondly, require controlling shareholders make the public commitment for the number of shares in right issues in order to provide a decision-making reference for other outside minority investors; thirdly, place a "way of the Securities Act of consignment issue," and cited failure of the system, that is, "at the consignment deadline, if amount of shares to be allotted by the original shareholders have not reached 70%, relevant issuer shall issue price plus interest rate on bank deposits for the same return to shareholders who has been subscribed. ", which changed the underwriting by the underwriters in the past with "strong sell" status, giving investors the right to vote really thus helping refinancing regulation by market means to reflect market demand and promote companies create a more rational idea towards right issues.

After the reform, "the Issuance of Securities by Listed Companies" additionally require "issue price shall be not less than the average price within 20 days before the prospectus announcement shares or average price on the previous trading day", which shows the transition of the pricing mechanism of public offerings from discounts to the premium. Premium public offerings not only help listed companies make raise more funds to support their development needs, but also help to protect the interests of existing shareholders of tradable shares and boost investor confidence, and to the investors who are promising with the refinancing projects and the company's future prospects. Under premium public offerings, if the investors are pessimistic towards the refinancing projects and the company's future prospects, the issuances will fail. Thus, the major shareholders and the underwriters will have to carefully consider the introduction of public offerings, which to some extent form a restraint mechanism of the market for the large shareholders and help to inhibit the large shareholders' motives of "money-enclosure".

In addition to raising more funds brought by premium public offerings, no large shareholders' commitments are required to participate in the public offerings is also one of the reasons for preferences in public offerings. Either from changes in the trend by year (as show in Table 1 and 2, also Fig.1 and 2) or the summary statistics before and after the reform (as show in Table 3 and 4), the public offerings has become the preferred way of refinancing. In addition, the private placement also contributes the dominate positions of public offerings. The private placement is an innovation of refinancing after the reform. Compared with the public offerings, the private placement lowers the requirements for issuance. Meanwhile, due to the introduction of strategic investors and other actions to achieve strategic intent, private placement has been very popular after the reform.

4 Conclusions

We study the relation between the non-tradable shares reform and the refinancing preferences. From the viewpoints of change in market and policy environments led by the reform, by taking the realities of China stock market into account we find that right issues dominate before the reform, however, public offerings (including private placement) dominate after reform, which could be attributed to more money encirclement induced by the shift of the public offering mechanism from in discount to in premium after reform and no requirements for large shareholders' participation commitments in public offerings.

References

1. Zhang, W.: Equity Refinancing Behavior and Efficiency Analysis of China's listed Companies with the Split Share Structure. Ph.D Dissertation, Huazhong University of Science and Technology (2005)
2. Wu, J., Ruan, T.: The structure of split shares and the preference of equity financing of China's listed companies. Journal of Financial Research 6, 56–67 (2004)
3. Du, M., Wang, L.: Operating performance of China's listed firms following rights issues: Causes and consequences. Management World 3, 114–121 (2006)
4. Li, L., Liao, L., Zhou, H.: On the Refinancing Performance Evaluation of Listed Companies in China. Reformation & Strategy 4, 74–78 (2008)
5. Deng, Q.: Comparative Research on the financing behavior of listed company before and after the reform of non-tradable shares. Master thesis, Southwestern University of Finance and Economics (2007)
6. He, Q.: Investigation of the behavior and market-place achievement about listed companies after the reform of split share structure in China. Master thesis, Sichuan University (2007)
7. Liu, L., Wang, D., Wang, Z.: Market reaction to seasoned equity offerings (SEO) and duality ownership structure theory of Chinese listed companies. Journal of Financial Research 8, 60–71 (2003)

Approximating Maximum Edge 2-Coloring in Simple Graphs

Zhi-Zhong Chen, Sayuri Konno, and Yuki Matsushita

Department of Mathematical Sciences, Tokyo Denki University, Hatoyama, Saitama 350-0394, Japan
zzchen@mail.dendai.ac.jp

Abstract. We present a polynomial-time approximation algorithm for legally coloring as many edges of a given simple graph as possible using two colors. It achieves an approximation ratio of roughly 0.842 and runs in $O(n^3 m)$ time, where n (respectively, m) is the number of vertices (respectively, edges) in the input graph. The previously best ratio achieved by a polynomial-time approximation algorithm was $\frac{5}{6} \approx 0.833$.

Keywords: Approximation algorithms, graph algorithms, edge coloring, NP-hardness.

1 Introduction

Given a graph G and a natural number t, the *maximum edge t-coloring problem* (called Max Edge t-Coloring for short) is to find a maximum-sized set F of edges in G such that F can be partitioned into at most t matchings of G. Motivated by call admittance issues in satellite based telecommunication networks, Feige et al. [3] introduced the problem and proved its APX-hardness. They also observed that Max Edge t-Coloring is a special case of the well-known maximum coverage problem (see [6]). Since the maximum coverage problem can be approximated by a greedy algorithm within a ratio of $1 - (1 - \frac{1}{t})^t$ [6], so can Max Edge t-Coloring. In particular, the greedy algorithm achieves an approximation ratio of $\frac{3}{4}$ for Max Edge 2-Coloring, which is the special case of Max Edge t-Coloring where the input number t is fixed to 2. Feige et al. [3] has improved the trivial ratio $\frac{3}{4} = 0.75$ to $\frac{10}{13} \approx 0.769$ by an LP approach.

The APX-hardness proof for Max Edge t-Coloring given by Feige et al. [3] indeed shows that the problem remains APX-hard even if we restrict the input graph to a simple graph and fix the input integer t to 2. We call this restriction (special case) of the problem Max Simple Edge 2-Coloring. Feige et al. [3] also pointed out that for Max Simple Edge 2-Coloring, an approximation ratio of $\frac{4}{5}$ can be achieved by the following *simple algorithm*: Given a simple graph G, first compute a maximum-sized subgraph H of G such that the degree of each vertex in H is at most 2 and there is no 3-cycle in H, and then remove one *arbitrary* edge from each odd cycle of H. This simple algorithm has been improved in [1,2,9]. The previously best ratio (namely, $\frac{5}{6}$) was given in [9]. In this

B. Chen (Ed.): AAIM 2010, LNCS 6124, pp. 78–89, 2010.

paper, we improve on both the algorithm in [1] and the algorithm in [9] to obtain a new approximation algorithm that achieves a ratio of roughly 0.842. Roughly speaking, our algorithm is based on local improvement, dynamic programming, and recursion. Its analysis is based on an intriguing charging scheme and certain structural properties of *caterpillar* graphs and *starlike* graphs (see Section 3 for definitions).

Kosowski et al. [10] also considered MAX SIMPLE EDGE 2-COLORING. They presented an approximation algorithm that achieves a ratio of $\frac{28\Delta-12}{35\Delta-21}$, where Δ is the maximum degree of a vertex in the input simple graph. This ratio can be arbitrarily close to the trivial ratio $\frac{4}{5}$ because Δ can be very large. In particular, this ratio is worse than our new ratio 0.842 when $\Delta \geq 4$. Moreover, when $\Delta = 3$, our algorithm indeed achieves a ratio of $\frac{6}{7}$, which is equal to the ratio $\frac{28\Delta-12}{35\Delta-21}$ achieved by Kosowski et al.'s algorithm [10]. Note that MAX SIMPLE EDGE 2-COLORING becomes trivial when $\Delta \leq 2$. Therefore, no matter what Δ is, our algorithm is better than or as good as all known approximation algorithms for MAX SIMPLE EDGE 2-COLORING.

Kosowski et al. [10] showed that approximation algorithms for MAX SIMPLE EDGE 2-COLORING can be used to obtain approximation algorithms for certain packing problems and fault-tolerant guarding problems. Combining their reductions and our improved approximation algorithm for MAX SIMPLE EDGE 2-COLORING, we can obtain improved approximation algorithms for their packing problems and fault-tolerant guarding problems immediately.

2 Basic Definitions

Throughout the remainder of this paper, a graph means a simple undirected graph (i.e., it has neither parallel edges nor self-loops).

Let G be a graph. We denote the vertex set of G by $V(G)$, and denote the edge set of G by $E(G)$. The *degree* of a vertex v in G, denoted by $d_G(v)$, is the number of vertices adjacent to v in G. A vertex v of G with $d_G(v) = 0$ is called an *isolated vertex*. For a subset U of $V(G)$, let $G[U]$ denote the graph (U, E_U) where E_U consists of all edges $\{u, v\}$ of G with $u \in U$ and $v \in U$. We call $G[U]$ the *subgraph of G induced by U*. For a subset U of $V(G)$, we use $G-U$ to denote $G[V(G) - U]$. G is a *star* if G is connected, G has at least three vertices, and there is a vertex u (called the *center* of G) such that every edge of G is incident to u. Each vertex of a star other than the center is called a *satellite* of the star.

A *cycle* in G is a connected subgraph of G in which each vertex is of degree 2. A *path* in G is a connected subgraph of G in which exactly two vertices are of degree 1 and the others are of degree 2. Each vertex of degree 1 in a path P is called an *endpoint* of P, while each vertex of degree 2 in P is called an *inner vertex* of P. An edge $\{u, v\}$ of a path P is called an *inner edge* of P if both u and v are inner vertices of P. The *length* of a cycle or path C is the number of edges in C. A cycle of odd (respectively, even) length is called an *odd* (respectively, *even*) cycle.

A *path-cycle cover* of G is a subgraph H of G such that $V(H) = V(G)$ and $d_H(v) \leq 2$ for every $v \in V(H)$. Note that each connected component of a path-cycle cover of G is a single vertex, path, or cycle. A path-cycle cover $\mathcal{C}$ of G is *triangle-free* if $\mathcal{C}$ does not contain a cycle of length 3. A path-cycle cover C of G is *maximum-sized* if the number of edges in C is maximized over all path-cycle covers of G.

G is *edge-2-colorable* if each connected component of G is an isolated vertex, a path, or an even cycle. Note that MAX SIMPLE EDGE 2-COLORING is the problem of finding a maximum-sized edge-2-colorable subgraph in a given graph.

3 Two Crucial Lemmas and the Outline of Our Algorithm

We say that a graph $K = (V_K, E_K \cup F_K)$ is a *caterpillar graph* if it satisfies the following conditions:

- The graph (V_K, E_K) has $h+1$ connected components $C_0, \ldots, C_h$ with $h \geq 0$.
- C_0 is a path while C_1 through C_h are odd cycles of length at least 5.
- F_K is a matching consisting of h edges $\{u_1, v_1\}, \ldots, \{u_h, v_h\}$.
- For each $i \in \{1, \ldots, h\}$, u_i is an inner vertex of path C_0 while v_i is a vertex of C_i.

We call the edges of F_K the *leg edges* of K, call path C_0 the *spine path* of K, and call cycles C_1 through C_h the *foot cycles* of K.

We say that a graph $K = (V_K, E_K \cup F_K)$ is a *starlike graph* if it satisfies the following conditions:

- The graph (V_K, E_K) has $h+1$ connected components $C_0, \ldots, C_h$ with $h \geq 2$.
- C_0 is a cycle of length at least 4 while C_1 through C_h are odd cycles of length at least 5.
- F_K is a matching consisting of h edges $\{u_1, v_1\}, \ldots, \{u_h, v_h\}$.
- For each $i \in \{1, \ldots, h\}$, u_i is a vertex of C_0 while v_i is a vertex of C_i.

We call the edges of F_K the *bridge edges* of K, call C_0 the *central cycle* of K, and call C_1 through C_h the *satellite cycles* of K.

Let r be the root of the quadratic equation $23r^2 - 55r + 30 = 0$ that is smaller than 1. Note that $r = 0.84176\ldots \approx 0.842$. The reason why we choose r in this way will become clear later in the proof of Lemma 7.

Lemma 1. *Suppose that K is a caterpillar graph such that each foot cycle of K is charged a penalty of $6 - 7r$. Let $p(K)$ be the total penalties charged to the foot cycles of K. Then, K has an edge-2-colorable subgraph K' such that $|E(K')| - p(K) \geq r|E_K|$, where E_K is the set of edges on the spine path or the foot cycles of K.*

Lemma 2. *Suppose that K is a starlike graph such that each satellite cycle of K is charged a penalty of $6 - 7r$. Let $p(K)$ be the total penalties charged to the satellite cycles of K. Then, K has an edge-2-colorable subgraph K' such that $|E(K')| - p(K) \geq r|E_K|$, where E_K is the set of edges on the central or satellite cycles of K.*

Based on Lemmas 1 and 2, we will design our algorithm roughly as follows: Given an input graph G, we will first construct a suitable maximum-sized triangle-free path-cycle cover $\mathcal{C}$ of G and compute a suitable set F of edges such that the endpoints of each edge in F fall into different connected components of $\mathcal{C}$ and each odd cycle of $\mathcal{C}$ has at least one vertex that is an endpoint of an edge in F. Note that $\mathcal{C}$ has at least as many edges as a maximum-sized edge-2-colorable subgraph of G. The edges in F will play the following role: we will break each odd cycle C in $\mathcal{C}$ by removing one edge of C incident to an edge of F and then this edge of F can possibly be added to $\mathcal{C}$ so that $\mathcal{C}$ becomes an edge-2-colorable subgraph of G. Unfortunately, not every edge of F can be added to $\mathcal{C}$ and we have to discard some edges from F, leaving some odd cycles of $\mathcal{C}$ *F-free* (i.e., having no vertex incident to an edge of F). Clearly, breaking an F-free odd cycle C of short length (namely, 5) by removing one edge from C results in a significant loss of edges from $\mathcal{C}$. We charge the loss to the non-F-free odd cycles (unevenly) as penalties. Fortunately, adding the edges of F to $\mathcal{C}$ will yield a graph whose connected components are caterpillar graphs, starlike graphs, or certain other kinds of graphs with good properties. Now, Lemmas 1 and 2 help us show that our algorithm achieves a ratio of r.

4 The Algorithm

Throughout this section, fix a graph G and a maximum-sized edge-2-colorable subgraph $\mathcal{B}$ (for "best") of G. Let n (respectively, m) be the number of vertices (respectively, edges) in G. Our algorithm starts by performing the following four steps:

1. If $|V(G)| \leq 2$, then output G itself and halt.
2. Compute a maximum-sized triangle-free path-cycle cover $\mathcal{C}$ of G. (*Comment:* This step can be done in $O(n^2 m)$ time [5].)
3. While there is an edge $\{u, v\} \in E(G) - E(\mathcal{C})$ such that $d_{\mathcal{C}}(u) \leq 1$ and v is a vertex of some cycle C of $\mathcal{C}$, modify $\mathcal{C}$ by deleting one (arbitrary) edge of C incident to v and adding edge $\{u, v\}$.
4. Construct a graph $G_1 = (V(G), E_1)$, where E_1 is the set of all edges $\{u, v\} \in E(G) - E(\mathcal{C})$ such that u and v appear in different connected components of $\mathcal{C}$ and at least one of u and v appears on an odd cycle of $\mathcal{C}$.

Hereafter, $\mathcal{C}$ always means that we have finished modifying it in Step 3. We give several definitions related to the graphs G_1 and $\mathcal{C}$. Let S be a subgraph of G_1. S *saturates* an odd cycle C of $\mathcal{C}$ if at least one edge of S is incident to a vertex of C. The *weight* of S is the number of odd cycles of $\mathcal{C}$ saturated by S. For convenience, we say that two connected components C_1 and C_2 of $\mathcal{C}$ are *adjacent* in G if there is an edge $\{u_1, u_2\} \in E(G)$ such that $u_1 \in V(C_1)$ and $u_2 \in V(C_2)$.

Lemma 3. *We can compute a maximum-weighted path-cycle cover in G_1 in $O(nm \log n)$ time.*

Our algorithm then proceeds to performing the following four steps:

5. Compute a maximum-weight path-cycle cover M in G_1.
6. While there is an edge $e \in M$ such that the weight of $M - \{e\}$ is the same as that of M, delete e from M.
7. Construct a graph $G_2 = (V(G), E(\mathcal{C}) \cup M)$. (*Comment:* For each pair of connected components of $\mathcal{C}$, there is at most one edge between them in G_2 because of Step 6.)
8. Construct a graph G_3, where the vertices of G_3 one-to-one correspond to the connected components of $\mathcal{C}$ and two vertices are adjacent in G_3 if and only if the corresponding connected components of $\mathcal{C}$ are adjacent in G_2.

Fact 1. *Suppose that C' is a connected component of G_3. Then, the following statements hold:*

1. *C' is a vertex, an edge, or a star.*
2. *If C' is an edge, then at least one endpoint of C' corresponds to an odd cycle of $\mathcal{C}$.*
3. *If C' is a star, then every satellite of C' corresponds to an odd cycle of $\mathcal{C}$.*

An *isolated odd-cycle* of G_2 is an odd cycle of G_2 whose corresponding vertex in G_3 is isolated in G_3. Similarly, a *leaf odd-cycle* of G_2 is an odd cycle of G_2 whose corresponding vertex in G_3 is of degree 1 in G_3. Moreover, a *branching odd-cycle* of G_2 is an odd cycle of G_2 whose corresponding vertex in G_3 is of degree 2 or more in G_3.

Lemma 4. *Let I be the set of isolated odd-cycles in G_2. Then, $|E(\mathcal{B})| \leq |E(\mathcal{C})| - |I|$.*

Proof. Let $C_1, \ldots, C_h$ be the odd cycles of $\mathcal{C}$ such that for each $i \in \{1, \ldots, h\}$, $\mathcal{B}$ contains no edge $\{u, v\}$ with $|\{u, v\} \cap V(C_i)| = 1$. Let $U_1 = \bigcup_{i=1}^{h} V(C_i)$ and $U_2 = V(G) - U_1$. For convenience, let $C_0 = G[U_2]$. Note that for each $e \in E(\mathcal{B})$, one of the graphs $C_0, C_1, \ldots, C_h$ contains both endpoints of e. So, $\mathcal{B}$ can be partitioned into $h+1$ disjoint subgraphs $\mathcal{B}_0, \ldots, \mathcal{B}_h$ such that $\mathcal{B}_i$ is a path-cycle cover of $G[V(C_i)]$ for every $i \in \{0, \ldots, h\}$. Since $\mathcal{C}[U_2]$ must be a maximum-sized path-cycle cover of C_0, $|E(\mathcal{C}[U_2])| \geq |E(\mathcal{B}_0)|$. The crucial point is that for every $i \in \{1, \ldots, h\}$, $|E(\mathcal{B}_i)| \leq |V(C_i)| - 1 = |E(C_i)| - 1$ because $|V(C_i)|$ is odd. Thus, $|E(\mathcal{C})| = |E(\mathcal{C}[U_2])| + \sum_{i=1}^{h} |E(C_i)| \geq |E(\mathcal{B}_0)| + \sum_{i=1}^{h}(|E(\mathcal{B}_i)| + 1) = |E(\mathcal{B})| + h$.

Note that $(V(G), E(G_1) \cap E(\mathcal{B}))$ is a path-cycle cover in G_1 of weight $k - h$, where k is the number of odd cycles in $\mathcal{C}$. So, $k - h \leq k - |I|$ because M is a maximum-weight path-cycle cover in G_1 of weight $k - |I|$. So, by the last inequality in the last paragraph, $|E(\mathcal{B})| \leq |E(\mathcal{C})| - h \leq |E(\mathcal{C})| - |I|$. ∎

Some definitions are in order (see Figure 1 for an example). A *bicycle* of G_2 is a connected component of G_2 that consists of two odd cycles and an edge between them. Note that a connected component of G_3 is an edge if it corresponds to a

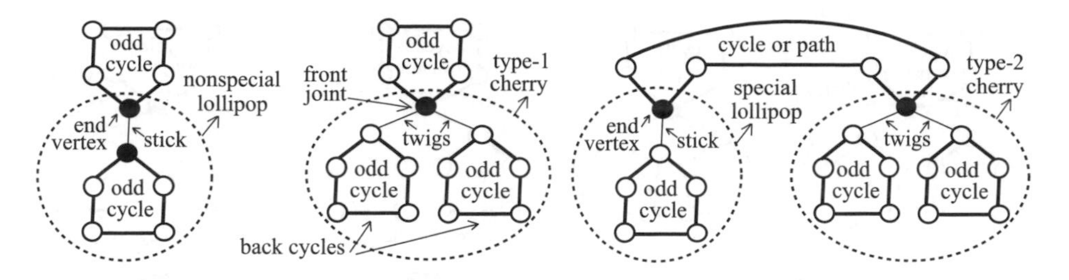

Fig. 1. An example of G_2, where the hollow vertices are free, the bold edges belong to $\mathcal{C}$, the left connected component is a bicycle, and the middle connected component is a tricycle

bicycle in G_2. A *tricycle* of G_2 is a connected component T of G_2 that consists of one branching odd-cycle C_1, two leaf odd-cycles C_2 and C_3, and two edges $\{u_1, u_2\}$ and $\{u_1, u_3\}$ such that $u_1 \in V(C_1)$, $u_2 \in V(C_2)$, and $u_3 \in V(C_3)$. For convenience, we call C_1 the *front cycle* of tricycle K, call C_2 and C_3 the *back cycles* of tricycle K, and call u_1 the *front joint* of tricycle K.

A *cherry* of G_2 is a subgraph Q of G_2 that consists of two leaf odd-cycles C_1 and C_2 of $\mathcal{C}$, a vertex $u \in V(G) - (V(C_1) \cup V(C_2))$, and two edges $\{u, v_1\}$ and $\{u, v_2\}$ such that $v_1 \in V(C_1)$ and $v_2 \in V(C_2)$. For convenience, we call edges $\{u, v_1\}$ and $\{u, v_2\}$ the *twigs* of cherry Q. By the construction of G_2, each pair of cherries are vertex-disjoint. Note that each odd cycle in a cherry of G_2 is a satellite of a star in G_3. We classify the cherries of G_2 into two types as follows. A cherry Q of G_2 is of *type-1* if Q is a subgraph of a tricycle of G_2. Note that the two odd cycles in a type-1 cherry of G_2 are the back cycles of a tricycle of G_2. A cherry of G_2 is of *type-2* if it is not of type-1. Further note that there is no edge $\{u, v\}$ in G such that u appears on an isolated odd-cycle of G_2 and v appears on an odd cycle in a cherry of G_2.

A *lollipop* of G_2 is a subgraph L of G_2 that consists of a leaf odd-cycle C of G_2, a vertex $u \notin V(C)$, and an edge $\{u, v\}$ with $v \in V(C)$. For convenience, we call edge $\{u, v\}$ the *stick* of lollipop L and call vertex u the *end vertex* of lollipop L. A lollipop of G_2 is *special* if it is neither a subgraph of a cherry of G_2 nor a subgraph of a bicycle of G_2. A vertex u of G_2 is *free* if no lollipop of G_2 has u as its end vertex. Because of Step 3, each vertex of degree at most 2 in G_2 is free.

We next define two types of operations that will be performed on G_2. An operation on G_2 is *robust* if it removes no edge of $\mathcal{C}$, creates no new odd cycle, and creates no new isolated odd-cycle of G_2.

Type 1: Suppose that C is an odd cycle of a cherry Q of G_2 and u is a free vertex of G_2 with $u \notin V(C)$ such that

- some vertex v of C is adjacent to u in G and
- if Q is a type-1 cherry of G_2, then u is not an endpoint of a twig of Q.

Then, a *type-1 operation* on G_2 using cherry Q and edge $\{u, v\}$ modifies G_2 by performing the following steps (see Figure 2 for example cases):

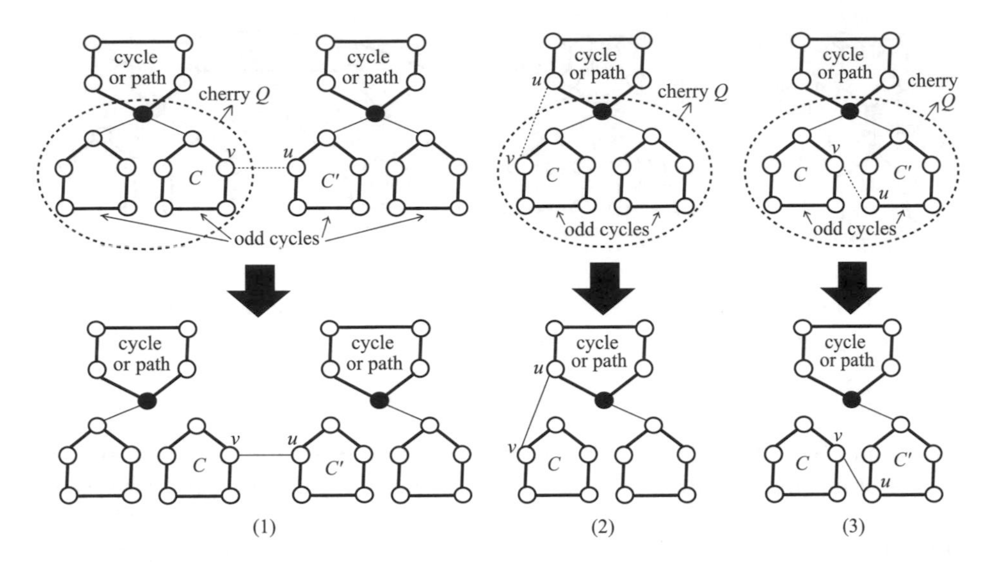

Fig. 2. Three example cases of a type-1 operation, where the dotted lines are edges in $E(G) - E(G_2)$

(1) If u appears on a leaf odd cycle C' of G_2 such that C' is not part of a bicycle of G_2 and Q is not a type-1 cherry of G_2 with $u \in V(Q)$, then delete the stick of the lollipop containing C' from G_2.

(2) Delete the twig of Q incident to a vertex of C from G_2.

(3) Add edge $\{u, v\}$ to G_2.

(*Comment:* A type-1 operation on G_2 is robust and destroys at least one cherry of G_2 without creating a new cherry in G_2.)

Type 2: Suppose that Q is a type-2 cherry of G_2, B is a bicycle of G_2, and $\{u, v\}$ is an edge in $E(G_1) - E(G_2)$ such that u appears on an odd cycle C of Q and v appears on an odd cycle of B. Then, a *type-2 operation* on G_2 using cherry Q, bicycle B, and edge $\{u, v\}$ modifies G_2 by deleting the twig of Q incident to a vertex of C and adding edge $\{u, v\}$.

(*Comment:* A type-2 operation on G_2 is robust. Moreover, when no type-1 operation on G_2 is possible, a type-2 operation on G_2 destroys a type-2 cherry of G_2 and creates a new type-1 cherry in G_2.)

Now, Step 9 of our algorithm is as follows.

9. While a type-1 or type-2 operation on G_2 is possible, perform the following step:
   (a) If a type-1 operation on G_2 is possible, perform a type-1 operation on G_2; otherwise, perform a type-2 operation on G_2.

Fact 2. *After Step 9, the following statements hold:*

1. *There is no edge* $\{u, v\}$ *in* $E(G)$ *such that* u *appears on an odd cycle in a type-2 cherry of* G_2 *and* v *appears on another odd cycle in a type-2 cherry of* G_2.
2. *If* $\{u, v\}$ *is an edge of* G_1 *such that* u *appears on an odd cycle of a type-2 cherry of* G_2 *and no type-2 cherry of* G_2 *contains* v, *then* v *is the end vertex of a special lollipop or the front joint of a tricycle of* G_2.

Hereafter, G_2 always means that we have finished modifying it in Step 9. Now, the final three steps of our algorithm are as follows:

10. Let U be the set of vertices that appear in type-2 cherries of G_2.
11. If $U = \emptyset$, then perform the following steps:
    (a) For each connected component K of G_2, compute a maximum-sized edge-2-colorable subgraph of K. (*Comment:* Because of the simple structure of K, this step can be done in linear time by a standard dynamic programming.)
    (b) Output the union of the edge-2-colorable subgraphs computed in Step 11a, and halt.
12. If $U \neq \emptyset$, then perform the following steps:
    (a) Obtain an edge-2-colorable subgraph R of $G - U$ by recursively calling the algorithm on $G - U$.
    (b) For each type-2 cherry Q of G_2, obtain an edge-2-colorable subgraph of Q by removing one edge from each odd cycle C of Q that shares an endpoint with a twig of Q.
    (c) Let $\mathcal{A}_1$ be the union of R and the edge-2-colorable subgraphs computed in Step 12b.
    (d) For each connected component K of G_2, compute a maximum-sized edge-2-colorable subgraph of K. (*Comment:* Because of the simple structure of K, this step can be done in linear time by a standard dynamic programming.)
    (e) Let $\mathcal{A}_2$ be the union of the edge-2-colorable subgraphs computed in Step 12d.
    (f) If $|E(\mathcal{A}_1)| \geq |E(\mathcal{A}_2)|$, output $\mathcal{A}_1$ and halt; otherwise, output $\mathcal{A}_2$ and halt.

Lemma 5. *Assume that* G_2 *has no type-2 cherry. Then, the edge-2-colorable subgraph of* G *output in Step 11b contains at least* $r|E(\mathcal{B})|$ *edges.*

Proof. Let $\mathcal{C}_2$ be the graph obtained from G_2 by removing one edge from each isolated odd-cycle of G_2. By Lemma 4, $|E(\mathcal{C}_2) \cap E(\mathcal{C})| \geq |E(\mathcal{B})|$. Consider an arbitrary connected component K of $\mathcal{C}_2$. To prove the lemma, it suffices to prove that K has an edge-2-colorable subgraph K' with $|E(K')| \geq r|E(K) \cap E(\mathcal{C})|$. We distinguish several cases as follows:

Case 1: K is a bicycle of $\mathcal{C}_2$. To obtain an edge-2-colorable subgraph K' of K, we remove one edge e from each odd cycle of K such that one endpoint of e is of degree 3 in K. Note that $|E(K')| = |E(K)| - 2 = |E(K) \cap E(\mathcal{C})| - 1$. Since $|E(K) \cap E(\mathcal{C})| \geq 10$, $|E(K')| \geq \frac{9}{10}|E(K) \cap E(\mathcal{C})| > r|E(K) \cap E(\mathcal{C})|$.

Case 2: K is a tricycle of $\mathcal{C}_2$. To obtain an edge-2-colorable subgraph K' of K, we first remove one edge e from each back odd-cycle of K such that one endpoint of e is of degree 3 in K, and then remove the two edges of the front odd-cycle incident to the vertex of degree 4 in K. Note that $|E(K')| = |E(K)| - 4 = |E(K) \cap E(\mathcal{C})| - 2$. Since $|E(K) \cap E(\mathcal{C})| \geq 15$, $|E(K')| \geq \frac{13}{15}|E(K) \cap E(\mathcal{C})| > r|E(K) \cap E(\mathcal{C})|$.

Case 3: K is neither a bicycle nor a tricycle of $\mathcal{C}_2$. If K contains no odd cycle of $\mathcal{C}$, then K itself is edge-2-colorable and hence we are done. So, assume that K contains at least one odd cycle of $\mathcal{C}$. Then, K is also a connected component of G_2. Moreover, the connected component K'' of G_3 corresponding to K is either an edge or a star.

Case 3.1: K'' is an edge. To obtain an edge-2-colorable subgraph K' of K, we start with K, delete the edge in $E(K) - E(\mathcal{C})$, and delete one edge from the unique odd cycle of K. Note that $|E(K')| = |E(K)| - 2 = |E(K) \cap E(\mathcal{C})| - 1$. Moreover, $|E(K) \cap E(\mathcal{C})| \geq 7$ because of Step 3 and the robustness of Type-1 or Type-2 operations. Hence, $|E(K')| \geq \frac{6}{7}|E(K) \cap E(\mathcal{C})| > r|E(K) \cap E(\mathcal{C})|$.

Case 3.2: K'' is a star. Let C_0 be the connected component of $\mathcal{C}$ corresponding to the center of K''. Let $C_1, \ldots, C_h$ be the odd cycles of $\mathcal{C}$ corresponding to the satellites of K''. If C_0 is a path, then K is a caterpillar graph and we are done by Lemma 1; otherwise, K is a starlike graph and we are done by Lemma 2. ∎

Corollary 1. *If the maximum degree Δ of a vertex in G is at most 3, then the ratio achieved by the algorithm is at least $\frac{6}{7}$.*

Proof. When $\Delta \leq 3$, G_2 has no cherry because of Step 3. Moreover, Lemmas 1, 2, and 5 still hold even when we replace the ratio r by $\frac{6}{7}$. ∎

In order to analyze the approximation ratio achieved by our algorithm when G_2 has at least one type-2 cherry after Step 9, we need to define several notations as follows:

- Let s be the number of special lollipops in G_2.
- Let t be the number of tricycles in G_2.
- Let c be the number of type-2 cherries in G_2.
- Let ℓ be the total number of vertices that appear on odd cycles in the type-2 cherries in G_2.

Lemma 6. *Let $E(\mathcal{B}_2)$ be the set of all edges $e \in E(\mathcal{B})$ such that at least one endpoint of e appears in a type-2 cherry of G_2. Then, $|E(\mathcal{B}_2)| \leq \ell + 2s + 2t$.*

Proof. $E(\mathcal{B}_2)$ can be partitioned into the following three subsets:

- $E(\mathcal{B}_{2,1})$ consists of those edges $e \in E(\mathcal{B})$ such that at least one endpoint of e is the vertex of a type-2 cherry of G_2 that is a common endpoint of the two twigs of the cherry.

– $E(\mathcal{B}_{2,2})$ consists of those edges $e \in E(\mathcal{B})$ such that each endpoint of e appears on an odd cycle of a type-2 cherry of G_2.
– $E(\mathcal{B}_{2,3})$ consists of those edges $\{u, v\} \in E(\mathcal{B})$ such that u appears on an odd cycle of a type-2 cherry of G_2 and no type-2 cherry of G_2 contains v.

Obviously, $|E(\mathcal{B}_{2,1})| \leq 2c$. By Statement 1 in Fact 2, $|E(\mathcal{B}_{2,2})| \leq \ell - 2c$ because for each odd cycle C, $\mathcal{B}_{2,2}$ can contain at most $|V(C)| - 1$ edges $\{u, v\}$ with $\{u, v\} \subseteq V(C)$. By Statement 2 in Fact 2, $|E(\mathcal{B}_{2,3})| \leq 2s + 2t$. So, $|E(\mathcal{B}_2)| \leq \ell + 2s + 2t$. ∎

Lemma 7. *The ratio achieved by the algorithm is at least r.*

Proof. By induction on $|V(G)|$, the number of vertices in the input graph G. If $|V(G)| \leq 2$, then our algorithm outputs a maximum-sized edge-2-colorable subgraph of G. So, assume that $|V(G)| \geq 3$. Then, after our algorithm finishes executing Step 10, the set U may be empty or not. If $U = \emptyset$, then by Lemma 5, the edge-2-colorable subgraph output by our algorithm has at least $r|E(\mathcal{B})|$ edges and we are done. So, suppose that $U \neq \emptyset$.

First consider the case where $s + t \leq \frac{1-r}{2r}\ell$. In this case, $\frac{\ell + r|E(\mathcal{B}_1)|}{\ell + 2s + 2t + |E(\mathcal{B}_1)|} \geq r$, where $\mathcal{B}_1$ is a maximum-sized edge-2-colorable subgraph of $G - U$. Moreover, by the inductive hypothesis, $|E(\mathcal{A}_1)| \geq \ell + r|E(\mathcal{B}_1)|$. Furthermore, by Lemma 6, $|E(\mathcal{B})| \leq \ell + 2s + 2t + |E(\mathcal{B}_1)|$. So, the lemma holds in this case.

Next consider the case where $s + t > \frac{1-r}{2r}\ell$. Let $\mathcal{C}_2$ be the graph obtained from G_2 by removing one edge from each isolated odd-cycle of G_2. By Lemma 4, $|E(\mathcal{C}_2) \cap E(\mathcal{C})| \geq |E(\mathcal{B})|$. Let $\mathcal{C}_3$ be the graph obtained from $\mathcal{C}_2$ by removing one twig from each type-2 cherry. Note that there are exactly c isolated odd-cycles in $\mathcal{C}_3$. Moreover, since the removed twig does not belong to $E(\mathcal{C})$, we have $|E(\mathcal{C}_3) \cap E(\mathcal{C})| \geq |E(\mathcal{B})|$. Consider an arbitrary connected component K of $\mathcal{C}_3$. To prove the lemma, we want to prove that K has an edge-2-colorable subgraph K' with $|E(K')| \geq r|E(K) \cap E(\mathcal{C})|$. This goal can be achieved because of Lemma 5, when K is not an isolated odd-cycle. On the other hand, this goal can not be achieved when K is an isolated odd-cycle (of length at least 5). Our idea behind the proof is to charge the deficit in the edge numbers of isolated odd-cycles of $\mathcal{C}_3$ to the other connected components of K because they have surplus in their edge numbers.

The deficit in the edge number of each isolated odd-cycle of $\mathcal{C}_3$ is at most $5r - 4$. So, the total deficit in the edge numbers of the isolated odd-cycles of $\mathcal{C}_3$ is at most $(5r - 4)c$. We charge a penalty of $6 - 7r$ to each non-isolated odd-cycle of $\mathcal{C}_3$ that is also an odd cycle in a type-2 cherry of G_2 or is also the odd cycle in a special lollipop of G_2. We also charge a penalty of $\frac{6-7r}{3}$ to each odd cycle of $\mathcal{C}_3$ that is part of a tricycle of G_2. Clearly, the total penalties are $(6 - 7r)c + (6 - 7r)(s + t) > (6 - 7r)c + \frac{(6-7r)(1-r)}{2r}\ell$. Note that $\ell \geq 10c$. The total penalties are thus at least $(6 - 7r)c + \frac{5(6-7r)(1-r)}{r}c = \frac{30-59r+28r^2}{r}c \geq (5r - 4)c$, where the last inequality follows from the equation $23r^2 - 55r + 30 = 0$. So, the total penalties are at least as large as the total deficit in the edge numbers of the isolated odd-cycles of $\mathcal{C}_3$. Therefore, to prove the lemma, it suffices to prove that for every connected component K of $\mathcal{C}_3$, we can compute an edge-2-colorable subgraph K' of K such that $|E(K')| - p(K) \geq r|E(K) \cap E(\mathcal{C})|$, where $p(K)$ is the total penalties of the odd cycles in K. As in the proof of Lemma 5, we distinguish several cases as follows:

Case 1: K is a bicycle of $\mathcal{C}_2$. In this case, $p(K) = 0$. Moreover, we can compute an edge-2-colorable subgraph K' of K such that $|E(K')| \geq \frac{9}{10}|E(K) \cap E(\mathcal{C})|$ (cf. Case 1 in the proof of Lemma 5). So, $|E(K')| - p(K) \geq r|E(K) \cap E(\mathcal{C})|$ because $r \leq \frac{9}{10}$.

Case 2: K is a tricycle of $\mathcal{C}_2$. In this case, $p(K) = 6 - 7r$. Moreover, we can compute an edge-2-colorable subgraph K' of K such that $|E(K')| = |E(K) \cap E(\mathcal{C})| - 2$ (cf. Case 2 in the proof of Lemma 5). So, $|E(K')| - p(K) \geq r|E(K) \cap E(\mathcal{C})|$ because $|E(K) \cap E(\mathcal{C})| \geq 15$ and $r \leq \frac{7}{8}$.

Case 3: K is neither a bicycle nor a tricycle of $\mathcal{C}_2$. We may assume that K contains at least one odd cycle of $\mathcal{C}$. Then, K is also a connected component of G_2. Moreover, the connected component K'' of G_3 corresponding to K is either an edge or a star.

Case 3.1: K'' is an edge. In this case, $p(K) \leq 6-7r$. Moreover, we can compute an edge-2-colorable subgraph K' of K such that $|E(K')| = |E(K) \cap E(\mathcal{C})| - 1$ (cf. Case 3.1 in the proof of Lemma 5). So, $|E(K')| - p(K) \geq r|E(K) \cap E(\mathcal{C})|$ because $|E(K) \cap E(\mathcal{C})| \geq 7$.

Case 3.2: K'' is a star. Let C_0 be the connected component of $\mathcal{C}$ corresponding to the center of K''. Let $C_1, \ldots, C_h$ be the odd cycles of $\mathcal{C}$ corresponding to the satellites of K''. If C_0 is a path, then K is a caterpillar graph and we are done by Lemma 1; otherwise, K is a starlike graph and we are done by Lemma 2. ∎

Clearly, each step of our algorithm except Step 12a can be implemented in $O(n^2m)$ time. Since the recursion depth of the algorithm is $O(n)$, it runs in $O(n^3m)$ total time. In summary, we have shown the following theorem:

Theorem 3. *There is an $O(n^3m)$-time approximation algorithm for* Max Simple Edge 2-Coloring *that achieves a ratio of roughly* 0.842.

5 An Application

Let G be a graph. An *edge cover* of G is a set F of edges of G such that each vertex of G is incident to at least one edge of F. For a natural number k, a *$[1,\Delta]$-factor k-packing* of G is a collection of k disjoint edge covers of G. The *size* of a $[1,\Delta]$-factor k-packing $\{F_1, \ldots, F_k\}$ of G is $|F_1| + \cdots + |F_k|$. The problem of deciding whether a given graph has a $[1,\Delta]$-factor k-packing was considered in [7,8]. In [10], Kosowski *et al.* defined the *minimum $[1,\Delta]$-factor k-packing problem* (Min-k-FP) as follows: Given a graph G, find a $[1,\Delta]$-factor k-packing of G of minimum size or decide that G has no $[1,\Delta]$-factor k-packing at all.

According to [10], Min-2-FP is of special interest because it can be used to solve a fault tolerant variant of the guards problem in grids (which is one of the art gallery problems [11,12]). Indeed, they proved the NP-hardness of Min-2-FP and the following lemma:

Lemma 8. *If* MAX SIMPLE EDGE 2-COLORING *admits an approximation algorithm A achieving a ratio of α, then* MIN-2-FP *admits an approximation algorithm B achieving a ratio of $2 - \alpha$. Moreover, if the time complexity of A is $T(n)$, then the time complexity of B is $O(T(n))$.*

So, by Theorem 3, we have the following immediately:

Theorem 4. *There is an $O(n^3 m)$-time approximation algorithm for* MIN-2-FP *achieving a ratio of roughly* 1.158*, where n (respectively, m) is the number of vertices (respectively, edges) in the input graph.*

6 Open Problems

One obvious open question is to ask whether one can design a polynomial-time approximation algorithm for MAX SIMPLE EDGE 2-COLORING that achieves a ratio significantly better than 0.842. The APX-hardness of the problem implies an implicit lower bound of $1 - \epsilon$ on the ratio achievable by a polynomial-time approximation algorithm. It seems interesting to prove an explicit lower bound significantly better than $1 - \epsilon$.

References

1. Chen, Z.-Z., Tanahashi, R.: Approximating Maximum Edge 2-Coloring in Simple Graphs via Local Improvement. Theoretical Computer Science (special issue on AAIM 2008) 410, 4543–4553 (2009)
2. Chen, Z.-Z., Tanahashi, R., Wang, L.: An Improved Approximation Algorithm for Maximum Edge 2-Coloring in Simple Graphs. Journal of Discrete Algorithms 6, 205–215 (2008)
3. Feige, U., Ofek, E., Wieder, U.: Approximating Maximum Edge Coloring in Multigraphs. In: Jansen, K., Leonardi, S., Vazirani, V.V. (eds.) APPROX 2002. LNCS, vol. 2462, pp. 108–121. Springer, Heidelberg (2002)
4. Gabow, H.: An Efficient Reduction Technique for Degree-Constrained Subgraph and Bidirected Network Flow Problems. In: Proceedings of the 15th Annual ACM Symposium on Theory of Computing (STOC 1983), pp. 448–456. ACM, New York (1983)
5. Hartvigsen, D.: Extensions of Matching Theory. Ph.D. Thesis, Carnegie-Mellon University (1984)
6. Hochbaum, D.: Approximation Algorithms for NP-Hard Problems. PWS Publishing Company, Boston (1997)
7. Jacobs, D.P., Jamison, R.E.: Complexity of Recognizing Equal Unions in Families of Sets. Journal of Algorithms 37, 495–504 (2000)
8. Kawarabayashi, K., Matsuda, H., Oda, Y., Ota, K.: Path Factors in Cubic Graphs. Journal of Graph Theory 39, 188–193 (2002)
9. Kosowski, A.: Approximating the Maximum 2- and 3-Edge-Colorable Subgraph Problems. Discrete Applied Mathematics 157, 3593–3600 (2009)
10. Kosowski, A., Malafiejski, M., Zylinski, P.: Packing $[1, \Delta]$-Factors in Graphs of Small Degree. Journal of Combinatorial Optimization 14, 63–86 (2007)
11. O'Rourke, J.: Art Gallery Theorems and Algorithms. Oxford University Press, Oxford (1987)
12. Urrutia, J.: Art Gallery and Illumination Problems. In: Handbook on Computational Geometry. Elsevier Science, Amsterdam (2000)

A Linear Kernel for Co-Path/Cycle Packing

Zhi-Zhong Chen[1], Michael Fellows[2], Bin Fu[3], Haitao Jiang[5], Yang Liu[3], Lusheng Wang[4], and Binhai Zhu[5]

[1] Department of Mathematical Sciences, Tokyo Denki University, Hatoyama, Saitama 350-0394, Japan
zzchen@mail.dendai.ac.jp
[2] The University of New Castle, Callaghan, NSW 2308, Australia
michael.fellows@newcastle.edu.au
[3] Department of Computer Science, University of Texas-American, Edinburg, TX 78739-2999, USA
binfu,yliu@panam.edu
[4] Department of Computer Science, City University of Hong Kong, Kowloon, Hong Kong
cswangl@cityu.edu.hk
[5] Department of Computer Science, Montana State University, Bozeman, MT 59717-3880, USA
htjiang,bhz@cs.montana.edu

Abstract. Bounded-Degree Vertex Deletion is a fundamental problem in graph theory that has new applications in computational biology. In this paper, we address a special case of Bounded-Degree Vertex Deletion, the Co-Path/Cycle Packing problem, which asks to delete as few vertices as possible such that the graph of the remaining (residual) vertices is composed of disjoint paths and simple cycles. The problem falls into the well-known class of 'node-deletion problems with hereditary properties', is hence NP-complete and unlikely to admit a polynomial time approximation algorithm with approximation factor smaller than 2. In the framework of parameterized complexity, we present a kernelization algorithm that produces a kernel with at most $37k$ vertices, improving on the super-linear kernel of Fellows *et al.*'s general theorem for Bounded-Degree Vertex Deletion. Using this kernel, and the method of bounded search trees, we devise an FPT algorithm that runs in time $O^*(3.24^k)$. On the negative side, we show that the problem is APX-hard and unlikely to have a kernel smaller than $2k$ by a reduction from Vertex Cover.

1 Introduction

In computational biology, a fundamental problem is to build up phylogenetic networks (trees) for various species, some of which are possibly extinct. A basic problem along this line is to construct ancestral genomes from the genomes of currently living species. Recently, Chauve and Tannier proposed the use of PQ-trees, where each leave represents a gene marker, to represent possible ancestral genomes [3]. This approach raises a natural question: given two PQ-trees over

B. Chen (Ed.): AAIM 2010, LNCS 6124, pp. 90–102, 2010.

the same set of markers, how do we compare their similarity? For example, do they generate the same sequence?

In [9], the above problem is shown to be NP-complete. A natural extension of the problem is to delete the minimum number of common markers so that the two resulting PQ-trees can generate the same sequence. Modeling the markers in the PQ-trees as a hyper-graph, this is exactly the problem of deleting the minimum number of markers so that the resulting graph is composed of a set of paths; and, when circular genomes are allowed (as in [13]), a set of paths and cycles. We call this the Co-Path/Cycle Packing problem.

This problem belongs to the family of problems concerned with deleting a minimum number of vertices to obtain a graph belonging to a hereditary class of graphs (the well-known Vertex Cover problem is another example) [11]. A general polynomial time 2-approximation algorithm is known for this family of problems [5], and therefore Co-Path/Cycle Packing can be approximated within a factor of 2 in polynomial time. By a reduction from Vertex Cover, we show that an α-approximation algorithm for Co-Path/Cycle Packing yields an α-approximation algorithm for Vertex Cover. By the recent results of Khot and Regev [10], it follows that Co-Path/Cycle Packing does not admit a polynomial time approximation algorithm with performance factor $2 - \epsilon$, unless the Unique Games Conjecture fails.

In the parameterized framework, Fellows *et al.* recently considered the Bounded-Degree Vertex Deletion (d-BDD) problem, that of deleting the minimum number k of vertices (the parameter) so that the resulting graph has maximum degree d [6]. When $d = 0$, this is the Vertex Cover problem; 2-BDD is exactly our Co-Path/Cycle Packing problem. Fellows *et al.* presented a generalized Nemhauser-Trotter Theorem that implies that the d-BDD problem admits a kernel with a linear number of vertices for $d \leq 1$ and that d-BDD admits an $O(k^{1+\epsilon})$ kernel for $d \geq 2$ [6]. (We comment that in the conference version of their paper, the claimed linear kernel bound for all d was not correct; the result described is in the journal version.)

Here we present a $37k$ kernel for Co-Path/Cycle Packing. This is the first vertex linear kernel 2-BDD problem. Our approach here is similar to that of Fellows *et al.* Roughly speaking, using the fact that a path/cycle packing cannot contain any 3-star, we compute a *proper* maximal 3-star packing in the input graph and use them to compute a *triple crown decomposition*, and subsequently obtain the linear kernel. Using the $37k$ vertex kernel, we describe an FPT algorithm that runs in time $O^*(3.24^k)$, based on a bounded search tree approach.

This paper is organized as follows. In Section 2, we give some definitions. In Section 3, we show the kernelization algorithm and prove the lower bound. In Section 4, we present the bounded search tree algorithm. In Section 5, we conclude the paper with several open questions.

2 Preliminaries

We begin this section with some basic definitions and notations of graph theory.

Given an undirected graph $G = (V, E)$, for a vertex subset $S \subseteq V$, let $G[S]$ be the subgraph induced by S and $G - S = G[V/S]$. Similarly, for an edge set $P \subseteq E$, $G - P = (V, E - P)$. The *neighborhood* of a vertex v and a vertex set S is denoted as $N(v) = \{u \in V | \{u, v\} \in E\}$ and $N(S) = \{u \in V - S | v \in S, \{u, v\} \in E\}$, respectively. $N[S] = N(S) \cup S$. $d(v) = |N(v)|$ is the *degree* of a vertex v. The graph $K_{1,s} = (\{u, v_1, \ldots, v_s\}, \{(u, v_1), \ldots, (u, v_s)\})$ is called an s-star. u is the *center* and v_i's are the leaves. An s-star packing is a collection of vertex-disjoint s-stars. A *path/cycle packing* for an undirected graph $G = (V, E)$ is a vertex set whose induced subgraph is composed of disjoint paths and simple cycles. (An isolated vertex is also considered as a path, of length zero.) For convenience, we also call the corresponding subgraph *path/cycle set*. A *co- path/cycle packing* for an undirected graph $G = (V, E)$ is a vertex set whose deletion results in a graph which is a path/cycle set. We now make the following formal definition for the problem involved in this paper.

Minimum Co-Path/Cycle Packing:
Input: An undirected graph $G = (V, E)$, integer k.
Question: Does there exist a vertex set $S \subseteq V$ of size at most k such that the induced subgraph $G - S$ is a path/cycle set?

We now present some definitions regarding FPT algorithms. Given a parameterized problem instance (I,k), an FPT (Fixed-Parameter Tractable) algorithm solves the problem in $O(f(k)n^c)$ time (often simplified as $O^*(f(k))$), where f is a function of k only, n is the input size and c is some fixed constant (i.e., not depending on k or n). A useful technique in parameterized algorithmics is to provide polynomial time executable data-reduction rules that lead to a *problem kernel.* A data-reduction rule replaces (I,k) by an instance (I',k') in polynomial time such that: (1) $|I'| \leq |I|$, $k' \leq k$, and (2) (I,k) is a Yes-instance if and only if (I',k') is a Yes-instance. A set of polynomial-time data-reduction rules for a problem are applied to an instance of the problem to achieve a *reduced* instance termed the *kernel.* A parameterized problem is FPT if and only if there is a polynomial time algorithm applying data-reduction rules that reduce any instance of the problem to a kernelized instance of size $g(k)$. More about parameterized complexity can be found in the monographs [4,7,12].

3 A Linear Kernel

In this section, we describe a polynomial time data-reduction rule and show that it yields a kernel having at most $37k$ vertices for the Co-Path/Cycle Packing problem, improving the super-linear kernel for 2-BDD problem by Fellows *et al.* [6].

As a common trick in data reduction, we remove vertices of high degree and remove useless vertices connecting vertices of degree at most two. The corresponding rules are summarized in the following lemmas.

Lemma 1. *Let G be a graph such that there exists $v \in V$ with $d(v) > k+2$, then G has a k-co-path/cycle packing iff $G - v$ has a $(k-1)$-co-path/cycle packing.*

Lemma 2. *Let G be a graph such that there exists an edge $e = (u, v)$ and $d(u) = d(v) \leq 2$, then G has a minimum co-path/cycle packing that does not contain u and v.*

Proof. W.l.o.g., we only consider the case when u is in some minimum co-path/cycle packing. Assume that W is a minimum co-path/cycle packing with $u \in W$, then $G - W$ contains a path ending at v and a path containing x (the other neighbor of u). We can construct another co-path/cycle packing W' from W by replacing u with x. Symmetrically, we just add e at the end of some path in $G - W$. Therefore, W' is a minimum co-path/cycle packing. □

Lemma 3. *Let G be a graph such that there exists a path $P=\langle v_1, v_2, \ldots, v_t\rangle$ where $t \geq 3$, and $d(v_i) \leq 2$ for all $1 \leq i \leq t$. Let Z denote the vertex set in the middle of the path, i.e., $Z = \{v_2, v_3, ..., v_{t-1}\}$, and let G' be the graph constructed by adding a new edge $e = (v_1, v_t)$ to $G - M$, then G has a k-co-path/cycle packing iff G' has a k-co-path/cycle packing.*

Proof. The 'only if' part is trivially true, so we will focus on the 'if' part. Following Lemma 2, there is a k-co-path/cycle packing W in G' which does not contain v_1 and v_t. So we will not create any vertex with degree more than two in $G - W$ by inserting the vertices in Z between v_1 and v_t. □

Lemma 3 basically implies that we can contract a vertex of degree at most two to either one of its neighbors, as long as its neighbors also have degrees at most two. From now on, we assume that an input graph is already preprocessed by Lemma 3.

We next review a famous structure in parameterized complexity, the *crown decomposition*, which was used to obtain a small kernel for Vertex Cover [1,2].

Definition 1. A *crown decomposition* (H,C,R) in a graph $G = (V, E)$ is a partition of V into three sets H, C and R which have the following properties:

(1) H (the head) is a separator in G such that there are no edges between the vertices in C and the vertices in R.

(2) $C = C_u \cup C_m$ (the crown) is an independent set in G.

(3) $|C_m| = |H|$, and there is a perfect matching between C_m and H.

We modify and generalize the crown decomposition to handle our particular problem. The variation is called *triple crown decomposition*, where each vertex in H has three vertices in C matched to it. We elaborate the details as follows.

Definition 2. A *triple crown decomposition* (H,C,L,R) in a graph $G = (V, E)$ is a partition of the vertices in V into four sets H,C,L and R which have the following properties:

(1) H (the head) is a separator in G such that there are no edges between the vertices in C and the vertices in R.

(2) $L = N(C)/H$ (the neighbor), $G[L \cup C]$ is a path/cycle set, and $|N(l) \cap (R \cup C)| \leq 2$ for all $l \in L$.

(3) $C = C_u \cup C_{m_1} \cup C_{m_2} \cup C_{m_3}$ (the crown). $|C_{m_i}| = |H|$, and there is a perfect matching between every C_{m_i} and H, for all $i \in \{1, 2, 3\}$.

Based on the triple crown decomposition, we describe the critical data reduction rule in this paper through the following lemma.

Lemma 4. *A graph $G = (V, E)$ which admits a triple crown decomposition (H, C, L, R) has a k-co-path/cycle packing iff $G - H$ has a $(k - |H|)$-co-path/cycle packing.*

Proof. The 'if' part is easy to prove because any co-path/cycle packing of $G - H$ together with H is certainly a solution for G.

We now prove the other direction. Let G have a co-path/cycle packing of size k. First, we can see that any co-path/cycle packing for $G[H \cup C]$ contains at least $|H|$ vertices. Since (H, C, L, R) is a *triple crown decomposition*, there are $|H|$ vertex-disjoint 3-stars in $G[H \cup C]$, at least one vertex of every star should be deleted in order to obtain a path/cycle set. Moreover, every vertex belonging to $C \cup L$ has degree at most two and every vertex belonging to C has no neighbor outside of L in $G - H$. There is no minimum co-path/cycle packing W for $G - H$ such that $v \in W \cap C$, otherwise v connects at most two paths in $G - H - W$, and $W - v$ is hence a smaller co-path/cycle packing. So the size of the minimum co-path/cycle packing for G is at least $|W| + |H|$, which means $|W| \leq k - |H|$. Hence the 'only if' part holds. □

Now, our main idea of the kernelization algorithm is to search for *triple crown decomposition* in the graph iteratively. When we cannot find that structure at all, we can conclude that the graph has bounded size. At the beginning, the algorithm computes a maximal 3-star packing in a greedy fashion. Note that the maximal 3-star packing thus found is a co-path/cycle packing, i.e., if we delete all these 3-stars the resulting graph has no vertex of degree greater than two. Then we refine the maximal 3-star packing such that the following lemma is satisfied. After that, the algorithm tries to search a *triple crown decomposition* with H being a subset of the star centers and C belonging to the path/cycle set.

We first summarize the method to obtain/refine a proper maximal 3-star packing in the following lemma.

Lemma 5. *Given a graph $G = (V, E)$, we can produce a maximal 3-star packing W such that every 3-star $P \in W$ falls into one of the three cases,*

1. *if the center u of P has at least 4 neighbors in $G - W$, then every leaf of P has at most two neighbors in $G - W$, all of them are of degree one in $G - W$.*
2. *if the center u of P has one to three neighbors in $G - W$, then any 3-star Q composed of one leaf v of P and three other vertices in $G - W$ contains all neighbors of u in $G - W$.*
3. *if the center u has no neighbor in $G - W$, then each leaf of P has at most two distinct neighbors, for a total of at most 6, in $G - W$.*

Proof. We prove the three cases respectively.

1. Suppose on the contrary that there exists a leaf x of P which is incident to some vertex y of degree-2 in $G - W$. Then we have a 3-star P' centered at y, with x being one of its leaves. Therefore, we can obtain one extra 3-star besides P': just modify P by replacing x with v as a new leaf. (v must exist due to that u has at least 4 neighbors in $G - W$.) This contradicts the optimality of W.
2. Otherwise, we can swap v with some neighbor x of u in $G - W$, with $x \notin Q$. We thus obtain one more 3-star. Again, this contradicts the optimality of W.
3. Let the leaves of P be $\{v_1, v_2, v_3\}$. Assume on the contrary that there exists a leaf v_1 of P which has more than three neighbors different from the neighbors of v_2 and v_3 in $G - W$. We can replace P with a new 3-star composed of v and its three neighbors. □

The algorithm *Proper Maximal 3-Star Packing* follows directly from Lemma 5. In the following box, we describe the corresponding algorithm with pseudo-code.

Algorithm *Proper Maximal 3-Star Packing*
Input: *Graph G=(V,E)*
Output: *A proper maximal 3-star packing W*
1 Compute a maximal 3-star packing W greedily.
2 For every $P = \langle u, \{x, y, z\}\rangle \in W$, check the following properties iteratively until W fulfills Lemma 5
2.1 if there are two or three disjoint 3-stars Q,R (or S) each contains only one leaf of P,
then $W = W - P + Q + R$ $(+S)$.
2.2 if there is a 3-star Q which contains only one leaf x and u has a neighbor v distinct from Q in $G - W$,
then replace x with v in P, $W = W + Q$.
2.2 if there is a 3-star Q which contains only one leaf x, and y and z have no neighbor in Q,
then $W = W - P + Q$.

Since the maximum 3-star packing in a graph is bounded by $O(n)$, any violation of the conditions (1) and (2) leads to a larger 3-star packing and each 3-star can be computed in $O(n)$ time. So we can find such a proper maximal 3-star packing fulfilling Lemma 5 in $O(n^2)$ time.

In order to describe the main algorithm concisely, we make use of some notations for some vertex sets in the graph. Let W be the maximal 3-star packing we obtain (after running Lemma 5). L_W and C_W denotes all the leaves and centers in W respectively. Then, $D_W^L = N(L_W) \cap (G - W)$, $D_W^C = (N(C_W) - D_W^L) \cap (G - W)$, $F(L_W) = N[D_W^L] \cap (G - W)$, and for every $v \in C_W$, $F(v) = N[N(v)] \cap (G - W - F(L_W))$. $S \subseteq C_W$, $F(S) = \bigcup_{v \in S} F(v)$.

The next procedure computes a *triple crown decomposition*, if it exists.

Algorithm *Triple Crown Decomposition*
Input: *Graph G=(V,E), proper maximal 3-star packing W*
Output: *Triple crown decomposition (H,C,L,R)*
1 X is the case-1 centers in C_W, $Y = G - W - F(L_W) - F(C_W - X)$.
2 Construct a bipartite graph $J = (X, Y, T)$,
where $T = \{(u, v) \in E | u \in X, v \in Y\}$.
3 Replicate every vertex $u \in X$ twice such that u and its two copies have the same neighbor in J.
Let the new graph be J' and let u^* denote the original vertex in X for both of u's copies.
4 Compute the maximum matching M' in J' and for every edge $e = (u, v) \in M'$, construct $(u^*, v) \in M \subseteq T$
5 Repeat the following two steps with $C_0 = Y - M$ until $C_i = C_{i+1}$.
5.1 $H_i = N_J(C_i)$;
5.2 $C_{i+1} = N_M(H_i) \cup C_i$;
6 If $H_i = X$, then we get a triple crown decomposition (H, C, L, R) where $H = H_i$, $C = C_i$, $L = N_G(C) - H$, $R = G - H - C - L$.
7 Else F_X=$X - H_i$, $F_Y = F(X - H_i)$, $X = X - F_X$, $Y = Y - F_Y$.
8 If $H_i \neq \phi$, goto step 2; else exit.

We have the following lemmas regarding the above algorithm.

Lemma 6. *$(H \cup C) \cap M$ is a 3-star packing such that all centers are in H and all leaves are in C.*

Proof. If there exists a vertex $v \in (H \cap M)$ which has fewer than three neighbors in $(C \cap M)$. From the way we get v, there is an M-augmenting path from some vertex not in M to v. (An M-augmenting path is a path where the edges in M and edges not in M alternate.) Then M cannot be maximum since both the start and end edges of the M-augmenting path are not in M. □

Lemma 7. *If v connects a vertex $h \notin H$, then v is matched to some vertex not in H.*

Proof. If v is not in M, then $v \in C$ and $N_J(v) \subseteq H$, which contradicts the condition in the lemma. Therefore $v \in M$. If v is matched to some vertex in H, then $v \in C$ and $h \in H$, also contradicts the condition in the lemma. □

Note that every vertex in Y either belongs to C or belongs to $N(F_X)$.

Lemma 8. *If the procedure does not find a triple crown decomposition, then $|F_Y| \leq 9|F_X|$.*

Proof. From Lemma 7, every vertex in F_Y is either matched to some vertex in F_X or is a neighbor of a matched vertex in F_Y. Since every vertex in F_X has at most three neighbors in M, $G - W$ is a path/cycle set. Therefore, $|F_Y| \leq 3 \cdot (2 + 1)|F_X| = 9|F_X|$. □

If the procedure cannot find a triple crown decomposition, which means $F_X = X$, we then refer T to the vertices in $Y - F_Y$. The next lemma shows properties of vertices in T.

Lemma 9. *For every $u \in T$, u fulfills the following two properties:*

1. *u has no neighbor in X.*
2. *There exists a vertex $v \in N(u)$, v is included in $F_Y \cup F(L_W) \cup F(C_W - X)$ and v is connected to another vertex $w \in N(W)$.*

Proof. From the procedure, we can see that any vertex in $Y = G - W - F(L_W) - F(C_W - X)$ which has some neighbor in X or has some neighbor in $N(F_X)$ is included in F_Y. Therefore, if $u \notin F_Y$, both of u's neighbors either belong to T or belong to $N(N(W))$. From Lemma 3, there are at most two consecutive vertices whose degrees are at most two. So at least one of u's neighbors belongs to $N(N(W))$. □

It is easy to verify that any single vertex or two consecutive vertices in T lie between vertices in $N(N(W))$. The following lemma bounds the cardinality of T.

Lemma 10. $|T| \leq 12|C_W|$.

Proof. Assume that there are r_i 3-stars of case-i where $i \in \{1, 2, 3\}$, we define an *interval* as a maximal path whose endpoints are in $N(N(W))$ and other intermediate vertices are not in $N(N(W))$. The proof of this lemma is based on computing the number of intervals between the vertices in $N(N(W))$. From Lemma 8, there are at most $3r_1$ distinct vertices in $N(F_X)$. From Lemma 5, every leaf of a case-1 3-star connects at most two vertices in $N(W)$, both of which are of degree one in $G - W$. Also, there are at most $3r_2$ distinct vertices which are the neighbors of the centers of case-2 3-stars. We conclude that after deleting neighbors of its corresponding center, every leaf of a case-2 3-star connects at most two vertices in $N(W)$, both of which are of degree one. Besides, there are at most $6r_3$ distinct vertices who are neighbors of the leaves of case-3 3-stars. For any vertex u in $N(W)$, when we contract the vertices in $(N[N(u)])$ in $G - W$ to a single vertex, there are at most $6r_1 + 6r_2 + 6r_3 - m/2$ intervals between those resulting vertices, where m is the number of tails (endpoints) of paths that are not in $N(W)$. Since there are at most two consecutive vertices in T lying in each interval and at most one vertex lies in each of the m tails, $|T| \leq 2 * (6r_1 + 6r_2 + 6r_3 - m/2) + m = 12|C_W|$. □

Theorem 1. *The Co-Path/Cycle Packing problem has a linear kernel of size $37k$.*

Proof. First of all, note that $V = W \cup F(L_W) \cup F(C_W - X) \cup F_Y \cup T$. Assume that there are r_i 3-stars of case-i where $i \in \{1, 2, 3\}$, from Lemma 5, every case-3 3-star corresponds to at most 18 vertices in $F(L_W) \cup F(C_W - X)$, every case-2 3-star corresponds to at most 21 vertices in $F(L_W) \cup F(C_W - X)$ and every leaf of case-1 3-star corresponds to at most 12 vertices in $F(L_W)$. Therefore, $|F(L_W) \cup F(C_W - X)| \leq (12r_1 + 21r_2 + 18r_3)$. From Lemma 8, every center of case-1 3-star corresponds to at most 9 vertices in F_Y. Then, $|F_Y| \leq 9r_1$. Following Lemma 10, $|T| \leq 12|C_W| = 12r_1 + 12r_2 + 12r_3$. In addition, $|W| = 4(r_1 + r_2 + r_3)$. Consequently, $|V| = |W| + |F(L_W) \cup F(C_W - X)| + |F_Y| + |T| \leq (37r_1 + 37r_2 + 34r_3) \leq 37(r_1 + r_2 + r_3) \leq 37k$, since we have $r_1 + r_2 + r_3 \leq k$. □

For completeness, we show below that the Co-Path/Cycle Packing problem is at least as hard as Vertex Cover, in terms of designing both approximation and FPT algorithms.

Theorem 2. *The Co-Path/Cycle Packing problem is APX-hard and cannot have a linear kernel of size smaller than $2k$.*

Proof. We prove the theorem by a simple reduction from *Vertex Cover*, which is APX-hard. Moreover, there is a famous conjecture that Vertex Cover cannot be approximated within a factor smaller than 2 [10]; consequently, no kernel smaller than $2k$ exists (unless the conjecture is disproved). The detailed reduction is as follows. Given an instance of *Vertex Cover*, $G = (V, E)$, we construct an instance of Co-Path/Cycle Packing $G^{'} = (V^{'}, E^{'})$. Let V_1 and V_2 be two vertex set each has $|V|$ isolated vertices. For each $v_i \in V$, add an edge between v_i and an isolated vertices in V_1 and V_2 respectively, so the set of new edges is defined as $E^* = \{(v_i, x_i)|v_i \in V, x_i \in V_1\} \cup \{(v_i, y_i)|v_i \in V, y_i \in V_2\}$. Then, $V^{'} = V \cup V_1 \cup V_2$, $E^{'} = E \cup E^*$. It is easily seen that any Vertex Cover in G is also a co-path/cycle packing in $G^{'}$, and any co-path/cycle packing corresponds to a Vertex Cover with at most the same size (since we can replace the vertices in the solution from $V_1 \cup V_2$ with their neighbors in V). Then the two instances could have the same optimal solution. Hence, the Co-Path/Cycle Packing problem cannot achieve a smaller approximation factor than that for Vertex Cover, and is unlikely to have a kernel of size smaller than $2k$, unless the conjecture by Khot and Regev is disproved. □

4 The FPT Algorithm

In this section, we elaborate the FPT algorithm which runs in $O^*(3.24^k)$ time. The key idea of the algorithm is bounded search tree. (For a survey, please refer to the paper by Fernau and Raible [8].) We implement it by some involved analysis. The following lemmas are critical while applying the bounded search tree method.

Lemma 11. *Given a graph $G = (V, E)$, if there exists a vertex v with $d(v) \geq 3$, then any minimum co-path/cycle packing for G either contains v or contains all but at most two of its neighbors.*

Lemma 12. *Given a graph $G = (V, E)$, if there exists an edge (u, v) with $d(u) = d(v) = 3$ and $N(u) \cap N(v) = \phi$, then any minimum co-path/cycle packing for G either contains one of u and v or contains at least one neighbor of u and one neighbor of v.*

Lemma 13. *Given a graph $G = (V, E)$, if there exists three edges (u, v), (u, w) and (v, w) with $d(u) = d(v) = 3$ and $N(\{u, v, w\}) = \{x, y, z\}$, then any a minimum co-path/cycle packing either contains one of u, v, w or contains all of x, y, z.*

The above three lemmas are easy to check, since otherwise there will be some vertex with degree greater than two left.

Lemma 14. *Given a graph $G = (V, E)$, if there exists an edge (u, v) with $d(u) = d(v) = 3$, $x \in (N(u) \cap N(v))$, and $d(x) = 2$, then there exists a minimum co-path/cycle packing which does not contain x.*

Proof. When a minimum co-path/cycle packing W contains x, $W - x + u$ remains a minimum co-path/cycle packing. □

Lemma 15. *Given a graph $G = (V, E)$, if there exists an edge (u, v) with $d(u) = d(v) = 3$ and $N(u) \cap N(v) = \{x, y\}$, then there exists a minimum co-path/cycle packing which does not contain v.*

Proof. When a minimum co-path/cycle packing W contains v, $W - v + u$ remains a minimum co-path/cycle packing. □

Lemma 16. *Given a graph $G = (V, E)$, if there exists a 3-star with the center being of degree three and every leaf of degree at most two in G, then there is a minimum co-path/cycle packing for G which only contains one leaf in the star.*

Proof. Any minimum co-path/cycle packing should contain at least one vertex from any 3-star. From Lemma 2 and Lemma 3, the deletion of a leaf results in the reservation of the other three vertices (in an optimal solution). So we just prove the existence of a minimum co-path/cycle packing which does not contain the center. Suppose on the contrary that the center is in a minimum co-path/cycle packing W, then we can modify W by replacing the center with any leaf. Obviously, only two paths are connected together in $G - W$ as a result of the replacement. Hence, the lemma holds. □

The detailed algorithm based on the above lemmas is presented at the end of the paper.

Theorem 3. *Algorithm $CPCP(G, k)$ solves the Co-Path/Cycle Packing problem correctly in $O^*(3.24^k)$ time.*

Proof. Step 1 deals with the boundary cases for the k Co-Path/Cycle Packing problem: if $k < 0$, then no co-path/cycle packing of size at most k can be found, thus the algorithm returns 'NO'. If $k \geq 0$ and G consists of only paths and cycles, then there is no need to remove any vertex from G, thus ϕ can be returned safely.

Step 2 considers the case when there is a vertex v of degree greater than 3. Following Lemma 11, it returns a k co-path/cycle packing correctly.

Step 3 deals with the case when there is an edge (u, v) such that u and v are of degree-3. From Lemma 12, Step 3.1 returns a k co-path/cycle packing correctly if it exists. From Lemma 14, Step 3.2.1 returns a k co-path/cycle packing correctly if it exists. From Lemma 13, Step 3.2.2 returns a k co-path/cycle packing correctly if it exists. From Lemma 15, Step 3.3 returns a k co-path/cycle packing correctly if it exists.

After Step 1, we have that $k \geq 0$ and G cannot consist of paths and cycles only, which implies that there is at least a vertex of degree greater than 2. After Step 2, all the vertices have degree less than or equal to 3 in G. After Step

3, no two degree-3 vertices are adjacent. It follows that all three neighbors of any degree-3 vertex v must be of degree less than 2. From Lemma 16, Step 4 returns a k co-path/cycle packing correctly if it exists. After that, every vertex has degree at most two. This completes the correctness proof of the algorithm.

Now we analyze the running time. As a convention, we take the notation $O^*(g(k))$ to refer to $O(g(k)n^c)$, where c is a constant independent of k and n. Step 1 takes time of $O(|G|)$. Let $f(k)$ be the time complexity for the k Co-Path/Cycle Packing problem. Step 2 has recurrence

$$f(k) = f(k-1) + \binom{i}{2} f(k-(i-2)), \quad \text{where } i \geq 4.$$

Step 3 has recurrence

$$f(k) \leq \begin{cases} 2f(k-1) + 4f(k-2), \\ 2f(k-1) + 2f(k-2), \\ 3f(k-1) + f(k-3), \\ 3f(k-1). \end{cases}$$

Step 4 has recurrence

$$f(k) = 3f(k-1).$$

We can verify that $f(k) \leq 3.24^k$, which comes from $f(k) = 2f(k-1) + 4f(k-2)$ at Step 3. Actually, we can show that $f(k) \leq 3^k$ by induction for the recurrence at Step 2. This concludes the proof. □

5 Concluding Remarks

In this paper, we present a $37k$ kernel for the minimum Co-Path/Cycle Packing problem. With the bounded search technique, we obtain an FPT algorithm which runs in $O^*(3.24^k)$ time. This problem is a special case of the Bounded-Degree Vertex Deletion (BDD) problem (when $d = 2$). So our result is the first linear kernel for BDD when $d = 2$. Previously, a linear kernel is only known for BDD when $d = 0$ (Vertex Cover) or $d = 1$ [6]. An interesting question is whether the bounds can be further improved. Another interesting question is to disallow cycles when some vertices are deleted (in the ancestral genome reconstruction problem, that means that no circular genome is allowed). In this latter case, we only have an $O(k^2)$ kernel.

Acknowledgments

This research is partially supported by NSF Career Award 084537, NSF grant DMS-0918034, and by NSF of China under grant 60928006. LW is fully supported by a grant from the Research Grants Council of the Hong Kong SAR, China [Project No. CityU 121207].

Algorithm *Co-Path/Cycle Packing (CPCP)*
Input: *a graph G and an integer k*
Output: *a co-path/cycle packing S for G such that* $|S| \leq k$*, or report 'NO'*
1 **if** $k < 0$, return 'NO'
  if G consists of only paths and simple cycles, return ϕ
2 pick a vertex v of degree greater than 3
  2.1 $S \leftarrow CPCP(G - v, k - 1)$
    if S is not 'NO', return $S + v$
  2.2 for every two neighbors w, z of v, G' is the graph after removing all neighbors of v except w, z from G.
    $S \leftarrow CPCP(G', k - (i - 2))$ ($i \geq 4$ is the number of neighbors of v).
    if S is not 'NO', return $S+$ all neighbors of v other than w, z
3 pick an edge $e = (u, v)$ such that both u and v are of degree 3.
  3.1 u and v have no common neighbor. let u_1, u_2 (v_1, v_2) be the other two neighbors of u (v).
    3.1.1 $S \leftarrow CPCP(G - u, k - 1)$ **if** S is not 'NO', return $S + u$
    3.1.2 $S \leftarrow CPCP(G - v, k - 1)$ **if** S is not 'NO', return $S + v$
    3.1.3 $S \leftarrow CPCP(G - u_i - v_j, k - 2)$, for each u_i, v_j ($i, j \in \{1, 2\}$), **if** S is not 'NO', return $S + u_i - v_j$
  3.2 u and v have only one common neighbor w. let x (y) be the other neighbors of u (v).
    3.2.1. **if** $d(w) = 2$
      3.2.1.1 $S \leftarrow CPCP(G - u, k - 1)$ **if** S is not 'NO', return $S + u$
      3.2.1.2 $S \leftarrow CPCP(G - v, k - 1)$ **if** S is not 'NO', return $S + v$
      3.2.1.3 $S \leftarrow CPCP(G - x - y, k - 2)$ **if** S is not 'NO', return $S + x + y$
    3.2.2. **if** $d(w) = 3$, let z be the other neighbors of w and $z \neq x, z \neq y$
      3.2.2.1 $S \leftarrow CPCP(G - u, k - 1)$ **if** S is not 'NO', return $S + u$
      3.2.2.2 $S \leftarrow CPCP(G - v, k - 1)$ **if** S is not 'NO', return $S + v$
      3.2.2.3 $S \leftarrow CPCP(G - w, k - 1)$ **if** S is not 'NO', return $S + w$
      3.2.2.4 $S \leftarrow CPCP(G - x - y - z, k - 3)$ **if** S is not 'NO', return $S + x + y + z$
    3.2.3 **if** $d(w) = 3$, let z be the other neighbors of w and $z \in \{x, y\}$
      3.2.3.1 **if** $z = x$, goto step3 with $e = (u, w)$.
      3.2.3.2 **if** $z = y$, goto step3 with $e = (v, w)$.
  3.3 u and v have two neighbors w_1, w_2in common, let x (y) be the other neighbor of w_1 (w_2)
    3.3.1 $S \leftarrow CPCP(G - u, k - 1)$ **if** S is not 'NO', return $S + u$
    3.3.2 $S \leftarrow CPCP(G - w_1, k - 1)$ **if** S is not 'NO', return $S + w_1$
    3.3.3 $S \leftarrow CPCP(G - w_2, k - 1)$ **if** S is not 'NO', return $S + w_2$
4 pick a degree-3 vertex v with neighbors v_1, v_2, v_3, **for each** v_i ($i \in \{1, 2, 3\}$)
  $S \leftarrow CPCP(G - v_i, k - 1)$, **if** S is not 'NO', return $S + v_i$.
5 return 'NO'

References

1. Abu-Khzam, F.N., Fellows, M.R., Langston, M.A., Suters, W.H.: Crown structures for vertex cover kernelization. Theory Comput. Syst. 41(3), 411–430 (2007)
2. Chen, J., Kanj, I.A., Jia, W.: Vertex cover: Further observations and further improvements. J. Algorithms 41(2), 280–301 (2001)
3. Chauve, C., Tannier, E.: A methodological framework for the reconstruction of contiguous regions of ancestral genomes and its application to mammalian genome. PLoS Comput. Biol. 4, e1000234 (2008)
4. Downey, R., Fellows, M.: Parameterized Complexity. Springer, Heidelberg (1999)
5. Fujito, T.: Approximating node-deletion problems for matroidal properties. J. Algorithms 31, 211–227 (1999)
6. Fellows, M., Guo, J., Moser, H., Niedermeier, R.: A generalization of Nemhausser and Trotter's local optimization theorem. In: Proc. STACS 2009, pp. 409–420 (2009)
7. Flum, J., Grohe, M.: Parameterized Complexity Theory. Springer, Heidelberg (2006)
8. Fernau, H., Raible, D.: Search trees: an essay. In: Proc. TAMC 2009, pp. 59–70 (2009)
9. Jiang, H., Chauve, C., Zhu, B.: Breakpoint distance and PQ-trees. In: Javed, A. (ed.) CPM 2010. LNCS, vol. 6129, pp. 112–124. Springer, Heidelberg (2010)
10. Khot, S., Regev, O.: Vertex cover might be hard to approximate to within $2 - \epsilon$. J. Comput. System Sci. 74, 335–349 (2008)
11. Lewis, J., Yannakakis, M.: The node-deletion problem for hereditary properties is NP-complete. J. Comput. System Sci. 20, 425–440 (1980)
12. Niedermeier, R.: Invitation to fixed-parameter algorithms. Oxford University Press, Oxford (2006)
13. Tannier, E., Zheng, C., Sankoff, D.: Multichromosomal median and halving problems under different genomic distances. BMC Bioinformatics 10, 120 (2009)

A VaR Algorithm for Warrants Portfolio

Jun Dai[1], Liyun Ni[1], Xiangrong Wang[2], and Weizhong Chen[1]

[1] School of Economics and Management, Tongji University, Shanghai, 200092, China
[2] College of Info Sci & Engi, Shandong University of Science & Technology, Qingdao, 266510, China
shdaijun@gmail.com

Abstract. Based on Gamma Vega-Cornish Fish methodology, this paper propose the algorithm for calculating VaR via adjusting the quantile under the given confidence level using the four moments (e.g. mean, variance, skewness and kurtosis) of the warrants portfolio return and estimating the variance of portfolio by EWMA methodology. Meanwhile, the proposed algorithm considers the attenuation of the effect of history return on portfolio return of future days. Empirical study shows that, comparing with Gamma-Cornish Fish method and standard normal method, the VaR calculated by Gamma Vega-Cornish Fish can improve the effectiveness of forecasting the portfolio risk by virture of considering the Gamma risk and the Vega risk of the warrants. The significance test is conducted on the calculation results by employing two-tailed test developed by Kupiec. Test results show that the calculated VaRs of the warrants portfolio all pass the significance test under the significance level of 5%.

Keywords: Warrant VaR, Gamma Risk, Vega Risk, Gamma Vega-Cornish Fish, Gamma-Cornish Fish.

1 Introduction

With the rapid development of the financial market, financial institutions and business face serious financial risk because of financial market showing unprecedented volatility. Financial risk management has become the necessity of business and financial institutions. The quantitative analysis and assessment of risk, namely risk measurement, is the core and foundation of risk management. With the increase of the complexity of financial transaction and financial market, the financial theory has developed continuously. Risk measurement method has developed from the nominal method, sensitivity analysis method and fluctuation method to complex VaR, pressure test and extreme value theory. Since G30 suggested that the institution with derivatives should employ VaR model as the specific measure of market risk in the research reports of derivatives in July 1997, the conception of VaR model has been accepted by market gradually and become the mainstream of risk measurement method in financial market.

Classical VaR calculation method supposes that the returns of portfolio satisfy gaussian distribution (normal distribution) wherein the symmetricity of the returns of portfolios is hidden. Thus, we can calculate the portfolio's VaR via the second moment (variance) of the distribution of the earnings rate at most. However, since its

B. Chen (Ed.): AAIM 2010, LNCS 6124, pp. 103–111, 2010.

asymmetry, the distribution of the earnings rate of the portfolio which contains option assets is with skewed at least. That is to say, we will need at least third moment of the distribution of the earnings rate to obtain the VaR of the portfolio, which is an evident difference from standard normality.

Furthermore, the proportion of Vega risk may be very big in the portfolio including options. Especially, it is necessary to consider Vega risk when analyzing the characteristics of the portfolio which includes options and employs Delta spot hedging strategy, straddle type price spread strategy and Strangle type spread strategy etc [1].

The remainder of this paper is organized as follows. Section 2 introduces the Gamma Vega-CF method briefly. Section 3 elaborates the proposed VaR calaulation algorithm for the warrants portfolio. Section 4 gives empirical results to illustrate the effectiveness and efficiency of the proposed method. Finally, concluding remarks are drawn in Section 5.

2 Gamma Vega-CF Method

Delta model which employs linear form, simplifies VaR calculation. The weakness of Delta model is that it unable to identify non-linear risks. Warrant is nonlinear securities,which is with convexity risk. Therefore, this paper introduces Gamma model as the VaR calculation method to identify the convexity or Gamma risk .

Gamma normal model is similar to Delta normal model as they both assume that the change of market factor is subject to normal distribution. The difference is that Gamma normal model uses the value function, which employs second order Taylor expansion to describe portfolio, to better capture the nonlinear characteristics of the changes of portfolio's price. Suppose that the standard risk measurement methods can fully embody the risk of portfolio. The second-order Taylor expansion for the return of portfolio is as follows:

$$dP = \frac{\partial P}{\partial S} dS + \frac{1}{2}\frac{\partial^2 P}{\partial S^2}(dS)^2 + \frac{\partial P}{\partial Vol_{stock}} dVol_{stock} + \frac{\partial P}{\partial r} dr + \frac{\partial P}{\partial y} dy + \frac{\partial P}{\partial t}(T-t)\,dt \tag{1}$$

where P is the value of portfolio and dP descibes the change of portfolio's value. S is the price of the object's stock corresponding to the warrant. Vol_{stock} is the volitility of the stock and r is the return rate of the object's stock. y is the exercise price, T is the duration, and t is current time.

Because the option price is not the linear function of the above parameters, its VaR can not estimate via linear method merely. Thus, considering both the Gamma risk and the Vega risk, the approximate expression of the portfolio return is modified as following: (taking on Equation (1)'s preceding three terms)

$$dP = \delta dS + \frac{1}{2}\Gamma(dS)^2 + \kappa d\sigma \tag{2}$$

where σ is the implied volatility. Suppose that the implied volatility takes on log-normal distribution, we have

$$\ln\left(\frac{\sigma_{t+1}}{\sigma_t}\right) \sim N\left(0, Vol_\sigma^2\right) \tag{3}$$

where Vol_σ is the standard deviation of implied volatility.

For the approximate expression of dP, that is Equation (2), we have

$$E(dP)=\frac{1}{2}\Gamma S^2 E\left[\left(\frac{dS}{S}\right)^2\right]=\frac{1}{2}\Gamma S^2 Vol_{stock}^2 \tag{4}$$

$$V(dP)=S^2\delta^2 V\left(\frac{dS}{S}\right)+S^4\left(\frac{1}{2}\Gamma\right)^2 V\left[\left(\frac{dS}{S}\right)^2\right]+\kappa^2\sigma^2 V\left(\frac{d\sigma}{\sigma}\right)+2\left(S^3\delta\frac{1}{2}\Gamma\right)Cov\left[\frac{dS}{S},\left(\frac{dS}{S}\right)^2\right] \tag{5}$$
$$+2S\delta\kappa\sigma Cov\left(\frac{dS}{S},\frac{d\sigma}{\sigma}\right)+2S^2\frac{1}{2}\Gamma\kappa\sigma Cov\left[\left(\frac{dS}{S}\right)^2,\frac{d\sigma}{\sigma}\right]$$

where we suppose that $Cov\left[\left(\frac{dS}{S}\right)^2,\frac{d\sigma}{\sigma}\right]=0$.

Because Gamma normal model considers the influence of the Gamma risk, it is superior to Delta model in calculating the convexity assets such as options etc. Influenced by the Gamma risk, the distribution of the portfolio return rate including options often occurs offset instead of obeying normal distribution. Therefore, Morgan JP introduced Gamma-Cornish Fish model into his RiskMetrics[2].

The fundamental principle of Gamma-CF model is that it can adjust the confidence interval parameters to correct the influence of Gamma risk on the skewness of normal distribution. Cornish-Fisher expansion formula is based on the statistics principle: any distribution (e.g., chi-square distribution) can be considered as the function of other distribution (such as normal distribution), which can be expressed by the parameters of other distribution[3]. For any distribution, based on the normal distribution, the expansion of CF is defined as

$$F_{\Delta p}(\alpha)=z_\alpha+(z_\alpha^2-1)\rho_3/6+(z_\alpha^2-3z_\alpha)\rho_4/24-(2z_\alpha^3-5z_\alpha)\rho_3^2/36 \tag{6}$$

where $F_{\Delta p}(\alpha)$ is the distribution function of the portfolio return ΔP, z_α is the quantile of standard normal distribution. ρ_3 and ρ_4 are the cubic and quartic cumulants of ΔP, respectively. ρ_3 is the measure of the distribution skewness, ρ_4 is the measure of the distribution kurtosis. Thus, VaR can be estimated by replacing α with confidence parameter $-(\alpha-s_\alpha)$ and approximately yeilding quantile under the normal distribution.

3 Algorithm

3.1 Hypotheses

In order to give the algorithm, we propose the following hypotheses.

1) Now hold n warrants. The price of each warrant is $C_i(S_i,\sigma_i,r,X_i,T_i-t)$, $i=1,2....,n$, where S_i is the price of the object's stock corresponding to a warrant, σ_i is the volatility of an object's stock, X_i is the strike price of a warrant, and T_i is the maturity date of a warrant.
2) The volume of holding of each warrant is π_i, $i=1,2....,n$.

3) The day return rate of the object's stock corresponding to a warrant is r_i, $i = 1,2....,n$. The logarithmic return rate of a stock obeys normal distribution $N(\mu, Vol_{stock}^2)$. Then, we have

$$\begin{aligned} dS &= \mu S_t + Vol_{stock} S_t dw \\ dw &= \eta_1 \sqrt{dt} \\ \eta_1 &\sim N(0,1) \end{aligned}.$$

Therefore, the value of the holding position is

$$P(t_0, C) = \sum_{i=1}^{n} \pi_i C_i (S_i, \sigma_i, r, X_i, T_i - t). \tag{7}$$

4) The differential form of the variation of the holding position value is

$$dP = \sum_{i=1}^{n} \pi_i (\delta_i dS_i + \frac{1}{2}\Gamma_i (dS_i)^2 + \kappa_i d\sigma_i) = \sum_{i=1}^{n} \pi_i (\delta_i S_i \frac{dS_i}{S_i} + \frac{1}{2}\Gamma_i S_i^2 \left(\frac{dS_i}{S_i}\right)^2 + \kappa_i \sigma_i \frac{d\sigma_i}{\sigma_i}). \tag{8}$$

The variation value is

$$\begin{aligned} &\Delta P = \sum_{i=1}^{n} \pi_i (\delta_i S_i r_i + \frac{1}{2}\Gamma_i S_i^2 (r_i)^2) \quad , R_P^G = \frac{\Delta P}{P} = \sum_{i=1}^{n} (k_i r_i + f_i r_i^2) \quad , \\ &k_i = \pi_i \delta_i S_i / P, f_i = \frac{1}{2} \pi_i \Gamma_i S_i^2 / P \end{aligned} \tag{9}$$

where $\delta_i = N_i\left(d_1^i\right)$, $\Gamma_i = \dfrac{N_i'\left(d_1^i\right)}{S_i \sigma_i \sqrt{T_i - t_i}}$, $\kappa_i = S_i \sqrt{T_i - t_i} N_i'\left(d_1^i\right)$,

$$N_i'(x) = \frac{1}{\sqrt{2\pi}} \exp(-\frac{x^2}{2}),\ d_1^i = \frac{\ln(S_i / X_i) + (r + \sigma_i^2 / 2)(T_i - t_i)}{\sigma_i \sqrt{T_i - t_i}}.$$

3.2 Algorithm Description

Based on the above analysis and hypotheses, we propose the algorithm for calculating VaR of warrants portfolio. Our entire algorithm implements following the next steps:

Step 1) Select the past data quantity t. Let $r_i(t)$ indicate the *t-th* day return rate of the object's stock corresponding to warrant i. $t = 0$ indicate the current day return rate.

Step 2) Select window time T_0, e.g., 100.

Step 3) Calculate the day return rate of the object's stock, which is given by

$$r_{i,-t} = \ln\left(\frac{S_{i,-t}}{S_{i,-t-1}}\right) = \ln\left(S_{i,-t}\right) - \ln\left(S_{i,-t-1}\right),\ i = 1,2,\hbar\ ,n\ ,\ 1 \le t \le T_0. \tag{10}$$

Step 4) Acquire the warrant price historical data $C_{i,t}\left(S_{i,t}, X_i, r, T_t, \sigma_{i,t}\right)$ in the warrants portfolio. Given $S_{i,t}, X_i, r, T_t$, obtain $\sigma_{i,t}$, $i = 1, 2, \hbar\ , n$, $1 \le t \le T_0$ calculated by dichotomy.

Step 5) Calculate the change rate of implied volatility as follows:

$$r_{i,t}^{\sigma} = \ln(\frac{\sigma_{i,t}}{\sigma_{i,t-1}}),\ i = 1,2,\hbar\ ,n\ ,\ 1 \le t \le T_0\ . \tag{11}$$

Step 6) Calculate the mean and variance of the change rate of implied volatility.
The mean of the change rate of implied volatility is

$$\mu_i^{\sigma} = \frac{1}{T_0 - 2}\sum_{i=2}^{T_0} r_{i,t}^{\sigma}\ ,\ i = 1,2,\hbar\ ,n\ . \tag{12}$$

The variance of the change rate of implied volatility is

$$Vol_{i,\sigma} = \frac{1}{T_0 - 2}\sum_{i=2}^{T_0} (r_{i,t}^{\sigma} - \mu_i^{\sigma})^2\ ,\ i = 1,2,\hbar\ ,n\ . \tag{13}$$

Step 7) Calculate the mean, variance, skewness and kurtosis of the portfolio return rate.
The mean of the portfolio return rate is

$$\mu = \frac{1}{T_0 - 1}\sum_{t=1}^{T_0 - 1} R_{p,-t}^{G} = \frac{1}{T_0 - 1}\sum_{t=1}^{T_0 - 1}\sum_{i=1}^{n} (k_i r_{i,-t} + f_i r_{i,-t}^2)\ . \tag{14}$$

The variance of the portfolio return rate is $\sigma^2 = \frac{1}{T_0 - 1}\sum_{t=1}^{T_0 - 1}\left(R_{p,-t}^{G} - \mu\right)^2$. (15)

The skewness of the portfolio return rate is

$$\rho_3 = E\left(R_p^G - ER_p^G\right)^3 / \sigma^3 = \frac{1}{T_0 - 1}\sum_{t=1}^{T_0 - 1}\left(R_{p,-t}^{G} - \mu\right)^3 / \sigma^3\ . \tag{16}$$

The kurtosis of the portfolio return rate is

$$\rho_4 = E\left(R_p^G - ER_p^G\right)^4 / \sigma^4 - 3 = -3 + \frac{1}{T_0 - 1}\sum_{t=1}^{T_0 - 1}\left(R_{p,-t}^{G} - \mu\right)^4 / \sigma^4\ . \tag{17}$$

Step 8) Calculate the quantile cv_{α} of the corresponding change.
Based on normal distribution, the expansion of the Cornish-Fisher is

$$cv_{\alpha} = z_{\alpha} + (z_{\alpha}^2 - 1)\rho_3 / 6 + (z_{\alpha}^3 - 3z_{\alpha})\rho_4 / 24 - (2z_{\alpha}^3 - 5z_{\alpha})\rho_3^2 / 36 \tag{18}$$

where cv_{α} is the quantile of portfolio return rate R_p^G in the confidence level α, z_{α} is the quantile of in the standard normal distribution.

Step 9) Calculate the risk value of the portfolio value in one future day VaR.

The EWMA variance[4][5] is $\sigma_P^2 = (1 - \lambda)\sum_{t=1}^{T_0 - 1}\lambda^{T_0 - t}\left(R_{p,-t}^{G} - \mu\right)^2$. (19)

Thus, we have $VaR = P(t_0, C)(\sigma_P cv_{\alpha} + \mu)$. (20)

4 Empirical Results

4.1 Data and Empirical Tests

Considering both the Gamma risks and Vega risks, this paper calculates the VaR values of CITIC Guoan warrants to subscribe GAC1, steel vanadium warrants to subscribe GFC1, Shenfa warrants to subscribe SFC2, respectively, employing Gamma Vega-CF, Gamma-CF and standard normal distribution methods. Furthermore, we test the VaR calculation results via posteriori test method. Meanwhile, we calculate the risk values of portfolio of GFC1 and SFC2. The basic data for empirical research is given in Tabel 1.

Table 1. Data for VaR calculation and test

Type	Name	Code	Duration	Samples Time Period	Posteriori Test Time Period
Single Warrant	GAC1	031005	730	2007/9/25~ 2008/05/12	2007/12/25~ 2008/05/12
Single Warrant	GFC1	031002	730	2006/12/12~ 2008/05/12	2007/04/13~ 2008/05/12
Single Warrant	SFC2	031004	365	2007/06/29~ 2008/05/12	2007/10/29~ 2008/05/12
Portfolio Warrants	GFC1+SFC2	NA	NA	2007/06/29~ 2008/05/12	2007/10/29~ 2008/05/12

Our empirical research calculates the volatility of the object's stock and the return rate distribution characteristics of the warrants portfolio employing the historical data of 80 day. The risk-free rate, confidence level and decay factor are set as 0.03, 0.95 and 0.94, respectively. For the warrants portfolio, the configuration proportion of GFC1 and SFC2 is 2:1. Tabel 2 gives the empirical results of calculation. Note that all the data in the table except "samples number" and "exception" is mean.

Table 2. Empirical results of VaR calculation

Name	Samples number	Gamma Vega-Cornish Fish			Gamma-Cornish Fish			Standard Normal		
		Lower 5% (VaR)	Upper 5%	Exception	Lower 5%(VaR)	Upper 5%	Exception	Lower 5%(VaR)	Upper 5%	Exception
GAC1	89	-5.75%	7.40%	9*	-8.02%	8.11%	2*	-6.46%	6.62%	10
GFC1	202	-7.99%	9.49%	11*	-8.67%	8.77%	10*	-9.05%	10.09%	10*
SFC2	130	-7.77%	7.33%	10*	-9.14%	9.63%	8*	-7.09%	6.86%	15
GFC1+ SFC2	130	-6.08%	6.01%	11*	-7.22%	7.68%	8*	-5.29%	5.91%	14

4.2 Validity Tests

1) Performance Assessment. Based on the performance assessment method proposed by Morgan J.P. in [4], we can see the effect of the risk control intuitively through the comparison between the profit and loss of the portfolio and the VaR value of the next day via calculation. Fig. 1. (a) to (d) give the concrete results where the vertical axis denotes the profit and loss of the portfolio and the horizontal axis denotes the observed number.

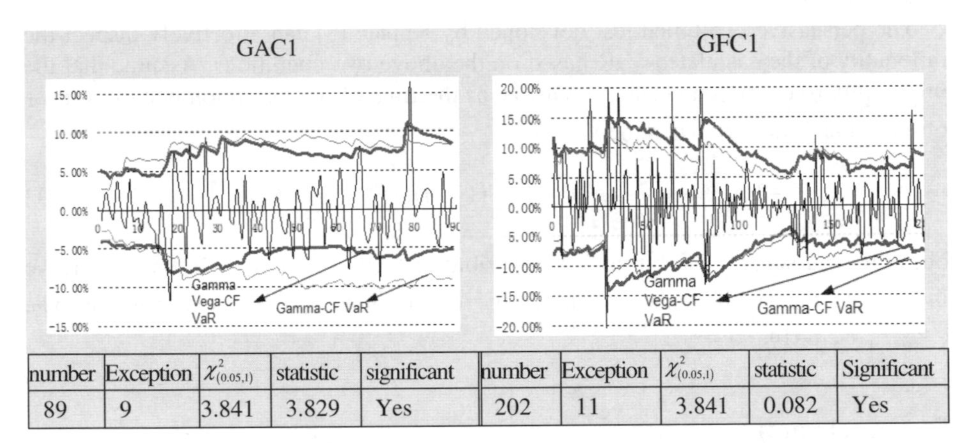

number	Exception	$\chi^2_{(0.05,1)}$	statistic	significant
89	9	3.841	3.829	Yes

number	Exception	$\chi^2_{(0.05,1)}$	statistic	Significant
202	11	3.841	0.082	Yes

(a) VaR values of CITIC Guoan warrants (b) VaR values of steel vanadium warrants

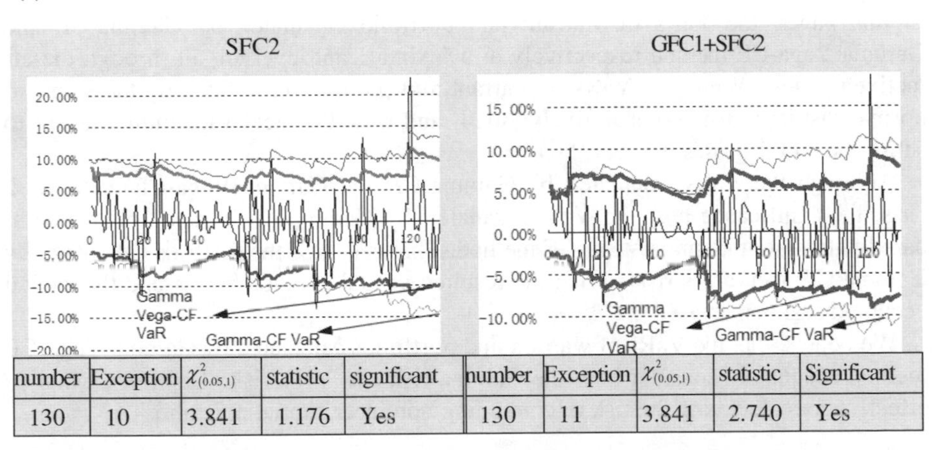

number	Exception	$\chi^2_{(0.05,1)}$	statistic	significant
130	10	3.841	1.176	Yes

number	Exception	$\chi^2_{(0.05,1)}$	statistic	Significant
130	11	3.841	2.740	Yes

(c) VaR values of Shenfa warrants (d) VaR values of portfolio of GFC1 and SFC2

Fig. 1. Comparison of the profit and loss of the portfolio and the VaR values of the next day via calculation employing Gamma Vega-CF and Gamma-CF, respectively

2)Significance Test. Regardless of which method employed to calculate VaR, we must compare VaR with the actual situation of the portfolio, which is called a back test. In back test, we must find the calculated VaR under the circumstances of the times of the portfolio loss a day surpass a given confidence level of 95%. The situation of VaR which the actual loss is more than 95% is called *exception*.

For the confidence level of 95%, if the exception days are around 5% of the total days, it shows that the model is credible. However, if the exception days far outweigh 5%, e.g. 10%, we have reason to believe that the VaR estimates much lower. From the perspective of the regulation, the capital of the VaR will be too low. On the other hand, if the frequency of the exception occurs far less than 5%, e.g. 1%, we have reason to believe that the estimated VaR is much higher and the resulted capital is much high too.

The persuasive two-tailed test developed by Kupiec [6] can effectively inspect the rationality of the calculated VaR based on the above two conditions. Assume that the probability of exception occurs in VaR is p, the time of the exception occurred in n observation days is m. Then the variable

$$-2\ln\left[(1-p)^{n-m}p^{m}\right]+2\ln\left[(1-m/n)^{n-m}(m/n)^{m}\right] \tag{21}$$

obeys the chi-square distribution (χ^2 distribution) with freedom 1. In this paper, we investigate the significance test under the circumstances of the exception is 5%, ($\chi^2_{(0.05,1)}=3.841$).

5 Conclusion

In this paper, the VaRs of warrant and portfolio calculated by Gamma-CF and Gamma Vega-CF method respectively at 5% significant level are all through the effectiveness test. While the VaRs of warrant and portfolio calculated by the standard normal distribution is comparatively small, and only the steel vanadium warrant to subscribe GFC1 passes the test.

Although the VaRs calculated by Gamma-CF and Gamma Vega-CF method all pass the significance test, the VaR calculated by Gamma-CF is much bigger compared with that of Gamma Vega-CF method. That is to say, the the VaR calculated by Gamma-CF overvalues risk so that the resulted capital is much higher and the capital occupied cost is increased.

We believe that the VaRs of warrant and portfolios calculated by Gamma Vega-CF method adequately consider both the Gamma risk and Vega risk of options, which is effective for calculating the risk of containing nonlinear captial portfolio.

Acknowledgment

This work is supported by the National Natural Science Foundation of China (NSFC), under Grant No. 70971079 and the Research Project of "SUST Spring Bud" under Grant No. 2008AZZ051.

References

1. Smith, T.F., Waterman, M.S.: Identification of Common Molecular Subsequences. J. Mol. Biol. 147, 195–197 (1981)
2. Malz, A.M.: Vega Risk and the Smile, Working Paper. The Risk Metrics Group, New York (2000)
3. Zangari, P.: A VaR Methodology for Portfolios that Including Options, New York (1996)
4. Johnson, N.L., Kotz, S.: Distributions in Statistics: Continuous Univariate Distributions. John Wiley and Sons, New York (1972)
5. Morgan, J.P.: Risk Metrics- Technical Document, 4th edn. New York, pp. 77–88, 219–220 (December 1996)
6. Zhou, D., Zhou, D.: Risk Management Frontier - the Theory and Application of Risk Value. Renmin University of China, Beijing (2004)
7. Kupiec, P.: Techniques for Verifying the Accuracy of Risk Management Models. Journal of Derivatives 3, 73–84 (1995)

Some Results on Incremental Vertex Cover Problem*

Wenqiang Dai

School of Management and Economics, University of Electronic Science and Technology of China, Chengdu, Sichuan, 610054, P. R. China
wqdai@uestc.edu.cn

Abstract. In the classical k-vertex cover problem, we wish to find a minimum weight set of vertices that covers at least k edges. In the incremental version of the k-vertex cover problem, we wish to find a sequence of vertices, such that if we choose the smallest prefix of vertices in the sequence that covers at least k edges, this solution is close in value to that of the optimal k-vertex cover solution. The maximum ratio is called *competitive ratio*. Previously the known upper bound of competitive ratio was 4α, where α is the approximation ratio of the k-vertex cover problem. And the known lower bound was 1.36 unless $P = NP$, or $2-\varepsilon$ for any constant ε assuming the Unique Game Conjecture. In this paper we present some new results for this problem. Firstly we prove that, without any computational complexity assumption, the lower bound of competitive ratio of incremental vertex cover problem is ϕ, where $\phi = \frac{\sqrt{5}+1}{2} \approx 1.618$ is the golden ratio. We then consider the restricted versions where k is restricted to one of two given values(Named 2-IVC problem) and one of three given values(Named 3-IVC problem). For 2-IVC problem, we give an algorithm to prove that the competitive ratio is at most $\phi\alpha$. This incremental algorithm is also optimal for 2-IVC problem if we are permitted to use non-polynomial time. For the 3-IVC problem, we give an incremental algorithm with ratio factor $(1+\sqrt{2})\alpha$.

1 Introduction

The classical VERTEX COVER problem has been widely studied in discrete optimization, see, e.g.[2,3,5,10,11,17,18]. In the standard weighted version, we are given an undirected graph $G = (V, E)$ with weights function on the vertices $w : V \rightarrow (R^+ \cup \infty)$, and are required to find a set $C \subseteq V$ with minimum total weight such that all edges in E are *covered* by having at least one endpoint in C. The problem is NP-hard and was one of the first problems shown to be NP-hard in Karp's seminal paper [13]. However, several different approximation algorithms have been developed for it [11,18]. The best of these achieve a performance ratio

* This work is supported in part by the NSF of China under Grant No.70901012, the Specialized Research Foundation for the Doctoral Program of Higher Education of China, Grant No. 200806141084, and Science Foundation of UESTC for Youths(JX0869).

B. Chen (Ed.): AAIM 2010, LNCS 6124, pp. 112–118, 2010.

of 2. Johan Håstad has proved in [8] that a performance ratio less than $\frac{7}{6}$ is not possible unless $P = NP$. This lower bound was improved to $10\sqrt{5}-21 \approx 1.36067$ under the same complexity assumption[6]. Recently there is another result, based on a stronger complexity assumption (Unique Game Conjecture) saying that it is hard to approximate to within $2 - \varepsilon$, for any constant ε[14].

Recently several generalizations of the vertex cover problem have been considered. Bshouty and Burroughs [4] were the first to study the k-vertex cover problem. They gave a 2-approximation algorithm for this problem. Subsequently, combinatorial algorithms with the same approximation guarantee were proposed, see, e.g.[12,1,7,16]. In this generalization we are not required to cover all edges of the graph, any edges may be left uncovered. More formally, in the k-vertex cover problem, we wish to find a minimum-weight set of vertices that covers at least k edges.

In this paper, we consider the incremental version of k-vertex cover problem where k is not specified in advance, which we call Incremental Vertex Cover Problem. Instead, authorizations for additional facilities arrive over time. Given an instance of the k-vertex cover problem, an algorithm produces an incremental sequence of vertices sets $\hat{S} = (S_1, S_2, \cdots, S_n)$ such that (i) $S_1 \subseteq S_2 \subseteq \cdots \subseteq S_n \subseteq V$; (ii) for any $k = 1, 2, \cdots, n$, S_k covers at least k edges, that is, any S_k is a feasible set of k-vertex cover problem; (iii) for any k, $cost(S_k) \leq c \cdot opt_k$, where opt_k denotes the corresponding optimum cost of k-vertex cover problem and c is called *competitive ratio.*

For this incremental problem, Lin et.al[15] described a competitive algorithm with competitive ratio 4α, where α is the approximation ratio of k-vertex cover problem. Thus by the known best 2-approximation algorithms, their algorithm presents the competitive ratio with 8 in polynomial time, and 4 in non-polynomial time. We also note that, the known lower bound is the same as the lower bound of approximation algorithm for k-vertex cover problem, and thus for vertex cover problem, that is, it is $10\sqrt{5} - 21 \approx 1.36067$ unless $P = NP$[6], or $2 - \varepsilon$ assuming the unique games conjecture[14].

In this paper, we will improve the lower bound of the competitive ratio. Indeed, we will show that no incremental sequence can have competitive ratio better than ϕ, where $\phi = \frac{\sqrt{5}+1}{2} \approx 1.618$ is the golden ratio, and this lower bound holds without any computational complexity assumption.

Motivated by the analysis process of lower bound, we then consider the restricted version of above incremental problem where k is restricted to one of two given values (Named 2-IVC problem). We give a simply competitive algorithm for 2-IVC problem to prove that the competitive ratio is at most $\phi\alpha$. This algorithm for 2-IVC problem is also optimal if we are permitted to use non-polynomial time by the proof process of the lower bound. An restricted version where k is restricted to one of three given values (Named 3-IVC problem) is also studied in this paper. For this problem, we give an incremental algorithm with ratio factor $(1 + \sqrt{2})\alpha$.

The rest of this paper is organized as follows. Section 2 gives the improved lower bound. Section 3 and 4 present the analysis of 2-IVC and 3-IVC problem as well as competitive ratio analysis, respectively. The final section, section 5, concludes the paper and describes future research.

2 Lower Bound Analysis

Theorem 1. *The competitive ratio of the incremental vertex cover problem is at least* $\phi = \frac{\sqrt{5}+1}{2} \approx 1.618$.

Proof. Consider the vertex weighted graph on five nodes shown in Fig. 1. In this graph, each vertex in A, B, C, D has a weight of $x = \phi - 1 \approx 0.618$, and vertex E has a weight of 1. Let us analysis any incremental algorithms.

Firstly, for the first one incremental request, any algorithm must choose one node in set $\{A, B, C, D\}$, otherwise, the weight of the 1-vertex cover subgraph chosen is 1, which is already off by the factor $\frac{1}{x} = \frac{1}{\phi-1} = \frac{\sqrt{5}+1}{2}$ from the optimal value of x (which is attained by the solution $\{A\}$ or $\{B\}$, or $\{C\}$, or $\{D\}$). Say, by symmetry, this step chooses node A. Now for the forth incremental request, the optimal incremental choose is E, resulting in an incremental solution $\{A, E\}$ with weight $1 + x$. However, the optimal choice for 4-vertex cover problem is set $\{E\}$, with a weight of 1.Thus, the solution chosen by the incremental algorithm is again off from the optimal value by a factor of $\frac{1+x}{1} = \phi = \frac{\sqrt{5}+1}{2}$. This completes the proof. □

Note that the lower bound of ϕ holds without any restriction on the amount of time that an algorithm may use to process each request, thus this lower bound is true for any algorithm, even if we are permitted to use non-polynomial time.

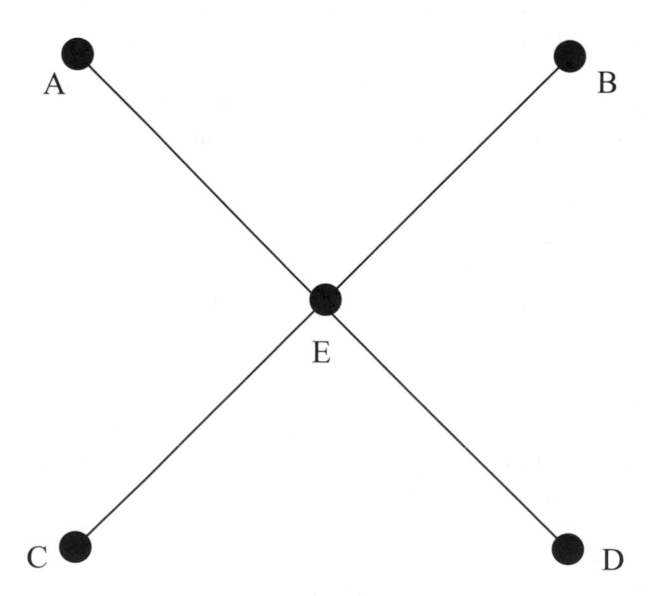

Fig. 1. Illustration used for the proof of Theorem 1. *Note*: vertex E has a weight of 1, and each vertex A, B, C, D has a weight of $x = \phi - 1$.

3 Incremental Algorithm for 2-IVC Problem

Motivated by the analysis process of lower bound, in this section, we will consider the restricted incremental vertex cover problem where k is restricted to one of two given values (Named 2-IVC problem), that is, for any given $1 \leq k < l \leq n$, we need to compute two vertex sets $S_k \subseteq S_l$ with each S_i covers at least i edges, minimizing the competitive ratio

$$c = max\{\frac{w(S_k)}{opt_k}, \frac{w(S_l)}{opt_l}\}$$

where $w(S_i) = \sum_{v \in S_i} w(v)$ for $i = k, l$.

The idea of algorithm and its analysis come from [9]. We now sketch the detail analysis as following for clarity and completeness, and for our later use.

Incremental Algorithm 1

Step 1. For each $i = k, l$, using an approximation algorithm of i-vertex cover problem to produce α-approximation solution sets F_k and F_l.

Step 2. Consider whether $w(F_k) \leq w(F_l)$. If it is not, then we simplify let $F_k = F_l$. Now two new approximation sets F_k,F_l with the conditions that $w(F_k) \leq w(F_l)$ is satisfied can be obtained.

This is because, if $w(F_l) < w(F_k)$, then the new set $F_k(new) = F_l$ satisfies it covers at least $l > k$ edges, and $w(F_k(new)) = w(F_l)) < w(F_k) \leq \alpha \cdot opt_k$.

Step 3. If $w(F_l) \geq \phi w(F_k)$, then $S_k = F_k$, $S_l = F_k \cup F_l$.

Step 4. Otherwise, we have $w(F_l) < \phi w(F_k)$, then $S_k = S_l = F_l$.

Firstly note that the incremental solution of above algorithm satisfy $S_k \subseteq S_l$ and for each $i = k, l$, S_i is a feasible solution for the i-vertex cover problem, that is, S_i covers at least i edges. And, the time consumption of above incremental algorithm is the same as the approximation algorithm of k-vertex cover problem used in Step 1. Finally we prove the algorithm have a good performance guarantees.

Theorem 2. *The incremental algorithm for 2-IVC problem presents a competitive ratio at most $\phi\alpha$, where $\phi = \frac{\sqrt{5}+1}{2}$ and α is the known approximation ratio of the k-vertex cover problem.*

Proof. When we have approximation solutions F_k and F_l, we have $w(F_k) \leq \alpha opt_k$ and $w(F_l) \leq \alpha opt_l$.

Now, if $w(F_l) \geq \phi w(F_k)$, we have $S_k = F_k$, $S_l = F_k \cup F_l$, then

$$w(S_k) = w(F_k) \leq \alpha opt_k$$

and

$$w(S_l) = w(F_k \cup F_l) \leq w(F_k) + w(F_l) \leq (\frac{1}{\phi} + 1)w(F_l) \leq (\frac{1}{\phi} + 1)\alpha opt_l$$

Thus in this case, the competitive ratio is $(\frac{1}{\phi} + 1)\alpha = \phi\alpha$.

On the other hand, if $w(F_l) < \phi w(F_k)$, we have $S_k = S_l = F_l$, then

$$w(S_k) = w(F_l) \leq \phi w(F_k) \leq \phi \alpha opt_k$$

and

$$w(S_l) = w(F_l) \leq \alpha opt_k$$

So in this case, the competitive ratio is also $\phi\alpha$. □

Combing the above 2-approximation algorithm and the proof process of Theorem 1, we have

Corollary 1. *For the 2-IVC problem, there exist an incremental algorithm with competitive ratio $2\phi \approx 3.236$ in polynomial time, and $\phi \approx 1.618$ in non-polynomial time. And this competitive ratio is optimal if we are permitted to use the non-polynomial time.*

4 Incremental Algorithm for 3-IVC Problem

In this section, we consider the 3-restricted incremental vertex cover problem, where k is restricted to one of three given values, that is, for any given $1 \leq k < t < l \leq n$, we need to compute three vertex sets $S_k \subseteq S_t \subseteq S_l$ with each S_i covers at least i edges, minimizing the competitive ratio

$$c = max\{\frac{w(S_k)}{opt_k}, \frac{w(S_t)}{opt_t}, \frac{w(S_l)}{opt_l}\}$$

where $w(S_i) = \sum_{v \in S_i} w(v)$ for $i = k, t, l$.

we will give an incremental algorithm with performance guarantee competitive ratio. Firstly we can easily know the following fact.

Lemma 1. *Let opt_k be the optimal weight of k-vertex cover problem, receptively, for each $k = 1, 2, ..., n$. Then we have*

$$opt_1 \leq opt_2 \leq \cdots \leq opt_n$$

Proof. This is because for each feasible solution of k-vertex cover problem, $k = 2, 3, ..., n$, it must also be a feasible solution of $(k-1)$-vertex cover problem. □

Note that by this lemma, for the 3-IVC problem, obtaining $c = 3\alpha$ approximation is rather easy——firstly generating three α-approximation solutions F_k, F_t, F_l, and simply let $S_k = F_k$, $S_t = F_k \cup F_t$, $S_l = F_k \cup F_k \cup F_l$. The challenge in this section is to improve the approximation below 3α, and indeed we present an algorithm achieving factor $(1 + \sqrt{2})\alpha$.

The idea of our algorithm come from the analysis of Algorithm 1. Now for the obtained three approximation solutions F_k, F_t, F_l, after a similarly done with Step 2 of Algorithm 1, we can assure that there are three approximation solutions F_k, F_t, F_l such that $w(F_k) \leq w(F_t) \leq w(F_l)$. Now there are two cases to be considered with a parameter λ, $1 \leq \lambda \leq 2$ to be chosen later.

CASE 1. $w(F_l) \leq \lambda w(F_k)(< \lambda w(F_t))$.

For this case, let

$$S_k = S_t = S_l = F_l$$

Now for any i, S_i covers at least i edges. And, in this case, we have for $i = k, t, l$, $w(S_i) = w(F_l) \leq \lambda w(F_1) \leq \alpha\lambda opt_k$. Thus, the competitive ratio is $\alpha\lambda$.

CASE 2. $w(F_l) > \lambda w(F_k)$.

Subcase 2.1. If $w(F_l) > \lambda w(F_t)$. That is, $w(F_l) > \lambda w(F_t) > \lambda w(F_k)$.

Let $S_k = F_k$, $S_t = F_k \cup F_t$, $S_l = F_k \cup F_t \cup F_l$. For any i, S_i also cover at least i edges. And we have

$$w(S_k) = w(F_k) \leq \alpha opt_k$$

$$w(S_t) \leq w(F_k) + w(F_t) \leq 2\alpha opt_t$$

$$w(S_l) \leq w(F_k) + w(F_t) + w(F_l) \leq (1 + \frac{2}{\lambda})\alpha opt_l$$

Thus, the competitive ratio is $\alpha(1 + \frac{2}{\lambda})$.

Subcase 2.2. If $w(F_l) < \lambda w(F_t)$. That is, $\lambda w(F_k) < w(F_l) < \lambda w(F_t) < \lambda w(F_l)$.

Let $S_k = F_k$, $S_t = F_k \cup F_l$, $S_l = F_k \cup F_l$. We have for any i, S_i also cover at least i edges. And

$$w(S_k) = w(F_k) \leq \alpha opt_k$$

$$w(S_t) \leq w(F_k) + w(F_l) \leq (1 + \lambda)\alpha opt_t$$

$$w(S_l) \leq w(F_k) + w(F_l) \leq 2\alpha opt_l$$

Thus, the competitive ratio is $\alpha(1 + \lambda)$.

Now let $\frac{2}{\lambda} = \lambda$, we get the competitive ratio $(1 + \sqrt{2})\alpha \approx 2.414\alpha$.

Summarized, we have the following result.

Theorem 3. *The above algorithm for 3-IVC problem presents a competitive ratio at most* $(1 + \sqrt{2})\alpha$, *where* α *is the known approximation ratio of the k-vertex cover problem.*

Corollary 2. *For the 3-IVC problem, there exist an incremental algorithm with competitive ratio* 4.828 *in polynomial time, and* 2.414 *in non-polynomial time.*

5 Conclusion

This paper presents some results concerning the incremental vertex cover problem. We have given an improved lower bound of $\phi = \frac{\sqrt{5}+1}{2}$ for this problem without any complexity assumption, and then presented an upper bound of $\phi\alpha$ for the restricted version where k is permitted to use only two values, and $(1+\sqrt{2})\alpha$ for the the restricted version where k is permitted to use only three values, where α is the known approximation algorithm for the k-vertex cover problem.

There are still various open problems for future research. For example, the optimal lower and upper bound of the competitive ratio for these problems are still open.

References

1. Bar-Yehuda, R.: Using homogeneous weights for approximating the partial cover problem. Journal of Algorithms 39(2), 137–144 (2001)
2. Bar-Yehuda, R., Even, S.: A linear time approximation algorithm for the weighted vertex cover problem. J. of Algorithms 2, 198–203 (1981)
3. Bar-Yehuda, R., Even, S.: A local-ratio theorem for approximating the weighted vertex cover problem. Annals of Discrete Mathematics 25, 27–45 (1985)
4. Bshouty, N., Burroughs, L.: Massaging a linear programming solution to give a 2-approximation for a generalization of the vertex cover problem. In: Meinel, C., Morvan, M. (eds.) STACS 1998. LNCS, vol. 1373, pp. 298–308. Springer, Heidelberg (1998)
5. Clarkson, K.L.: A modification of the greedy algorithm for the vertex cover. Information Processing Letters 16, 23–25 (1983)
6. Dinur, I., Safra, S.: On the hardness of approximating minimum vertex cover. Annuals of Mathematics 162(1) (2005); Preliminary version in STOC 2002 (2002)
7. Gandhi, R., Khuller, S., Srinivasan, A.: Approximation algorithms for partial covering problems. Journal of Algorithms 53(1), 55–84 (2004)
8. Håstad, J.: Some optimal inapproximability results. J. of ACM 48(4), 798–859 (2001)
9. Hartline, J., Sharp, A.: An Incremental Model for Combinatorial Minimization (2006), http://www.cs.cornell.edu/~asharp
10. Hochbaum, D.S.: Efficient bounds for the stable set, vertex cover and set packing problems. Discrete Applied Mathematics 6, 243–254 (1983)
11. Hochbaum, D.S. (ed.): Approximation Algorithms for NP-hard Problems. PWS Publishing Company (1997)
12. Hochbaum, D.S.: The t-vertex cover problem: Extending the half integrality framework with budget constraints. In: Jansen, K., Rolim, J.D.P. (eds.) APPROX 1998. LNCS, vol. 1444, pp. 111–122. Springer, Heidelberg (1998)
13. Karp, R.M.: Reducibility among combinatorial problems. In: Complexity of Computer Computations, pp. 85–103. Plenum Press, New York (1972)
14. Khot, S., Regev, O.: Vertex cover might be hard to approximaite to within $2-\varepsilon$. In: Proceedings of the 18th IEEE Conference on Computational Complexity (2003)
15. Lin, G.L., Nagarajan, C., Rajamaran, R., Williamson, D.P.: A general approach for incremental approximation and hierarchical clustering. In: Proc. 17th Symp. on Discrete Algorithms (SODA). ACM/SIAM (2006)
16. Mestre, J.: A primal-dual approximation algorithm for partial vertex cover: making educated guesses. In: Chekuri, C., Jansen, K., Rolim, J.D.P., Trevisan, L. (eds.) APPROX 2005 and RANDOM 2005. LNCS, vol. 3624, pp. 182–191. Springer, Heidelberg (2005)
17. Nemhauser, G.L., Trotter Jr., L.E.: Vertex packings: Structural properties and algorithms. Mathematical Programming 8, 232–248 (1975)
18. Vazirani, V.V.: Approximation Algorithms. Springer, Heidelberg (2001)

Finding Good Tours for Huge Euclidean TSP Instances by Iterative Backbone Contraction

Christian Ernst[1,3], Changxing Dong[1], Gerold Jäger[1,2], Dirk Richter[1], and Paul Molitor[1]

[1] Martin-Luther-University Halle-Wittenberg, D-06120 Halle (Saale), Germany
[2] Christian-Albrechts-University Kiel, D-24118 Kiel, Germany
[3] GISA GmbH, D-06112 Halle (Saale), Germany

Abstract. This paper presents an iterative, highly parallelizable approach to find good tours for very large instances of the Euclidian version of the well-known Traveling Salesman Problem (TSP). The basic idea of the approach consists of iteratively transforming the TSP instance to another one with smaller size by contracting pseudo backbone edges. The iteration is stopped, if the new TSP instance is small enough for directly applying an exact algorithm or an efficient TSP heuristic. The pseudo backbone edges of each iteration are computed by a window based technique in which the TSP instance is tiled in *non-disjoint* sub-instances. For each of these sub-instances a good tour is computed, independently of the other sub-instances. An edge which is contained in the computed tour of *every* sub-instance (of the current iteration) containing this edge is denoted to be a pseudo backbone edge. Paths of pseudo-backbone edges are contracted to single edges which are fixed during the subsequent process.

1 Introduction

The Traveling Salesman Problem (TSP) is a well known and intensively studied problem [1,5,10,17] which plays a very important role in combinatorial optimization. It can be simply stated as follows. Given a set of cities and the distances between each pair of them, find a shortest cycle visiting each city exactly once. If the distance between two cities does not depend on the direction, the problem is called *symmetric*. The size of the problem instance is defined as the number n of cities. Formally, for a complete, undirected and weighted graph with n vertices, the problem consists of finding a minimal Hamiltonian cycle. In this paper we consider *Euclidean TSP* (ETSP) instances whose cities are embedded in the Euclidean plane[1].

Although TSP is easy to understand, it is hard to solve, namely $\mathcal{NP}$-hard. We distinguish two classes of algorithms for the symmetric TSP, namely heuristics

[1] However, the ideas presented in this paper can be easily extended to the case in which the cities are specified by their latitude and longitude, treating the Earth as a ball (see [19]).

B. Chen (Ed.): AAIM 2010, LNCS 6124, pp. 119–130, 2010.

and exact algorithms. For the exact algorithms the program package *Concorde* [1,20], which combines techniques of linear programming and branch-and-cut, is the currently leading code. Concorde has exactly solved many benchmark instances, the largest one has size 85,900 [2]. On the other hand, in the field of symmetric TSP heuristics, Helsgaun's code [6,7,8,21] (LKH), which is an effective implementation of the Lin-Kernighan heuristic [11], is one of the best packages. Especially for the most yet not exactly solved TSP benchmark instances [14,15,16,18,19] this code found the currently best tours.

An interesting observation [13] is that tours with good quality are likely to share many edges. Dong et al. [4] exploited this observation by first computing a number of good tours of a given TSP instance by using several different heuristical approaches, collecting the edges which are contained in each of these (not necessarily optimal) tours, computing the maximal paths consisting of only these edges, and contracting these maximal paths to single edges which are kept fixed during the following process. By the contraction step, a new TSP instance with *smaller* size is created which can be attacked more effectively. For some TSP benchmark instances of the VLSI Data Set [18] with sizes up to 47,608, this approach found better tours than the best ones so far reported.

The idea of fixing edges and reducing chains of fixed edges to single edges is not new. It has already been presented by Walshaw in his multilevel version of Helgaun's LKH [12]. Walshaw's process of fixing edges however is rather naive as it only matches vertices with their nearest unmatched neighbours instead of using more sophisticated edge measures.

An alternative to the approach would be fixing *without* backbone contraction. Thus the search space is considerably cut, although the size of the problem is not reduced. This basic concept of edge fixing was already used by Lin, Kernighan [11] and is implemented in LKH. The main difference between edge fixing without backbone contraction and the approaches presented in [4,12] is the reduction of the size by contracting. This reduction has great influence to the effectiveness of the approach. The reason is that all the edges incident to an inner vertex of the contracted paths do not appear in the new instance anymore. Another idea related to [4] is Cook and Seymour's tour merging algorithm [3], which merges a given set of starting tours to get an improved tour.

The bottleneck of the approach presented in [4] when applied to *huge* TSP instances is the computation of several good starting tours, i. e., tours of high quality, by using several different TSP methods. Using *different* TSP heuristics during the computation of the starting tours hopefully increases the probability that edges contained in each of the starting tours are edges which are also contained in optimal tours.

This paper focuses on TSP instances with very large sizes. Only a tiny part of the search space of such a huge TSP instance can be traversed in reasonable time. To overcome this problem, huge TSP instances are usually partitioned. In our new approach, which handles ETSP, this partitioning is done by moving a window frame across the bounding box of the vertices. The amount of the stepwise shift is chosen as a fraction $1/s$ of the width (height) of the window frame so that all vertices of the TSP instance but those located near the boundary are contained in exactly s^2 windows (see Figure 1(b) where the basic idea is illustrated for $s = 2$); parameter s determines the shift amount of the window frame. For the vertices of each window a good tour is computed by either an exact algorithm, e. g., Concorde, or some heuristics, e. g., by Helsgaun's LKH. If two vertices u and w which are contained in the same s^2 windows are neighbored in each of the s^2 tours constructed, the edge $\{u, w\}$ is assumed to have high probability to appear in an optimal tour of the original TSP instance, and we call it a *pseudo backbone edge*. As in [4,12], maximal paths of pseudo backbone edges are computed and contracted to single edges which are fixed during the following process.

Our experiments show that **(a)** actually most of the fixed edges are contained in an optimal tour, and **(b)** fixing edges and contracting chains of fixed edges considerably reduce the size of the original TSP instance. Because of (b), the width and height of the window frame applied in the next iteration can be increased so that larger sections of the bounding box will be considered by each window. The iteration stops, when the window frame is as large as the bounding box itself. In this case, LKH is directly applied to the remaining TSP instance. The experimental runs show that tours of high quality of huge TSP instances are constructed by this approach in acceptable runtime. For instance, for `World-TSP` the approach computes a tour of length 7,525,520,531, which is only 0,18 % above a known lower bound, within 45 hours on a standard personal computer, and a tour of length 7,524,796,079, which is 0,17 % above the lower bound, within 13 hours on a parallel computer with 32 processors. Similarly, it finds a tour for the DIMACS 3,162,278-sized TSP instance [15], whose length is only 0,0465 % larger than the best tour currently known for that TSP instance, within 6 days, and a tour for the DIMACS 10,000,000-sized TSP instance, whose length is

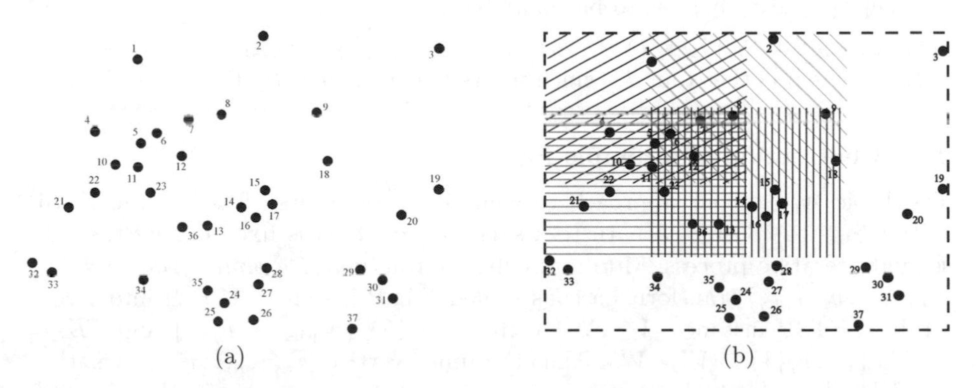

Fig. 1. (a) Vertices embedded in the Euclidean plane with $V = \{1, \ldots, 37\}$. **(b)** Four neighbored windows if the width (height) of the window frame is half the width (height) of the bounding box and parameter s is set to 2. The top left-hand (top middle-hand, middle left-hand, middle middle-hand) window is marked by slanted (back slanted, horizontal, vertical) lines. The vertices 5, 6, 7, 8, 11, and 12 are contained in each of the four windows.

only 0,0541 % larger, within 20 days. Moreover, we observe a high-level trade-off between tour length and runtime which can be controlled by modifying, e. g., parameter s or some other parameters.

The paper is structured as follows. Basic definitions with respect to TSP and our approach are given in Section 2. The overall algorithm together with a detailed illustration is described in Section 3. The parameters of the algorithm are listed in Section 4. Section 5 presents the experimental results. Finally, conclusions and suggestions for future work are given in Section 6.

2 Definitions

2.1 Basics

Let V be a set of n vertices embedded in the Euclidean plane (see Figure 1(a)) and let $dist(u, w)$ be the Euclidean distance between the vertices u and w. A sequence $p = (v_1, v_2, \ldots, v_q)$ with $\{v_1, v_2, \ldots, v_q\} \subseteq V$ is called a *path* of *length* q. The costs $dist(p)$ of such a path p is given by the sum of the Euclidean distances between neighboring vertices, i. e., $dist(p) := \sum_{i=1}^{q-1} dist(v_i, v_{i+1})$. The path is called *simple*, if it contains each vertex of V at most once, i. e., $v_i = v_j \Rightarrow i = j$ for $1 \le i \le q$ and $1 \le j \le q$. It is called *complete*, if the path is simple and contains each vertex of V exactly once. It is called *closed*, if $v_q = v_1$ holds. A complete path $p = (v_1, v_2, \ldots, v_n)$ can be extended to the closed path $T = (v_1, v_2, \ldots, v_n, v_1)$. Such a closed path of V is called a *tour*.

2.2 Euclidean Traveling Salesman Problem

ETSP is the problem of finding a tour with minimum costs for a given set V of vertices embedded in the Euclidean plane. An ETSP instance is *constrained* by a set $FE \subseteq \{\{v_i, v_j\};\ v_i, v_j \in V \text{ and } v_i \ne v_j\}$ of *fixed edges*, if a tour $T = (v_1, v_2, \ldots, v_n, v_1)$ has to be computed such that

(1) $(\forall \{u, w\} \in FE)(\exists i \in \{1, \ldots, n\})\ \{v_i, v_{(i \bmod n)+1}\} = \{u, w\}$
(2) there is no other tour T' which meets (1) such that $dist(T') < dist(T)$.

2.3 Contraction of a Simple Path

The basic step of our approach to compute good tours of very large ETSP instances is to contract a path to a single edge which is fixed during the subsequent iterative process. More formally, *contraction of a simple path* $p = (v_i, v_{i+1}, \ldots, v_{j-1}, v_j)$ transforms a constrained ETSP instance (V, FE) into a constrained ETSP instance (V', FE') with $V' = V \setminus \{v_{i+1}, \ldots, v_{j-1}\}$ and $FE' = (FE \cup \{\{v_i, v_j\}\}) \cap (V' \times V')$. Thus the inner vertices $v_{i+1}, \ldots, v_{j-1}$ of path p are deleted – only the boundary vertices v_i and v_j of p remain in the new constrained ETSP instance – and the size of the new constrained ETSP becomes smaller, unless $i + 1 = j$. (If $i + 1 = j$, the new instance is less complex than the current instance in the sense that the edge $\{v_i, v_j\}$ is fixed.) Figures 4(b) and 4(c) illustrate the contraction process before and after the contraction of the four simple paths $(5, 11)$, $(6, 12, 7, 8)$, $(9, 18)$, and $(23, 36, 13)$.

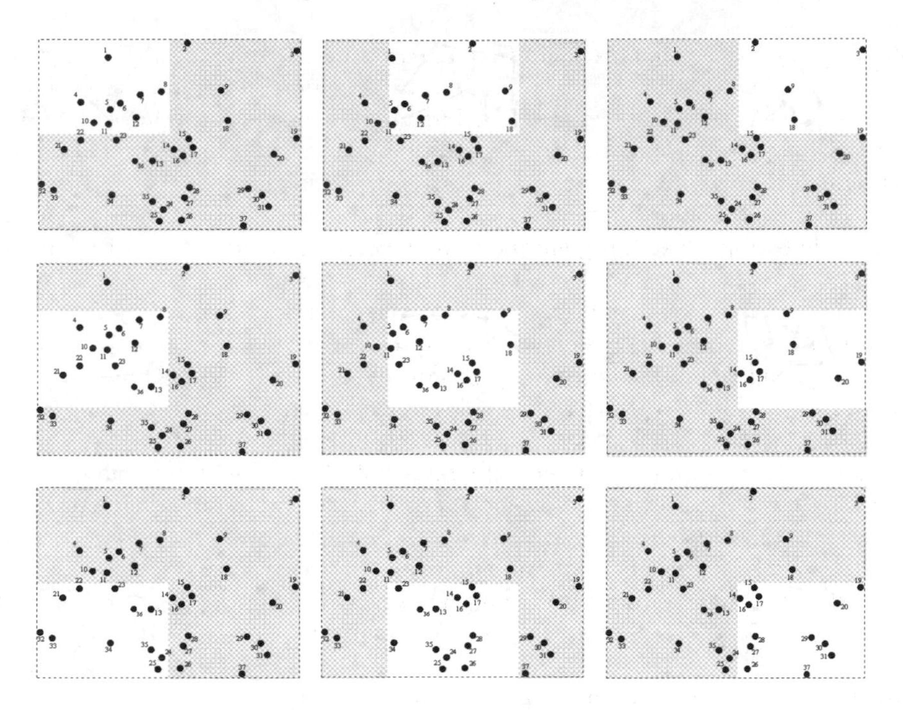

Fig. 2. Moving a window frame across the bounding box with `WINDOW_SCALE`= 2 and $s = 2$. Each window defines a sub-problem for which a good tour is computed, independently of the neighboring sub-problems.

3 The Overall Algorithm

To solve very large (constrained) ETSP instances, we apply a window based approach to iteratively find paths to be contracted. In each iteration a window frame is moved across the bounding box of the vertices. As illustrated in Figure 2, the windows considered are *not* disjoint. In fact, we move the window by the fraction $1/s$ of the width (height) of the window frame so that a vertex is contained in up to s^2 windows. To simplify matters, we shall illustrate our approach with respect to $s = 2$, although $s > 2$ might lead to better tours and actually does in some cases. The height and the width of the window frame are determined by dividing height and width of the bounding box by a parameter `WINDOW_SCALE` – you find more details on this parameter in Section 4 – which is chosen in such a way that the sub-problems induced by the windows have sizes which can be efficiently handled by Helsgaun's LKH [6,7,21] or by the exact solver Concorde [20]. Since LKH finds optimal solutions frequently for small instances and the sizes of the windows are rather small, we have not tried to solve them with Concorde.

The sub-problems should contain a number of vertices greater than a lower bound `MNL` (see Section 4) so that a corresponding good tour contains ample information on the original TSP instance. In the following, we call a TSP instance

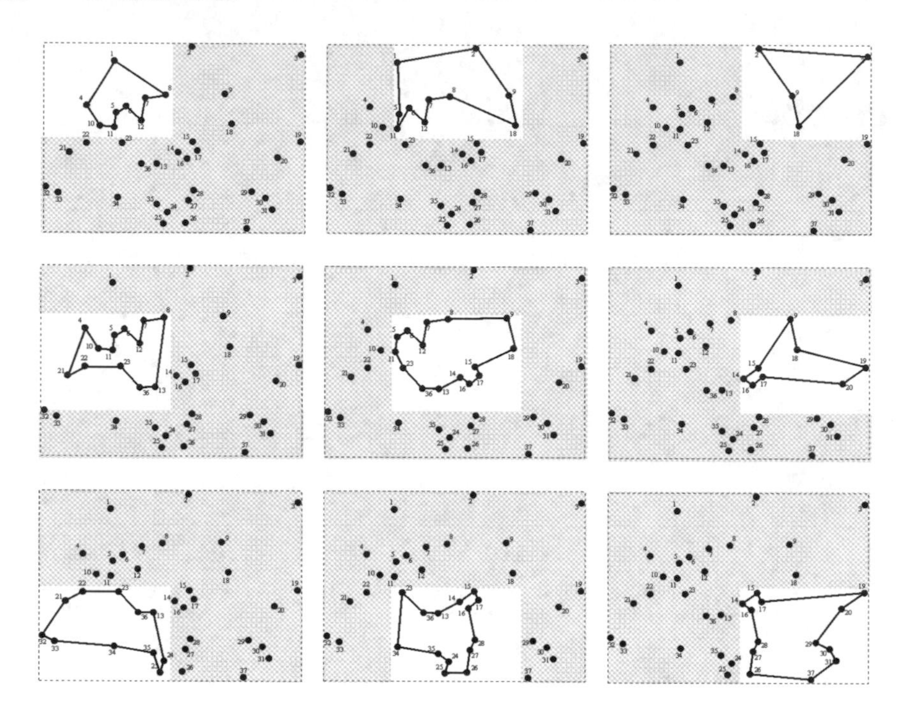

Fig. 3. A good tour is computed for each of the non-trivial sub-problems. In this example, parameter MNL has been set to 3, so that there are no trivial instances.

containing less than MNL vertices a *trivial* instance. A tour is computed for each of the non-trivial sub-problems. Figure 3 illustrates this step which is applied in every iteration.

Now, our approach is based on the assumption, that two vertices u and w which are contained in the same s^2 non-trivial windows and neighboring in each of the s^2 tours constructed have high probability to appear in an optimal tour for the original ETSP instance [2] – for $s = 2$, in some sense, the four windows together reflect the surrounding area of the common vertices with respect to the four directions. We call such an edge $\{u, w\}$ *pseudo backbone edge.* [3]

Figures 3 and 4(a) illustrate the notion of pseudo backbone edges. For this purpose, consider only the four top left-hand windows. Figure 3 shows a tour for each of these ETSP instances. Each of these four tours contain the edges $\{5, 11\}$, $\{6, 12\}$, $\{7, 12\}$, and $\{7, 8\}$. Thus, they are the pseudo backbone edges generated

[2] If one of the windows which contain u and w is trivial, the edge $\{u, w\}$ is not considered as pseudo-backbone.

[3] Note that edges at the boundary of the TSP instance cannot be contained in s^2 non-trivial windows and therefore cannot be pseudo backbone edges. This may cause a problem if there are a lot of vertices at the boundary because, in this case, the original TSP instance cannot be reduced to such a degree that the final TSP instance is small enough to be efficiently solved by LKH. To overcome this problem, we shift the window frame above the boundary line.

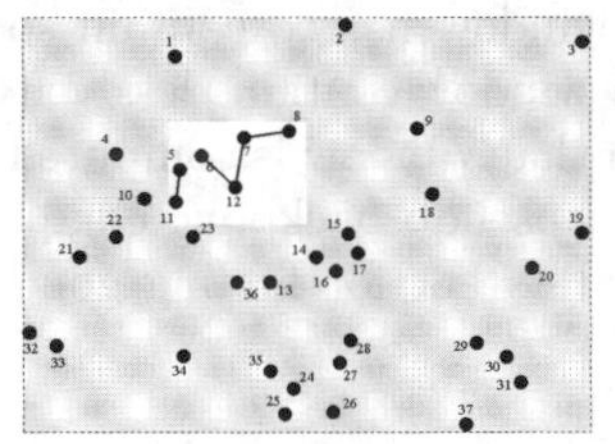

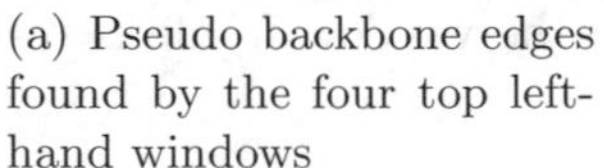

(a) Pseudo backbone edges found by the four top left-hand windows

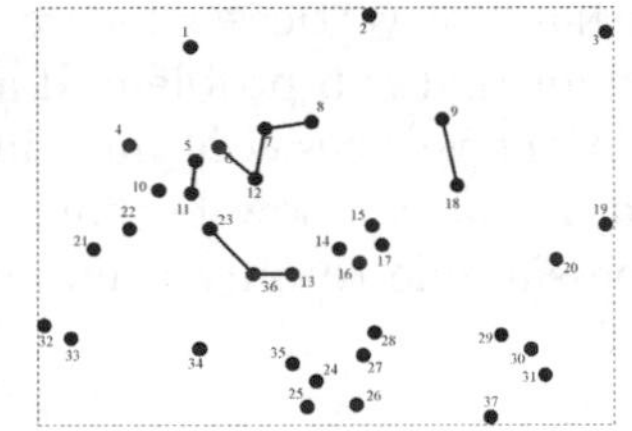

(b) Pseudo backbone edges found in the current iteration.

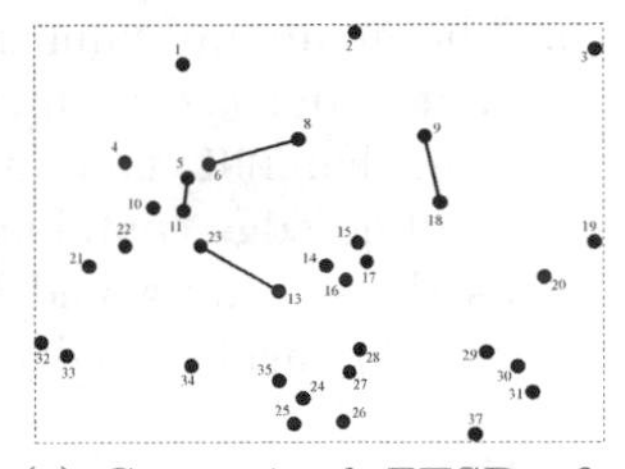

(c) Constrained ETSP after contraction of the paths shown in (b).

Fig. 4. Pseudo backbone contraction

by these four tours. The set of all pseudo backbone edges generated by the tours computed in this iteration (Figure 3) is shown in Figure 4(b). The pseudo backbone edges can be partitioned into a set of maximal paths. In our running example the set consists of four paths, namely $(5, 11)$, $(6, 12, 7, 8)$, $(23, 36, 13)$, and $(9, 18)$. Now, these maximal paths are contracted, as described in Section 2.3 which leads to a new constrained ETSP with smaller or equal size (see Figure 4(c)).

This process is iteratively repeated. The parameter `WINDOW_SCALE` which determines the width and height of the window frame is re-adjusted, i. e., decreased, in each iteration. The iteration stops if `WINDOW_SCALE` is set to 1, i. e., if the window frame spans the whole bounding box. In this case, LKH is directly applied to the current constrained ETSP instance. Finally the fixed edges are recursively re-substituted which results in a tour of the original ETSP instance.

4 The Algorithm's Parameters

The main parameters of the algorithm are the following

1. The scale parameter `IWS`[4] which determines the width and height of the window frame applied in the first iteration, namely

$$\text{width}_{\text{frame}} = \left\lceil \frac{\text{width}_{\text{bounding box}}}{\texttt{WINDOW\_SCALE}} \right\rceil \text{ and } \text{height}_{\text{frame}} = \left\lceil \frac{\text{height}_{\text{bounding box}}}{\texttt{WINDOW\_SCALE}} \right\rceil$$

   with `WINDOW_SCALE` := `IWS`.

2. `D`[5] which is the factor relative to the width (height) of the window frame by which the window frame is shifted, i. e., `D`$= 1/s$.

[4] `IWS` is an abbreviation for `INITIAL_WINDOW_SCALE`.

[5] `D` is an abbreviation for `DISPLACEMENT`.

3. The minimum number MNL [6] of vertices which a window has to contain so that a tour is computed for that sub-problem. If a window contains less vertices than MNL, no tour of the vertices contained in that window is computed (and no edge contained in that window becomes a pseudo backbone edge), as those tours would be suitable to only a limited extent for finding good pseudo backbone edges.

4. The parameter WGF [7] which is the factor by which the window frame is readjusted after each iteration, i. e.,

$$\mathtt{WINDOW\_SCALE}_{\mathrm{new}} = \frac{\mathtt{WINDOW\_SCALE}_{\mathrm{old}}}{\mathtt{WGF}} \, .$$

   Actually, the parameter WGF can be assigned values of the enumeration type {slow, medium, fast}. The default assignments to the values slow, medium, and fast are 1.2, 1.3, and 1.4.

5 Experimental Results

We performed the following three types of experiments:

1. We made an analysis in which we determined which of the edges that are fixed during our approach are actually present in an optimal solution.
2. We investigated the reduction rates reached by our approach.
3. We applied our approach to some huge TSP instances and compared costs (and running times, if possible) to Helsgaun's LKH.

Our first experiment deals with the question of how many edges are fixed by our approach and how many of these fixed edges are contained in at least one optimal tour. Unfortunately, we couldn't perform this experiment on huge TSP instances, but had to use middle-sized TSP instances consisting of "only" some thousands of vertices as we need to know the optimal solutions of the instances for being able to compute the number of fixed edges contained in optimal tours. Actually, as we also do not know *all* optimal solutions of middle-sized TSP instances, we counted the number of fixed edges contained in *one given* optimal solution, i. e., we computed a lower bound of the number of fixed edges contained in optimal solutions. Table 1 shows the result of this experiment with respect to the national TSP instances Greece (9,882 cities), Finland (10,639 cities), Italy (16,862 cities), Vietnam (22,775 cities), and Sweden (24,978 cities). The table gives us an impression of how many of the edges which are fixed by our approach are contained in an optimal tour, namely between 93 % and 96 %. Remember, this is only a lower bound; the ratio could be yet much higher. In the instances of Table 1 the ratio between the number of fixed edges and the size of the instance which is between 48 % and 66 % (see the second and third column of

[6] MNL is an abbreviation for MIN_NODE_LIMIT.

[7] WGF is an abbreviation for WINDOW_GROWTH_FACTOR.

Table 1. How many edges which are fixed are contained in a given optimal solution? The second column shows the number of cities of the instances, the third column the number of edges fixed by our approach, and the fourth one the number of fixed edges which are contained in a given optimal solution. The displacement D had been set to 1/2 during this experiment. The experiment has been performed on a standard personal computer.

instance	size	fixed edges	thereof optimal	fraction	runtime
Greece	9 882	4,739	4,436	93,61 %	475 s
Finland	10,639	4,840	4,655	96,18 %	376 s
Italiy	16,862	10,180	9,714	95,42 %	876 s
Vietnam	22,775	11,445	10,820	94,54 %	1,602 s
Sweden	24,978	16,523	15,470	93,63 %	1,502 s

Table 1) is unsatisfactory. Note that this unsatisfactory percentage is due to the fact that the TSP instances taken in this experiment are relative small and we set the parameter value MNL to about 1,000 which is a reasonable value for this parameter when the approach is applied to such middle-sized TSP instances. However, this yields comparatively many trivial windows so that comparatively many edges are excluded from becoming a pseudo backbone. Table 2 exemplarily shows how many cities are typically eliminated during each of the iterations when the approach is iteratively applied to huge TSP instances. (Note that the number of eliminated cities is a lower bound for the number of fixed edges.) For the World-TSP which consists of 1,904,711 cities we obtain a size reduction of 98 % from 1,904,711 vertices to 33,687 vertices.

Currently, the best known tour for World-TSP has been found by Keld Helsgaun using LKH. The length of this tour is 7,515,877,991 which is at most 0,0487% greater than the length of an optimal tour, as the currently best lower bound for the tour length of the World-TSP is 7,512,218,268 (see [19]). However, no overall running time comprising the running time of both, the computation of good starting tours *and* the iterative k-opt steps of LKH are reported. Helsgaun reports that by assigning the right values to the parameters, LKH computes a tour for the World-TSP in an hour or two [8] which is at most 0,6% greater than the length of an optimal tour [7]. By using sophisticated parameters, LKH even finds better tours (up to only 0.12% greater than the length of an optimal tour) in a couple of days [9].

We applied our iterative approach to the World-TSP instance, too. Using appropriate parameter values our approach constructs a tour for World-TSP of length 7,525,520,531 which is at most 0,18% above the lower bound in less than two days on a standard personal computer and a tour of length 7,530,566,694 which is at most 0,2442 % greater than the length of an optimal tour in about 7 hours; it computes a tour with length 7,524,796,079 (gap=0.1674 %) in less than 13 hours on a parallel computer with 32 Intel Xeon 2.4 GHz processors.

[8] The computation time of 1,500 seconds, Helsgaun states in [7, pages 92-93], does not include the computation time of the starting tour [9].

Table 2. How many edges are fixed by our approach when applied to a huge instance? Information about the reduction rates reached by our approach when applied to the World-TSP. After 14 iterations, LKH is directly applied to the remaining TSP instance consisting of 33,687 cities.

Iter-ation	Number of cities eliminated in this iteration	Number of paths contracted in this iteration	Size of the new TSP instance
1	103,762	20,811	1,800,949
2	228,887	56,007	1,572,062
3	325,034	87,511	1,247,028
4	327,625	100,443	919,403
5	271,871	95,234	647,532
6	194,280	80,672	453,252
7	148,191	62,884	305,061
8	94,678	47,089	210,383
9	70,599	34,432	139,784
10	43,061	24,673	96,723
11	21,082	19,292	75,641
12	18,830	14,574	56,811
13	15,265	11,286	41,546
14	7,859	9,496	33,687

Table 3. Trade-off between runtime and tour length. The table presents the best experimental results which we obtained by our approach applied to World-TSP. Column 6 gives the gaps between the lengths of the tours with respect to the lower bound 7,512,218,268 on the minimum tour length.

standard personal computer						
D	MNL	IWS	WGF	**tour length**	**gap** [%]	**runtime**
1/3	20,000	40	slow	7,520,207,210	0,1063 %	6 days 21 hours
1/3	10,000	30	slow	7,521,096,881	0,1181 %	5 days 12 hours
1/3	10,000	17	slow	7,522,418,605	0,1357 %	5 days 8 hours
1/2	20,000	18	medium	7,525,520,531	0,1770 %	1 day 21 hours
1/2	20,000	18	medium	7,528,686,717	0,2192 %	1 day 15 hours
1/2	5,000	40	medium	7,529,946,223	0,2359 %	1 day 10 hours
1/2	5,000	100	medium	7,533,272,830	0,2802 %	1 day 10 hours

parallel computer with 32 Intel Xeon 2.4 Ghz processors						
D	MNL	IWS	WGF	**tour length**	**gap** [%]	**runtime**
1/2	20,000	50	medium	7,524,796,079	0,1674 %	0 day 13 hours
1/2	20,000	15	medium	7,529,172,390	0,2256 %	0 day 11 hours
1/2	10,000	30	fast	7,530,566,694	0,2442 %	0 day 7 hours

Table 4. Best experimental results of our approach applied to the `DIMACS` instances E3M.0 and E10M.0 [15]. Column 8 gives the gap to the current best known tour [15].

standard personal computer								
	size	`D`	`MNL`	`IWS`	`WGF`	**tour length**	**gap** [%]	**runtime**
E3M.0	3,162,278	1/3	10,000	16	`slow`	1,267,959,544	0,0465 %	5.71 days
		1/3	10,000	20	`slow`	1,268,034,733	0,0525 %	5.46 days
		1/2	20,000	12	`slow`	1,268,280,730	0,0719 %	4.08 days
		1/2	10,000	30	`medium`	1,268,833,812	0,1156 %	1.62 days
E10M.0	10,000,000	1/3	10,000	30	`slow`	2,254,395,868	0,0541 %	19.67 days
		1/2	10,000	30	`medium`	2,256,164,398	0,1326 %	4.25 days
		1/2	5,000	40	`slow`	2,260,519,918	0,3259 %	2.62 days

Actually, we exploited one of the central properties of our iterative approach, namely the property that the approach can be highly parallelized, as the tours for the windows of an iteration can be computed in parallel.

Table 3 shows our experimental results with respect to `World-TSP` for both, runs on a standard personal computer and runs on the above mentioned parallel machine. Table 4 summarizes the results with respect to two instances of the DIMACS TSP Challenge which have sizes of 3,162,278 and 1,000,000 vertices, respectively [15]. Note the general high-level trade-off provided by the approach:

- the smaller the factor `D` by which the window frame is shifted, the better the tours and the worse the runtime, although the runtimes remain acceptable;
- the faster the increase of the window frame, the better the runtime and the worse the tours, although the tour lengths remain very good.

6 Future Work

One problem with the approach presented in this paper arises, if the vertices are non-uniformly distributed, i. e., if there are both, regions with very high densities and regions with very low densities. In order to partition the regions of high density in such a way that LKH can handle the windows efficiently with respect to both, tour quality and running time, the parameter `IWS` should be large. However this results in many trivial windows located in regions of low density. To overcome this problem, we reason about recursion which can be applied to windows containing too much vertices and about non-uniform clustering of dense regions.

Acknowledgement

The work presented in this paper was supported by the German Research Foundation (DFG) under grant MO 645/7-3.

References

1. Applegate, D.L., Bixby, R.E., Chvátal, V., Cook, W.J.: The Traveling Salesman Problem. In: A Computational Study. Princeton University Press, Princeton (2006)
2. Applegate, D.L., Bixby, R.E., Chvátal, V., Cook, W.J., Espinoza, D., Goycoolea, M., Helsgaun, K.: Certification of an Optimal Tour through 85,900 Cities. Operations Research Letters 37(1), 11–15 (2009)
3. Cook, W., Seymour, P.: Tour Merging via Branch-Decomposition. INFORMS Journal on Computing 15(3), 233–248 (2003)
4. Dong, C., Jäger, G., Richter, D., Molitor, P.: Effective Tour Searching for TSP by Contraction of Pseudo Backbone Edges. In: Goldberg, A.V., Zhou, Y. (eds.) AAIM 2009. LNCS, vol. 5564, pp. 175–187. Springer, Heidelberg (2009)
5. Gutin, G., Punnen, A.P. (eds.): The Traveling Salesman Problem and Its Variations. Kluwer Academic Publishers, Dordrecht (2002)
6. Helsgaun, K.: An Effective Implementation of the Lin-Kernighan Traveling Salesman Heuristic. European Journal Operations Research 126(1), 106–130 (2000)
7. Helsgaun, K.: An Effective Implementation of k-opt Moves for the Lin-Kernighan TSP Heuristic. In: Writings in Computer Science, vol. 109. Roskilde University (2006)
8. Helsgaun, K.: General k-opt submoves for the Lin-Kernighan TSP heuristic. Mathematical Programming Computation 1(2-3), 119–163 (2009)
9. Helsgaun, K.: Private Communication (March 2009)
10. Lawler, E.L., Lenstra, J.K., Rinnooy Kan, A.H.G., Shmoys, D.B. (eds.): The Traveling Salesman Problem - A Guided Tour of Combinatorial Optimization. John Wiley & Sons, Chicester (1985)
11. Lin, S., Kernighan, B.W.: An Effective Heuristic Algorithm for the Traveling Salesman Problem. Operations Research 21, 498–516 (1973)
12. Walshaw, C.: A Multilevel Lin-Kernighan-Helsgaun Algorithm for the Travelling Salesman Problem. TR 01/IM/80, Computing and Mathematical Sciences, University of Greenwich, UK (2001)
13. Zhang, W., Looks, M.: A Novel Local Search Algorithm for the Traveling Salesman Problem that Exploits Backbones. In: Proc. of the 19th Int'l Joint Conf. on Artificial Intelligence (IJCAI 2005), pp. 343–350 (2005)
14. TSPLIB Homepage, `http://elib.zib.de/pub/mp-testdata/tsp/tsplib/tsplib.html`
15. DIMACS Implementation Challenge, `http://www.research.att.com/~dsj/chtsp/`
16. National Instances, `http://www.tsp.gatech.edu/world/summary.html`
17. TSP Homepage, `http://www.tsp.gatech.edu/`
18. VLSI Instances from the TSP Homepage, `http://www.tsp.gatech.edu/vlsi/summary.html`
19. World-TSP from the TSP Homepage, `http://www.tsp.gatech.edu/world/`
20. Source Code of [1] (Concorde), `http://www.tsp.gatech.edu/concorde/index.html`
21. Source Code of [7] (LKH), `http://www.akira.ruc.dk/~keld/research/LKH/`

Point Location in the Continuous-Time Moving Network

Chenglin Fan[1,2] and Jun Luo[1]

[1] Shenzhen Institutes of Advanced Technology
Chinese Academy of Sciences, China
[2] School of Information Science and Engineering
Central South University, Changsha 410083, China
{cl.fan,jun.luo}@sub.siat.ac.cn

Abstract. We discuss two variations of the moving network Voronoi diagram. The first one addresses the following problem: given a network with n vertices and E edges. Suppose there are m sites (cars, postmen, *etc*) moving along the network edges and we know their moving trajectories with time information. Which site is the nearest one to a point p located on network edge at time t'? We present an algorithm to answer this query in $O(\log(mW \log m))$ time with $O(nmW \log^2 m + n^2 \log n + nE)$ time and $O(nmW \log m + E)$ space for preprocessing step, where E is the number of edges of the network graph (the definition of W is in section 3). The second variation views query point p as a customer with walking speed v. The question is which site he can catch the first? We can answer this query in $O(m + \log(mW \log m))$ time with same preprocessing time and space as the first case. If the customer is located at some node, then the query can be answered in $O(\log(mW \log m))$ time.

1 Introduction

Voronoi diagram is a fundamental technique in computational geometry and plays important roles in other fields such as GIS and physics [3]. The major goal of Voronoi diagram is to answer the *nearest-neighbor* query efficiently. Much has been written about variants of the Voronoi diagrams and the algorithms for computing the Voronoi diagrams in various fields. Many variants of the Voronoi diagrams give different definitions of distance in different fields, without limiting to the Euclidean distance [8].

The network Voronoi diagram [9, 7, 5, 2] divides a network (e.g. road network) into Voronoi subnetworks. A network Voronoi diagram is a specialization of a Voronoi diagram in which the locations of objects are restricted to the links that connect the nodes of the network. The distance between objects is defined as the length of the shortest network distance (e.g. shortest path or shortest time), instead of the Euclidean distance.

For network Voronoi diagram, any node located in a Voronoi region has a shortest path to its corresponding Voronoi site that is always less than that to any other Voronoi site. In this way, the entire graph is partitioned into several subdivisions as shown in figure 1. We can see that the network Voronoi edges intersect with the network edges in most cases. This means that a network edge may be divided into two parts and placed into two adjacent Voronoi regions.

To construct the network Voronoi diagram, we have to measure network distance between nodes and Voronoi sites. Therefore a shortest-path searching algorithm is required. In [5], Erwig presented a variation of Dijkstra's algorithm, the parallel Dijkstra

B. Chen (Ed.): AAIM 2010, LNCS 6124, pp. 131–140, 2010.

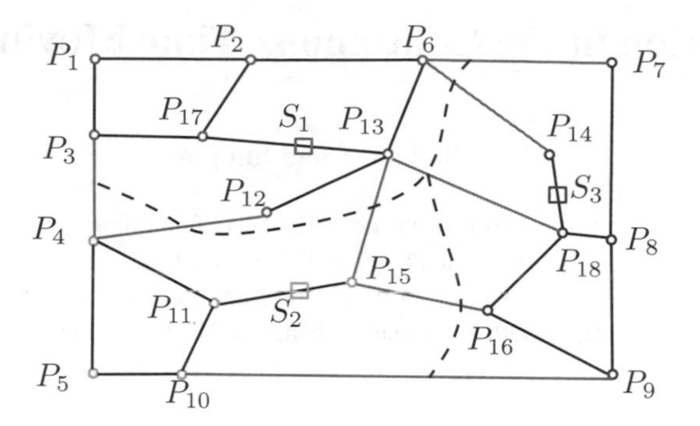

Fig. 1. The network Voronoi diagram. S_1, S_2 and S_3 are Voronoi sites and $P_1 - P_{18}$ are nodes.

algorithm, which can be used to compute the multiple source shortest paths, to construct the network Voronoi diagram.

There are two types of dynamic network Voronoi diagram. The first one is caused by the changes of edge weight. In [10], Ramalingam and Reps present an incremental search method, the Dynamic SWSF-FP algorithm, to handle "multiple heterogeneous modification" between updates. The input graph is allowed to be reconstructed by an arbitrary mixture of edge insertions, edge deletions, and edge-length changes. The second type of dynamic network Voronoi diagram is caused by moving sites. In paper [4], Devillers and Golin discuss the moving Voronoi diagram based on Euclidean distance on two-dimension. They assume that each site is moving along a line with constant speed. They give two types of the "closest", namely closest in static case and kinetic case. In the static case the meaning of "closest" is quite clear. The closest site is the nearest site as to a query point q. In kinetic case, the closest site is the site a customer (who starts from q at time t_0 with speed v) can reach quickest.

In this paper we study the problem of moving network Voronoi diagram when all sites $S_1, S_2, ..., S_m$ move on the edges of network continuously. Every edge of the network is endowed with a positive value which denotes the network length (may not satisfy the triangle inequality) of edge. We can imagine that the set of Voronoi sites is the set of public service platform (bus, postman, policeman, etc). We assume that each site S_i ($1 \leq i \leq m$) moves with constant velocity v_i and their trajectories with time information are also know in advance. n nodes are denoted as $P_1, P_2, ..., P_n$. Then the kinetic state of site S_i can be described as follows:

$$S_i(t) \in edge[P_{i0}, P_{i1}], t \in [t_0, t_1]$$

$$S_i(t) \in edge[P_{i1}, P_{i2}], t \in [t_1, t_2]$$

$$\vdots$$

$$S_i(t) \in edge[P_{i(w-1)}, P_{iw}], t \in [t_{w-1}, t_w]$$

where $S_i(t)$ is the point where the site S_i is located at time t. We use $length(P_a, P_b)$ to denote the length of edge$[P_a, P_b]$. We assume that all edges are undirected, so

$length(P_a, P_b) = length(P_b, P_a)$. When the site S_i moves on edge$[P_a, P_b]$ (from P_a to P_b) during time interval $[t_1, t_2]$, we have:

$$length(S_i(t), P_a) = length(P_a, P_b)(t - t_1)/(t_2 - t_1)$$

We use $d(P_a, P_b)$ to denote the shortest network distance between P_a and P_b. If the length of edge is not satisfied with the triangle inequality, the $d(P_a, P_b)$ may not equal $length(P_a, P_b)$.

In this paper we want to be able to answer the following two different types of queries in continuous-time moving network:

1. Static case query $StaticNearest(q, t', v)$: given a customer at location q (q is on edge of network), to find the nearest site at time t'. That is, given q, t', return i such that
$$d(q, S_i(t')) \leq d(q, S_j(t')), j = 1, \ldots, m$$
2. Kinetic case query $KineticNearest(q, t')$: The query inputs are q, t', and $v > 0$. They specify a customer located at q at time t' with walking speed v. The customer (who can only move on the edges of the network) wants to reach a site as soon as possible. The problem here is to find the site the customer can reach quickest. Set
$$t_j = min\{t \geq t' : (t_j - t')v = d(q, S_j(t))\}, j = 1, \ldots, m$$
be the first time that the customer can catch site S_j starting from q at time t'. Then the query return i such that $t_i = \min\{t_1, t_2, ..., t_m\}$.

The structure of the paper is as follows. Section 2 gives some definitions that we need in the whole paper. We discuss the algorithms for two queries $StaticNearest(q, t', v)$ and $KineticNearest(q, t', v)$ in section 3 and 4 respectively. Finally we give conclusions in section 5.

2 Preliminaries

In moving network Voronoi, sites move with constant velocity continuously. When a site moves on the edge during some time interval, for example in figure 2, S_i move from P_a to P_b during time interval $[t_1, t_2]$, for any other node P_j that is connected (we assume the graph is a connected graph) with P_a and P_b,

$$d(P_j, S_i(t)) = min\{d(P_j, P_a) + length(P_a, S_i(t)), d(P_j, P_b) + length(P_b, S_i(t))\}$$

where $t \in [t_1, t_2]$. Each site S_i moves on the preestablished path with constant velocity v_i (see figure 3). For a node P_j, the network distance between P_j and $S_i(t)$ can be described by segmented functions as follows:

$$d(P_j, S_i(t)) = \min\{d(P_j, P_{i0}) + v_i(t - t_0), d(P_j, P_{i1}) + v_i(t_1 - t)\}, t \in [t_0, t_1]$$

$$\ldots$$

$$d(P_j, S_i(t)) = \min\{d(P_j, P_{i(w-1)}) + v_i(t - t_{w-1}), d(P_j, P_{iw}) + v_i(t_w - t)\}, t \in [t_{w-1}, t_w]$$

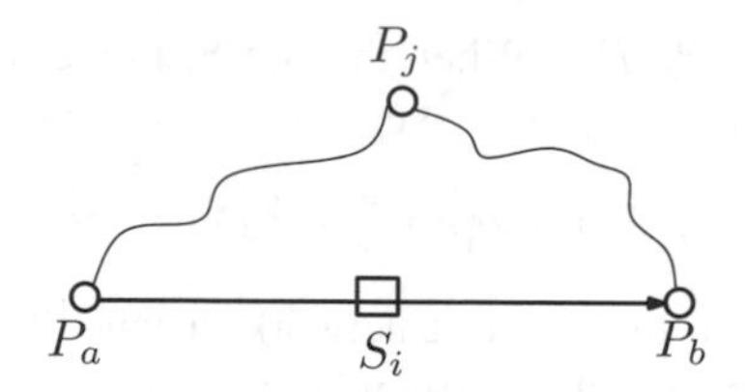

Fig. 2. A site S_i moves from P_a to P_b

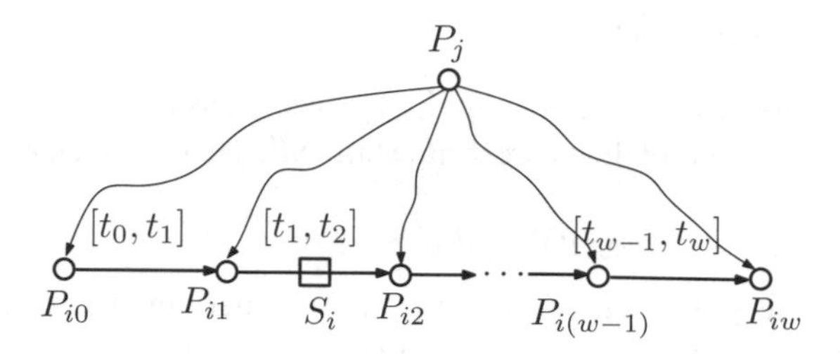

Fig. 3. A site S_i moves from P_{i0} to P_{iw}

Lemma 1. *All segmented functions* $d(P_j, S_i(t))$ *(*$1 \leq i \leq m, 1 \leq j \leq n$*) can be computed in* $O(n^2 \log n + nE + Y)$ *time and* $O(E + Y)$ *space, where* n *and* E *are the number of nodes and edges of the network respectively, and* Y *is the total number of segmented functions for all nodes.*

Proof. If we already know the shortest path tree rooted at P_j, then each $d(P_j, S_i(t))$ can be computed in constant time. Each shortest path tree rooted at P_j can be computed in $O(n \log n + E)$ time using Dijkstra's shortest path algorithm by implementing the priority queue with a Fibonacci heap. Since we have n nodes, we need $O(n^2 \log n + nE)$ time to compute all n shortest path trees. And we need extra Y time to compute all segmented functions. Therefore the total time is $O(n^2 \log n + nE + Y)$. For the space, we need $O(E)$ space to store the network graph and $O(Y)$ space to store all segmented functions. Therefore, the total space is $O(E + Y)$.

3 Point Location in Static Moving Network

As we described above, for an arbitrary node P_j, the network distance $d(P_j, S_i(t))$ is a piecewise linear function, and it consists of at most w_i positive slope and w_i negative slope line segments, where $i = 1, 2, \ldots, m$. Figure 4 shows an example of the network distance function diagram of $d(P_j, S_1(t))$, $d(P_j, S_2(t))$, $d(P_j, S_3(t))$. P_j's nearest site is S_3 during time intervals $[t_0, t_1]$, $[t_4, t_5]$, $[t_6, t_7]$. During time intervals $[t_2, t_3]$, $[t_8, t_9]$, P_j's nearest site is S_1. Hence we only need to compute the lower envelop of these piecewise linear functions $d(P_j, S_i(t))$, where $i = 1, 2, \ldots, m$.

We use the divide-and-conquer algorithm to compute the lower envelope of those segmented functions. To compute the lower envelop $h(t) = \min_{1 \leq i \leq m}\{d(P_j, S_i(t))\}$, we partition $\{d(P_j, S_1(t)), d(P_j, S_2(t)), \ldots, d(P_j, S_m(t))\}$ into two parts,

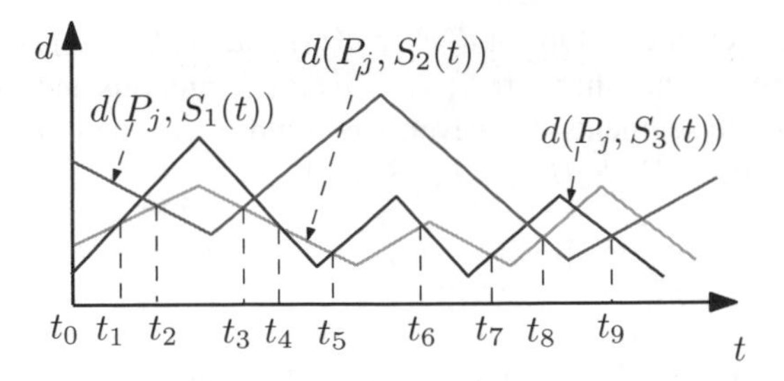

Fig. 4. The network distance function diagram of $d(P_j, S_i(t)), i = 1, 2, 3$

$\{d(P_j, S_1(t)), d(P_j, S_2(t)), \ldots, d(P_j, S_{\lceil m/2\rceil}(t))\}$ and $\{d(P_j, S_{\lceil m/2\rceil+1}(t)), d(P_j, S_{\lceil m/2\rceil+2}(t)), \ldots, d(P_j, S_m(t))\}$, then compute $\min\{d(P_j, S_1(t)), d(P_j, S_2(t)), \ldots, d(P_j, S_{\lceil m/2\rceil}(t))\}$ and $\min\{d(P_j, S_{\lceil m/2\rceil+1}(t)), d(P_j, S_{\lceil m/2\rceil+2}(t)), \ldots, d(P_j, S_m(t))\}$ recursively and finally merge them together to obtain $h(t)$.

Now we consider merging two distance functions $d(P_j, S_k(t))$ and $d(P_j, S_l(t))$ (see figure 5).

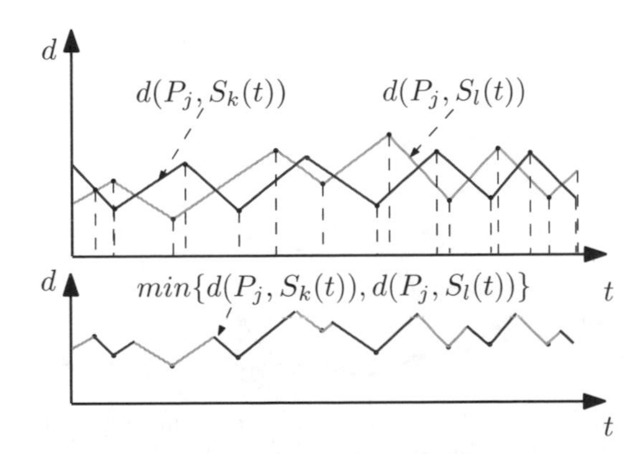

Fig. 5. The example of merging two functions: $d(P_j, S_k(t))and\ d(P_j, S_l(t))$

Lemma 2. *There are at most* $2(w_k+w_l)$–1 *intersections between functions* $d(P_j, S_k(t))$ *and* $d(P_j, S_l(t))$.

Proof. $d(P_j, S_k(t))$ and $d(P_j, S_l(t))$ are two continuous and piecewise linear polygonal curves. Suppose the function $d(P_j, S_k(t))$ is composed of x line segments ($1 \leq x \leq 2w_k$) and the time intervals for those line segments are $[t_{begin}, t_1]$, $[t_1, t_2]$, ..., $[t_{x-1}, t_{end}]$. Similarly suppose the function $d(P_j, S_l(t))$ is composed of y line segments ($1 \leq y \leq 2w_l$) and the time intervals for those line segments are $[t_{begin}, t'_1]$, $(t'_1, t'_2]$, ..., $(t'_{y-1}, t'_{end}]$. If we mix those two set of intervals, we get $x + y - 1$ new intervals. At each new interval, there is at most one intersection between two functions. Hence the total number of intersections is $x + y - 1 \leq 2(w_k + w_l) - 1$.

It seems that the lower envelope of $d(P_j, S_k(t))$ and $d(P_j, S_l(t))$ has at most $4(w_k + wl) - 2$ line segments since there are $2(w_k + wl) - 1$ intervals and each interval could have at most two line segments. However, we could prove there are fewer number of line segments of $\min\{d(P_j, S_k(t)), d(P_j, S_l(t))\}$.

Lemma 3. *The function* $\min\{d(P_j, S_k(t)), d(P_j, S_l(t))\}$ *is composed of at most* $3(w_k + w_l)$ *line segments.*

Proof. The function $d(P_j, S_k(t))$ is composed of x line segments ($1 \le x \le 2w_k$). $d(P_j, S_l(t))$ is composed of y line segments ($1 \le y \le 2w_l$). We define a *valley* as a convex chain such that if we walk on the convex chain from left to right, we always make left turn from one line segment to the next line segment (see Figure 6). Thus $d(P_j, S_k(t))$ consists of at most $x/2$ valleys and $d(P_j, S_l(t))$ consists of at most $y/2$ valleys. For each valley of $d(P_j, S_k(t))$, it can cut one line segment of $d(P_j, S_l(t))$ into three line segments, of which at most two could be in the lower envelope. It is similar for valleys of $d(P_j, S_l(t))$. Therefore, besides the original line segments of two polylines, $x/2 + y/2$ valleys could add at most $x/2 + y/2$ line segments for the lower envelope. The total number of line segments of the lower envelope is $\frac{3}{2}(x + y) \le 3(w_k + w_l)$.

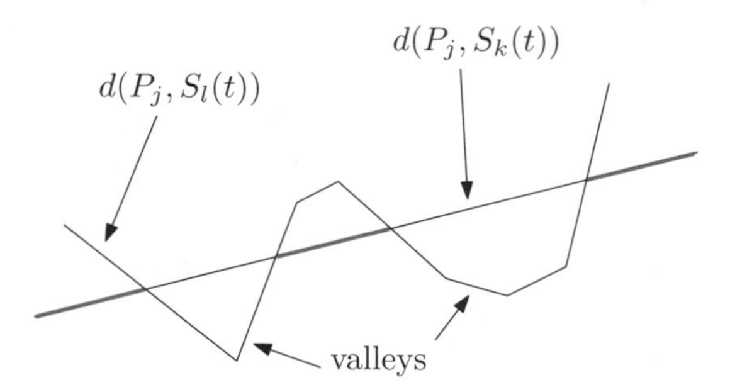

Fig. 6. One line segment of $d(P_j, S_k(t))$ is cut into 5 line segments by two valleys of $d(P_j, S_l(t))$, of which 3 line segments (red) are in the lower envelope of $d(P_j, S_k(t))$ and $d(P_j, S_l(t))$. Notice the number of valleys does not change after merging.

Lemma 4. *The lower envelope* $h(t) = \min\{d(P_j, S_1(t)), d(P_j, S_2(t)), \ldots, d(P_j, S_m(t))\}$ *can be computed in* $O(mW \log^2 m)$ *time and* $O(mW \log m)$ *space where* $W = \max\{2w_1, 2w_2, \ldots, 2w_m\}$ *and it consists of* $O(mW \log m)$ *line segments.*

Proof. The first step of our algorithm is to compute $\min\{d(P_j, S_{2a+1}(t)), d(P_j, S_{2a+2}(t))\}$ where $a = 0, \ldots, \frac{m}{2} - 1$. According to lemma 3, each function $\min\{d(P_j, S_{2a+1}(t)), d(P_j, S_{2a+2}(t))\}$ can be computed in at most $3W$ time, which is composed of at most $3W$ line segments. The second step is to compute $\min\{\min\{d(P_j, S_{4a+1}(t)), d(P_j, S_{4a+2}(t))\}, \min\{d(P_j, S_{4a+3}(t)), d(P_j, S_{4a+4}(t))\}$ where $a = 0, \ldots, \frac{m}{4} - 1$. It seems each $\min\{\min\{d(P_j, S_{4a+1}(t)), d(P_j, S_{4a+2}(t))\}, \min\{d(P_j, S_{4a+3}(t)), d(P_j, S_{4a+4}(t))\}$ consists of $9W$ line segments. However, we observe that the number of the valleys does not change as the number of line segments of lower envelope increases after merging (see Figure 6). That means each lower envelope after first step consists of at most W valleys and $3W$ line segments. After second step, each lower envelope consists of at most $2W$ valleys and $2 \times 3W + 2W = 8W$ line segments. Let $N(i)$ denote the number of line segments of each lower envelope after ith ($1 \leq i \leq \log m$) step. Then we have:

$$\begin{aligned}
N(1) &= 2 \times W + W = (2+1)W = 3W \\
N(2) &= 2 \times 3W + 2W = (3+1)2W = 4 \times 2W = 8W \\
N(3) &= 2 \times 8W + 4W = (4+1)4W = 5 \times 4W = 20W \\
N(4) &= 2 \times 20W + 8W = (5+1)8W = 6 \times 8W = 48W \\
&\dots \\
N(i) &= 2N(i-1) + 2^{i-1}W = (i+2)2^{i-1}W \\
&\dots \\
N(\log m) &= 2N(\log m - 1) + 2^{\log m - 1}W = (\log m + 2)2^{\log m - 1}W \\
&= mW(2 + \log m)/2 = O(mW \log m)
\end{aligned}$$

The computation time for ith step is $\frac{m}{2^i} N(i) = \frac{m}{2^i}(i+2)2^{i-1}W$. The total computation time is $\sum_{i=1}^{\log m} \frac{m}{2^i}(i+2)2^{i-1}W = O(mW \log^2 m)$. The space complexity is $O(mW \log m)$ since we need so much space to store $O(mW \log m)$ line segments of lower envelope. Note that Agarwal and Sharir [1] prove that for mW line segments, the size of the lower envelope is $\Theta(mW\alpha(mW))$ where $\alpha(mW)$ is the inverse of Ackermann's function and the lower envelope can be computed in $O(mW \log(mW))$ time. Our algorithm is better than the algorithm in paper [1] when W is much greater than m.

So far we compute the lower envelope for one node. The envelopes for all n nodes can be computed in $O(nmW \log^2 m)$ time. If the query point q is on some node P_j, we can use binary search for the query time t' on lower envelope associated with P_j to find the nearest site in $O(\log(mW \log m))$ time. If the query point q is on some edge $[P_a, P_b]$, we can just perform the same binary search on both lower envelopes associated with P_a and P_b and then plus the extra length from q to P_a and P_b respectively. The minimum one of those two values and its corresponding site is the answer to the query. Therefore we have following theorem:

Theorem 1. *For a static query of nearest site in moving network, we can answer it in* $\log(mW \log m)$ *time with* $O(nmW \log^2 m + n^2 \log n + nE)$ *time and* $O(nmW \log m + E)$ *space for preprocessing step.*

Note that the preprocessing time and space includes the time and space to compute segmented functions that are $O(n^2 \log n + nE + Y)$ and $O(E + Y)$ where $Y \leq mW$. Sometimes, the number of line segments X of lower envelope could be very small. Here we give an output sensitive algorithm to compute the lower envelope. The general idea of the algorithm is as follows: first, we find function $\min_{1 \leq k \leq m}\{d(P_j, S_k(t_0))\}$, which means the lowest line segment l_e at time t_0. Suppose l_e corresponds the site S_e. Then compute the minimum time intersection point of $d(P_j, S_k(t))$ ($1 \leq k \leq m$ and $k \neq e$) with l_e and get the next line segment $l_{e'}$ of lower envelope. We compute the intersection point from left to right. If the current line segment on $d(P_j, S_k(t))$ has no intersection with l_e then we try the next line segment until there is an intersection point or $d(P_j, S_k(t))$ is out of the time interval of l_e. The line segments of $d(P_j, S_k(t))$ without intersection with l_e will not be tested again later since they are above l_e and can not be the lower envelop. Let $l_e = l_{e'}$ and we perform above process repeatedly until the end of time. The time complexity of this output sensitive algorithm is $O(mX + Y) = O(mX + mW)$ and we need $O(X)$ space to store the lower envelope. Therefore if $X \leq W$, then the preprocessing time and space could be reduced to $O(n^2 \log n + nE + nWm)$ and $O(E + nmW)$ respectively.

4 Point Location in Kinetic Moving Network

We now consider kinetic case queries. Imagine that the moving sites are postmen. There is a customer at a query point on the road with walking speed v and searching for a postman that he can reach in minimum time for delivering his package. Both postmen and the customer can only move on the road.

These types of queries differ from the static case query in the previous section. It is possible that the customer might reach a postman further away (that is traveling toward it) quicker than a nearby postman (that is traveling away from it). The answer to the query depends strongly on v. Note that if we let v approach infinity, then the problem becomes the static query case.

We first consider the query $KineticNearest(P_j, t', v)$, which means the customer lies on the node P_j at t', it is clear that customer can reach the point P if $d(P_j, P) = v(t - t')$ at time t. So we add a line $d = v(t - t')$ over $d(P_j, S_i(t)), i = 1, 2, \ldots, m$ (see figure 7). If $d = v(t - t')$ intersects with $d(P_x, S_i(t)), i = 1, 2, \ldots, m$ at earliest time t'', then S_x which corresponds to the line segment intersecting with $d = v(t - t')$ at t'' would be the site the customer can reach first. Actually it equivalents to find the intersection point of $d = v(t - t')$ with lower envelope of $d(P_j, S_i(t)), i = 1, 2, \ldots, m$ (see figure 8). As we discussed in previous section, for each node P_j, the lower envelope $min\{d(P_j, S_i(t)), i = 1, 2, \ldots, m\}$ can be computed in $O(mW \log^2 m)$ times, which consists of $O(mW \log m)$ line segments. Then we only need to consider the intersection between function $d = v(t - t')$ and function $min\{d(P_j, S_i(t)), i = 1, 2, \ldots, m\}$ (see figure 8).

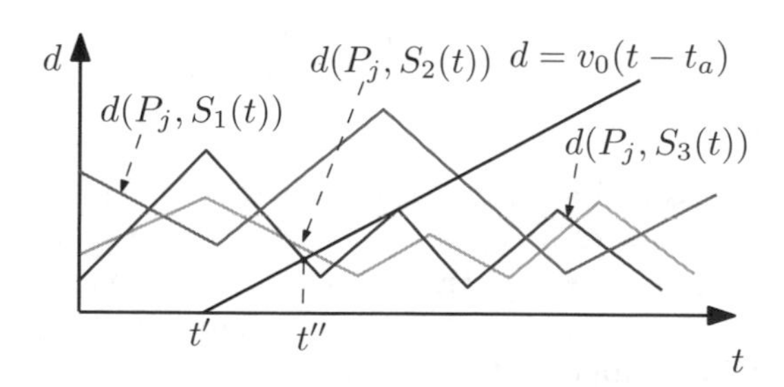

Fig. 7. The diagram of function $d(P_j, S_i(t)), i = 1, 2, 3$ and $d = v(t - t')$

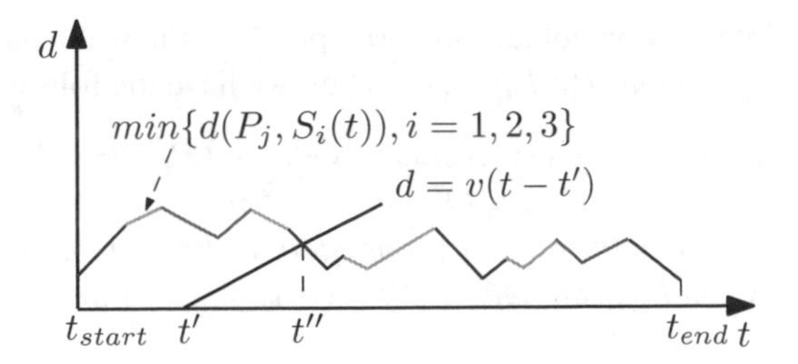

Fig. 8. The diagram of function $min\{d(P_j, S_i(t)), i = 1, 2, 3\}$ and $d = v(t - t')$

If we already compute the function $min\{d(P_j, S_i(t)), i = 1, 2, \ldots, m\}$ in preprocessing step, how to get the first intersection with $d = v(t - t')$. In [6], Leonidas Guibas gave an algorithm: Given a segment e inside P, preprocess P so that for each query ray r emanating from some point on e into P, the first intersection of r with the boundary of P can be calculated in $O(\log V)$ time(V denote the number of vertex of polygon P), after given a triangulation of a simple polygon P. Actually the lower envelope $min\{d(P_j, S_i(t)), i = 1, 2, \ldots, m\}$ with two vertical line segments which correspond to $t = t_{start}$ and $t = t_{end}$ (see figure 8) and the horizontal line segment of time axis can be treated as a simple polygon. The line $d = v(t - t')$ can be treated as a ray emanating from point $(t', 0)$ on horizontal line segment along the time axis. Thus the intersection point can be computed in $O(\log(mW \log m))$ time, which means the $KineticNearest(P_j, t', v)$ can be answered in $O(\log(mW \log m))$ time.

Above we assume that the customer locates exactly on some node. But the customer may locate in the middle of some edge. So the query becomes $KineticNearest(P_s \in edge[P_a, P_b], t', v)$. Let $length(P_s, P_a) = L_1$, $length(P_s, P_b) = L_2$, and $t_a = L_1/v$, $t_b = L_2/v$. If the customer meets some site on some edge other than $edge[P_a, P_b]$, then we know that the customer must walk to P_a or P_b first. Therefore the problem becomes to find than the first intersection between function $min\{d(P_a, S_i(t)), i = 1, 2, \ldots, m\}$ and function $d = v(t - t' - t_a)$ where $t \geq t' + t_a$, and the first intersection between function $min\{d(P_b, S_i(t)), i = 1, 2, \ldots, m\}$ and function $d = v(t - t' - t_b)$ where $t \geq t' + t_b$. This is exactly the same as the previous case. If the customer meets some site on edge $[P_a, P_b]$, then we know they must meet during the site travels through the edge $[P_a, P_b]$ at first time (note that one site could travel through the same edge many times). In preprocessing step, we can extract the line segment of $min\{d(P_a, S_i(t)), i = 1, 2, \ldots, m\}$ which corresponds to one site traveling through the edge $[P_a, P_b]$ from P_a to P_b at first time. Then the intersection point between the extracted line segment and the line segment $d = v(t' + t_a - t)$ where $(t' \leq t \leq t' + t_a)$ is the meeting time and position between the customer and that site. Note that there could be no intersection between two line segments. Since any of m sites could meet the customer on edge $[P_a, P_b]$, we just compute all intersection points in $O(m)$ time. Similarly we can compute the intersection point if the customer meet some site when the site travels through the edge $[P_a, P_b]$ from P_b to P_a at first time. Then we have $O(m)$ intersection points on $d = v(t' + t_a - t)$ where $(t' \leq t \leq t' + t_a)$, $O(m)$ intersection points on $d = v(t' + t_b - t)$ where $(t' \leq t \leq t' + t_b)$, one intersection on $d = v(t - t' - t_a)$ where $t \geq t' + t_a$, and the other intersection on $d = v(t - t' - t_b)$ where $t \geq t' + t_b$. The minimum time of all those $O(m)$ intersection points and corresponding site is the answer to the query $KineticNearest(P_s \in edge[P_a, P_b], t', v)$. Then we have the following theorem:

Theorem 2. *We can answer query $KineticNearest(P_s \in edge[P_a, P_b], t', v)$ in $O(m + \log(mW \log m))$ time with $O(nmW \log^2 m + n^2 \log n + nE)$ time and $O(nmW \log m + E)$ space for preprocessing step. If the customer is located at some node, then the query can be answered in $O(\log(mW \log m))$ time.*

5 Conclusions

In this paper, we consider two variants of the point location problem in moving network. For the static case query, we can answer it in $O(\log(mW \log m))$ time and for the kinetic case query we can answer it in $O(m + \log(mW \log m))$ time. Both need $O(nmW \log^2 m + n^2 \log n + nE)$ time and $O(nmW \log m + E)$ space for preprocessing step. If $X \leq W$, then the preprocessing time and space could be reduced to $O(n^2 \log n + nE + nWm)$ and $O(E + nmW)$ respectively. For both queries, we assume that the trajectories of m sites are known in advance. In the future work, it will be interesting to study the online version of this problem: we only know the trajectories of m sites up to now. How to maintain the Voronoi diagram such that we can answer the static case query efficiently? Note that we do not have kinetic case query for online problem since we don't know the trajectories of m sites in advance.

References

1. Agarwal, K.P., Sharir, M.: Davenport–Schinzel Sequences and Their Geometric Applications. Cambridge University Press, Cambridge (1995)
2. Bae, S.W., Kim, J.-H., Chwa, K.-Y.: Optimal construction of the city voronoi diagram. In: ISAAC, pp. 183–192 (2006)
3. de Berg, M., van Kreveld, M., Overmars, M., Schwarzkopf, O.: Computational Geometry Algorithms and Applications. Springer, Heidelberg (1997)
4. Devillers, O., Golin, M.J.: Dog bites postman: Point location in the moving voronoi diagram and related problems. In: Lengauer, T. (ed.) ESA 1993. LNCS, vol. 726, pp. 133–144. Springer, Heidelberg (1993)
5. Erwig, M.: The graph voronoi diagram with application. Networks 36, 156–163 (2000)
6. Guibas, L., Hershberger, J., Leven, D., Sharir, M., Tarjan, R.: Linear time algorithms for visibility and shortest path problems inside simple polygons. In: SCG 1986: Proceedings of the Second Annual Symposium on Computational Geometry, pp. 1–13. ACM, New York (1986)
7. Hakimi, S., Labbe, M., Schmeiche, E.: The voronoi partition of a network and its implications in location theory. INFIRMS Journal on Computing 4, 412–417 (1992)
8. Lee, D.T.: Two-dimensional voronoi diagrams in the lp-metric. Journal of the ACM 27(4), 604–618 (1980)
9. Okabe, A., Satoh, T., Furuta, T., Suzuki, A., Okano, K.: Generalized network voronoi diagrams: Concepts, computational methods, and applications. International Journal of Geographical Information Science 22(9), 965–994 (2008)
10. Ramalingam, G., Reps, T.: An incremental algorithm for a generalization of the shortest-path problem. Journal of Algorithms 21(2), 267–305 (1996)

Coordinated Scheduling of Production and Delivery with Production Window and Delivery Capacity Constraints

Bin Fu[1], Yumei Huo[2], and Hairong Zhao[3]

[1] Department of Computer Science,
University of Texas–Pan American, Edinburg, TX 78539, USA
binfu@cs.panam.edu

[2] Department of Computer Science,
College of Staten Island, CUNY, Staten Island, New York 10314, USA
yumei.huo@csi.cuny.edu

[3] Department of Mathematics, Computer Science & Statistics,
Purdue University, Hammond, IN 46323, USA
hairong@calumet.purdue.edu

Abstract. This paper considers the coordinated production and delivery scheduling problem. We have a planning horizon consisting of z delivery times each with a unique delivery capacity. Suppose we have a set of jobs each with a committed delivery time, processing time, production window, and profit. The company can earn the profit if the job is produced in its production window and delivered before its committed delivery time. From the company point of view, we are interested in picking a subset of jobs to process and deliver so as to maximize the total profit subject to the delivery capacity constraint. We consider both the single delivery time case and the multiple delivery times case.

Suppose the given set of jobs are k-disjoint, that is, the jobs can be partitioned into k lists of jobs such that the jobs in each list have disjoint production windows. When k is a constant, we developed a PTAS for the single delivery case. For multiple delivery times case, we also develop a PTAS when the number of delivery times is a constant as well.

1 Introduction

Under the current competitive manufacturing environment, companies tend to put more emphasis on the coordination of different stages of a supply chain, i.e. suppliers, manufacturers, distributors and customers. Among these four stages, the issue of coordinating production and distribution (delivery) has been widely discussed.

In the stage of distribution (delivery), the company may own transportation vehicles which deliver products at periodic or aperiodic times, or the company may rely on a third party to deliver, which picks up products at regular or irregular times. The products incurred by different orders may be delivered together if the destinations are close to each other, e.g. delivery to same countries by

B. Chen (Ed.): AAIM 2010, LNCS 6124, pp. 141–149, 2010.

ships, delivery to same states or cities by flights, or delivery to same areas by trucks. The delivery capacity may vary at different delivery times and is always bounded.

In the stage of production, each order may have a non-customer defined production window. For example, the company may rely on another company to complete a sub-process, or may rely on a manufacturer to make the products or semi-products. With some pre-scheduled jobs, the manufacturer can only provide partial production line, or in some cases, it provides a production window for each order where the windows of different orders may overlap. Another example of production window is that the company may have to wait for arrivals of raw materials to start the manufacturing process. If the raw materials are perishable, the company has to start the manufacturing process immediately. Given the arrival schedule of raw materials, the company creates a production window for each order.

In summary, in the production stage, the company has the constraint of production window, and in the delivery stage, the company has the constraint of delivery capacity and promised delivery date. The company has to decide which orders to accept based on these constraints and the potential profit of each order in order to maximize the total profit .

This paper addresses the problem faced by the company under the above scenarios. We consider the commit-to-ship model, i.e. if an order is accepted, the company guarantees the products be shipped to the customer before the committed time, we call this time the *committed delivery date*. We focus on the single machine production environment. We are interested in selecting a subset of orders in order to maximize the total profit. When orders are selected, both production schedule and delivery schedule should be considered simultaneously. Thus we face a "coordinated scheduling problem": generate a coordinated schedule, which consists of a production schedule and a delivery schedule subject to the production window, delivery date, and delivery capacity constraints.

Problem definition. Our problem can be formally defined as follows. We have a planning horizon consisting of z delivery times, $T = \{D^1, D^2, \cdots, D^z\}$. Each delivery time D^j is associated with a delivery capacity C^j. We have a set of jobs $J = \{J_1, J_2, \cdots, J_n\}$. Each job J_i has a promised delivery time $d_i \in T$, a processing time p_i, a production window $[l_i, r_i]$, a size c_i, and a profit f_i which can be earned if J_i is processed at or before r_i, and delivered before or at d_i. Without loss of generality, we assume that $p_i \leq r_i - l_i$ and $p_i \leq d_i$ for all jobs J_i. We also assume $r_i \leq d_i$ for all jobs J_i. The problem is to select a subset of jobs from $J = \{J_1, J_2, \cdots, J_n\}$, and generate a feasible coordinated production and delivery schedule S of these jobs so as to maximize the total profit. A feasible coordinated schedule S consists of a feasible production schedule and a feasible delivery schedule. A production schedule is feasible if all the jobs are processed within their production windows; and a delivery schedule is feasible if all jobs are delivered before the promised delivery time and the delivery capacities at all times are satisfied.

Literature review. In recent two decades, coordinated production and delivery scheduling problems have received considerable interest. However, most of the research is done at the strategic and tactical levels (see survey articles [10], [5], [7], [2], [3]). At the operational scheduling level, Chen [4] gives a state-of-the-art survey of the models and results in this area. Based on the delivery mode, he classified the models into five classes: (1) models with individual and immediate delivery; (2) models with batch delivery to a single customer by direct shipping method; (3) models with batch delivery to multiple customers by direct shipping method; (4) models with batch delivery to multiple customers by routing method (5) models with fixed delivery departure date. In the first model, jobs have delivery windows, and thus production windows can be incurred, however, due to the immediate delivery and no delivery capacity constraints considered, the problems under this model can be reduced to fixed-interval scheduling problems without the delivery, which can be solved as a min-cost network flow problem ([9]). For all other models, no production windows have been specially considered in the survey.

Several papers considered problems with time window constraints and/or delivery capacity constraints. Amstrong et al. ([1]) considered the integrated scheduling problem with batch delivery to multiple customers by routing method, subject to delivery windows constraints. The objective is to choose a subset of the orders to be delivered such that the total demand of the delivered orders is maximized. Garcia and Lozano ([6]) considered the production and delivery scheduling problems in which time windows are defined for the jobs' starting times. In their paper, orders must be delivered individually and immediately after they are manufactured, so delivery capacity is not an issue. In [8], Huo, Leung and Wang considered the integrated production and delivery scheduling problem with disjoint time windows where windows are defined for the jobs' completion times. In their paper, they assume a sufficient number of capacitated vehicles are available.

New contribution. Compared with existing models, our model is more practical and thus more complicated. The problem is NP-hard since the problem at each stage is already NP-hard by itself. So our focus is to develop approximation algorithms. Suppose a set of jobs are k-disjoint, that is, the jobs can be partitioned into k lists of jobs such that the jobs in each list have disjoint production windows. When k is constant and there is a single delivery time,we develop the first PTAS(Polynomial Time Approximation Scheme) for the coordinated production and delivery scheduling problem. For multiple delivery times, we also develop a PTAS when the number of delivery time is a constant as well.

The paper is organized as follows. In Section 2, we present an approximation scheme for single delivery time. In Section 3, we present an approximation scheme for multiple delivery times. In Section 4, we draw some conclusions.

2 Single Delivery

In this section, we study the coordinated production and delivery scheduling problem where the jobs have the same promised delivery time D which is associated with a delivery capacity C. In this case, all jobs will be delivered at same time, thus no delivery schedule is necessary as long as the delivery capacity constraint is satisfied by the selected jobs and the production schedule of the selected jobs is feasible. Therefore the problem in this case becomes selecting a subset of orders and generate a feasible production schedule subject to the capacity constraint. Given a constant ϵ, we develop an algorithm which generates a feasible production schedule of a subset of jobs subject to the capacity constraint, whose profit is at least $(1-\epsilon)$ times the optimal. Our algorithm is a PTAS when the set of jobs $J = \{J_1, J_2, \cdots, J_n\}$ is k-disjoint and k is a constant.

Our algorithm has four phases:

Phase I: select large jobs;
Phase II: schedule the large jobs selected from Phase I along with some small jobs selected in this phase;
Phase III: from the schedules generated in Phase II, search for the one with maximum total profit;
Phase IV: convert the schedule from phase III to a feasible schedule.

2.1 Phase I: Select Large Jobs

In this phase, we select large jobs for production and delivery without generating the production schedule. Let us define *large jobs* first. Given a constant parameter $0 < \delta < 1$ which depends on ϵ and will be determined later, a job is said to be *large* if its size is at least δ times the "available" delivery capacity; otherwise, it is a *small job*. By definition, we can see that a job may be "small" at the beginning and becomes "large" later as the available capacity becomes smaller due to more jobs are selected.

To select the large jobs, we use brute force. Specifically, we enumerate all the possible selections of large jobs subject to the available capacity constraint. We use A to denote the set of all possible selection. The jobs in each selection $A_p \in A$ are selected in $\lceil \frac{1}{\delta} \rceil$ iterations. For each A_p, at the beginning of each iteration, the current available capacity $\bar{C}_p$ is calculated and the set of large jobs (and so small jobs) from the remaining jobs is identified, and then a subset of large jobs is selected and added to A_p. If no large jobs is selected and added to A_p at certain iteration, we mark A_p as "finalized", which means no more large jobs will be selected and added to A_p in later iterations.

Phase1-Alg

Let A be the set of all possible selections of large jobs so far; $A = \{\emptyset\}$.
For $i = 1$ to $\lceil \frac{1}{\delta} \rceil$
  Let $A' = \emptyset$
  For each selection of large jobs $A_p \in A$
    If A_p is marked as "finalized", add A_p directly to A'
    Else

a. let $\bar{C}_p$ be the capacity available for jobs not in A_p, i.e. $\bar{C}_p = C - \sum_{J_i \in A_p} c_i$
b. from the jobs not selected in A_p, find the large jobs with respect to $\bar{C}_p$
c. generate all possible selections of these large jobs, say X, subject to the available capacity constraint
d. for each $X_j \in X$
generate a new large job selection $A_q = A_p \cup X_j$, and add A_q into A'
if $A_q = A_p$, mark A_q as "finalized".

$A = A'$

return A

Lemma 1. *There are at most $O(n^{O(1/\delta^2)})$ possible ways to select the large jobs, where $0 < \delta < 1$ is a constant.*

2.2 Phase II: Schedule Large and Small Jobs

From Phase I, we get a set of large job selections A without scheduling the jobs. However, for a job selection $A_p \in A$, it is possible no feasible production schedule exists for jobs in A_p. In this case, we say A_p is an infeasible job selection. In this phase, our goal is, to identify all feasible large job selections in A; and for each feasible selection A_p, find a feasible production schedule S that contains the large jobs in A_p, and some newly added "small" jobs, where the small jobs are identified at the beginning of the last iteration of Phase1-Alg, and each has a size less than $\delta\bar{C}_p$. Furthermore, the profit of S, denoted by $\text{Profit}(S) = \sum_{J_i \in S} f_i$, is close to the maximum among all feasible schedules whose large job selection is exactly A_p. On the other hand, the generated schedule S in this phase may violate the capacity constraint. Specifically, let $\widehat{C}_p$ be the *available capacity for small jobs*, i.e. $\widehat{C}_p = C - \sum_{i \in A_p} C_i$, and let $\text{Load}(S/A_p) = \sum_{J_i \in S \setminus A_p} c_i$ be the total size of the small jobs in S, it is possible that $\widehat{C}_p \leq \text{Load}(S/A_p) \leq (1+\delta)\widehat{C}_p$. In this case, we say S is a "valid" schedule.

Even though we know that the large job selection in a schedule S is exactly A_p, we still can not determine how to schedule the jobs in A_p due to the production window constraint of the jobs and the unknown small jobs in S. We build S using dynamic programming. We add jobs to the schedule one by one in certain order. For this, we assume the set of jobs $J = \{J_1, J_2, \cdots, J_n\}$ is k-disjoint, and J has been divided into k job lists $L_1, L_2, \cdots, L_k$ such that production windows of jobs in the same list are disjoint. Let n_u be the number of jobs in job list L_u $(1 \leq u \leq k)$. We relabel the jobs in each job list L_u in increasing order of their production windows' starting time, and we use $J_{u,v}$, $1 \leq v \leq n_u$, to denote the v-th job in the job list L_u. It is easy to see that in any feasible schedule, one can always assume the jobs in the same list are scheduled in the order they appear in the list. So in our dynamic programming, the jobs in each list are considered in this order. In case the lists $L_1, L_2, \cdots, L_k$ are not given, note that the problem can be solved greedily.

For a given large job selection produced from Phase I, $A_p \in A$, if we can find all schedules whose large job selection is exactly A_p, we can easily find the best schedule. However, that will be both time and space consuming. To reduce space and time, we find a subset of schedules to approximate all possible schedules so that no two jobs in the set are "similar". Let us formally define "similar" schedules. Given a schedule S with a large job selection A_p, let Profit(S/A_p) be the total profit of small jobs in S. For two schedules S_1 and S_2 that both have the same set of large jobs A_p, given a constant $\omega = 1 + \frac{\delta}{2n}$, we say they are *similar*, if Profit(S_1/A_p) and Profit(S_2/A_p) are both in $[\omega^x, \omega^{x+1})$, and Load(S_1/A_p) and Load$(S_2/A_p))$ are both in $[\omega^y, \omega^{y+1})$ for some integer x and y.

In the following, we use $T(A_p, n'_1, \cdots, n'_k)$ to denote a set of valid schedules such that (a) no two schedules in the set are similar; (b) only the first n'_u jobs from list L_u $(1 \le u \le k)$ can be scheduled; (c) and among these n'_u jobs in list L_u, all the jobs in A_p must be scheduled. For each schedule S, we use $C_{\max}(S)$ to represent the last job's completion time. In $T(A_p, n'_1, \cdots, n'_k)$, from each group of schedules that are similar to each other, we only keep the schedule with the smallest $C_{\max}(S)$ in the group.

Phase2-Alg(A_p, J, $\bar{C}_p$)

- Input: a set of jobs J, which has been divided into k disjoint job lists $L_1, L_2, \cdots, L_k$;
  A_p: a large job selection obtained from Phase I;
  $\bar{C}_p$: the available capacity at the beginning of last iteration in Phase I for obtaining A_p.
- Let $\widehat{C}_p = C - \sum_{i \in A_p} C_i$, i.e., the available capacity for small jobs
- Initialize $T(A_p, 0, \cdots, 0) = \{\emptyset\}$.
- Construct $T(A_p, n_1, \cdots, n_k)$ using dynamic programming
  To find the set $T(A_p, n'_1, \cdots, n'_k)$, do the following steps
  1. For $t = 1$ to k
     (a) consider job J_{t,n'_t}, let $[l_{t,n'_t}, r_{t,n'_t}]$ be its production window, p_{t,n'_t} be its processing time, c_{t,n'_t} be its size
     (b) If $J_{t,n'_t} \in A_p$
        For each schedule S in $T(A_p, n'_1, \cdots, n'_t - 1, \cdots, n'_k)$
          if $\max(C_{\max}(S), l_{t,n'_t}) + p_{t,n'_t} \le r_{t,n'_t}$
            get a schedule S' by adding J_{t,n'_t} to S and schedule it at $\max(C_{\max}(S), l_{t,n'_t})$;
            add S' to $T(A_p, n'_1, \cdots, n'_k)$;
     (c) Else
        For each schedule S in $T(A_p, n'_1, \cdots, n'_t - 1, \cdots, n'_k)$
          add S into $T(A_p, n'_1, \cdots, n'_k)$;
          if $c_{t,n'_t} < \delta \bar{C}_p$(i.e. a small job) and $\max(C_{\max}(S), l_{t,n'_t}) +$ $p_{t,n'_t} \le r_{t,n'_t}$ and Load$(S \cup \{J_{t,n'_t}\}/A_p) \le (1+\delta)\widehat{C}_p)$
            get a schedule$S'$by adding J_{t,n'_t} to S at $\max(C_{\max}(S), l_{t,n'_t})$;
            add S' into $T(A_p, n'_1, \cdots, n'_k)$;

2. From each group of schedules in $T(A_p, n'_1, \cdots, n'_k)$ that are similar to each other, delete all but the schedule S with minimum $C_{\max}(S)$ in the group

– return the schedule S from $T(A_p, n_1, \cdots, n_k)$ with maximum profit

One should note that in the algorithm $\bar{C}_p$ and $\widehat{C}_p$ are not exactly same. Since $\widehat{C}_p$ is available capacity for small jobs, while $\bar{C}_p$ is the available capacity at the beginning of last iteration of Phase1-Alg and some large jobs may be selected at the last iteration, we have $\bar{C}_p \geq \widehat{C}_p$. In particular, if A_p is marked "finalized" at the end of algorithm, then we have $\bar{C}_p = \widehat{C}_p$.

It is easy to see that if a large job selection $A_p \in A$ is not feasible, $T(A_p, n_1, \cdots, n_k)$ must be empty. For any feasible large job selection A_p, we have the following lemma.

Lemma 2. *Suppose $A_p \in A$ is a feasible large job selection obtained from Phase I, and let S' be the feasible schedule that has the maximum profit among all schedules whose large job selection is exactly A_p. Then Phase2-Alg(A_p, J, $\bar{C}_p$) returns a schedule S such that: the large job selection in S is A_p; S is valid,and* $\text{Profit}(S) \geq (1-\delta)\text{Profit}(S')$.

Lemma 3. *For a given A_p, the running time of Phase2-Alg(A_p, J, C) is $O(kn^k(\log_\omega \sum_{i=1}^n f_i) \log_\omega C)$.*

2.3 Phase III: Search for the Best Schedule

For each feasible large job selection $A_p \in A$ obtained from Phase I, the dynamic procedure of Phase II outputs a valid schedule whose large job selection is exactly A_p. In this phase, we find a good schedule to approximate the optimal schedule. This is done by selecting the schedule S with the maximum total profit among all the schedules generated in Phase II for all feasible A_p-s.

Lemma 4. *Let S be the schedule with the maximum profit among all schedules generated from Phase II. Then S must be valid and* $\text{Profit}(S) \geq (1-\delta)\text{Profit}(S^*)$, *where S^* is the optimal schedule. Furthermore, the total running time to obtain S is $O(\frac{k}{\delta^2} n^{O(k+1/\delta^2)}(\lg \sum_{i=1}^n f_i)(\lg C))$.*

2.4 Phase IV: Convert to a Feasible Schedule

From Phase III, we get the schedule S with the maximum total profit which is valid but may not be feasible, i.e. $\widehat{C}_p \leq \text{Load}(S/A_p) \leq (1+\delta)\widehat{C}_p$. To convert S into a feasible schedule, we have to delete some jobs carefully so that the total profit will not be affected greatly.

Phase4-Alg(S)

Input: S is the best schedule produced after Phase III, which is valid but may not be feasible

Let A_p be its corresponding large jobs selection in S.

Let $S' = S$
If Load(S') $> C$ (i.e. Load(S'/A_p) $> \widehat{C}_p$)
  If A_p is marked as "finalized"
    Delete a set of small jobs of total size of at most $2\delta\widehat{C}_p$
    and with the least possible profit from S'
  Else
    Delete the large job in A_p with least profit from S'
Return S'

Lemma 5. *Let δ be a constant of at most $\frac{\epsilon}{3}$. The schedule S' returned by Phase4-alg is feasible, and* Profit(S') $\geq (1-\epsilon)$Profit(S^*), *where S^* is an optimal schedule.*

For any constant ϵ, by Lemma 5 and 4, we have the following theorem.

Theorem 1. *For any coordinated production and delivery scheduling problem with production window and delivery capacity constraints, if there is only one delivery time, and the job set is k-disjoint, where k is a constant, there exists a polynomial time approximation scheme that runs in time $O(n^{O(k+1/\epsilon^2)}(\lg \sum_{i=1}^{n} f_i)(\lg C))$.*

3 Multiple Delivery

In this section, we study the coordinated production and delivery scheduling problem with multiple delivery times $D^1, D^2, \cdots, D^z$ which have delivery capacity of $C^1, C^2, \cdots, C^z$, respectively. Our goal is to find a feasible coordinated production and delivery schedule whose total profit is close to optimal. As the case of the single delivery time, a feasible production schedule is one that satisfies the production window constraint. A feasible delivery schedule, however, is more restricted than the single delivery time: for each selected job, we have to specify its delivery time which can not be later than its promised delivery date; and the delivery capacity constraint has to be satisfied for all delivery times.

When z is a constant, we develop a PTAS for this case which is based on the PTAS for the case of single delivery time. The structure of the two PTASes are similar, but the details are different. In particular, in Phase I and Phase II, we have to consider the delivery schedule of the jobs; and in Phase IV, we have to make sure the capacity is satisfied for all delivery times. Due to space limit, we will not given details of the algorithm.

Theorem 2. *For any coordinated production and multiple delivery scheduling problem with production window and delivery capacity constraints, when there are constant number of delivery times z, and the job set is k-disjoint where k is a constant, there exists a polynomial time approximation scheme. Furthermore, for any constant ϵ, the algorithm runs in time*

$$O(n^{O(\frac{z}{\epsilon^2}+k)}(\lg \sum_{i=1}^{n} f_i) \prod_{j=1}^{z} (\lg C^j)).$$

4 Conclusion

In this paper, we study the problem of coordinated scheduling problem with production window constraint and the delivery capacity constraint. When the jobs are k-disjoint and k is a constant, we develop a PTAS for the case of single delivery time. We then extend the PTAS to solve the problem with constant number of delivery times. One open question is to develop constant approximation algorithms for the problem.

References

1. Amstrong, R., Gao, S., Lei, L.: A zero-inventory production and distribution problem with a fixed customer sequence. Ann. Oper. Res. 159, 395–414 (2008)
2. Bilgen, B., Ozkarahan, I.: Strategic tactical and operational production-distribution models: A review. Internat. J. Tech. Management 28, 151–171 (2004)
3. Chen, Z.-L.: Integrated production and distribution operations: Taxonomy, models, and review. In: Simchi-Levi, D., Wu, S.D., Shen, Z.-J. (eds.) Handbook of Quantitative Supply Chain Analysis: Modeling in the E-Business Era. Kluwer Academic Publishers, Norwell (2004)
4. Chen, Z.-L.: Integrated production and outbound distribution scheduling: Review and extensions. In: Operations Research (2009) (to appear)
5. Erenguc, S.S., Simpson, N.C., Vakharia, A.J.: Integrated production/ distribution planning in supply chains: An invited review. Eur. J. Oper. Res. 115, 219–236 (1999)
6. Garcia, J.M., Lozano, S.: Production and delivery scheduling problem with time windows. Computers & Industrial Engineering 48, 733–742 (2005)
7. Goetschalckx, M., Vidal, C.J., Dogan, K.: Modeling and design of global logistics systems: A review of integrated strategic and tactical models and design algorithms. Eur. J. Oper. Res. 143, 1–18 (2002)
8. Huo, Y., Leung, J.Y.-T., Wang, X.: Integrated Production and Delivery Scheduling with disjoint windows. Discrete Applied Math. (2009) (to appear)
9. Kroon, L.G., Salomon, M., Van Wassenhove, L.N.: Exact and approximation algorithms for the operational fixed interval scheduling problem. Eur. J. Oper. Res. 82, 190–205 (1995)
10. Sarmiento, A.M., Nagi, R.: A review of integrated analysis of production-distribution systems. IIE Trans. 31, 1061–1074 (1999)

Inverse 1-median Problem on Trees under Weighted l_∞ Norm

Xiucui Guan[1,⋆] and Binwu Zhang[2]

[1] Department of Mathematics, Southeast University, Nanjing 210096, China
xcguan@163.com
[2] Department of Mathematics and Physics, Hohai University,
Changzhou 213022, China

Abstract. The inverse 1-median problem is concerned with modifying the weights of the customers at minimum cost such that a prespecified supplier becomes the 1-median of modified location problem. We first present the model of inverse 1-median problem on trees. Then we propose two algorithms to solve the problem under weighted l_∞ norm with bound constraints on modifications. Based on the approach of the unbounded case, we devise a greedy-type algorithm which runs in $O(n^2)$ time, where n is the number of vertices. Based on the property of the optimal solution, we propose an $O(n \log n)$ time algorithm using the binary search.

Keywords: Inverse 1-median problem; Tree; weighted l_∞ norm; Binary search.

1 Introduction

The inverse location problems have become an important aspect in the field of inverse optimization problems in recent years. Different from the classical location problem, a feasible solution for a location problem is first given in an inverse location problem. And we aim to modify parameters of the original problem at minimum total cost within certain modification bounds such that the given feasible solution becomes optimal with respect to the new parameter values.

The *1-median problem* can be stated as follows: let (X, d) be a metric space with distance function d. Let n vertices $v_1, v_2, \ldots, v_n$ (called customers) be given. Every customer v_i has a weight w_i. We assume throughout this paper that all given weights are positive. Find a new point s (the *1-median*, called the supplier) in the space X such that

$$\sum_{i=1}^{n} w_i d(v_i, s)$$

becomes minimum.

The inverse 1-median problem consists in changing the weights of the customers of a 1-median location problem at minimum cost such that a prespecified supplier becomes the 1-median of modified location problem. In this paper, we

⋆ Corresponding author.

B. Chen (Ed.): AAIM 2010, LNCS 6124, pp. 150–160, 2010.

consider the inverse 1-median problem where customers and supplier correspond to the vertices of a tree graph $T = (V, E)$ with vertex set V and edge set E, where the distance function d is given by the lengths of paths in the tree. Each vertex $v_i \in V$ has a positive weight w_i. Let s^* not be a 1-median of the tree T with respect to the given weights w. We aim to find new vertex weights w^* satisfying the bound constraint $0 \leq \underline{w} \leq w^* \leq \overline{w}$ such that s^* becomes a 1-median with respect to w^*. And the objective function is to minimize the adjustment of weights $\|w - w^*\|$ under some norm.

Cai et al. [8] proved that the inverse 1-center location problem with variable edge lengths on general unweighted directed graphs is NP-hard, while the underlying center location problem is solvable in polynomial time. Burkard et al. presented a greedy-type algorithm in $O(n \log n)$ time for the inverse 1-median problem with variable vertex weights if the underlying network is a tree or the location problem is defined in the plane [6] and an algorithm in $O(n^2)$ time on cycles [7]. Galavii [9] showed that the inverse 1-median problem can actually be solved in $O(n)$ time on a tree, or on a path with negative weights. Gassner [10] suggested an $O(n \log n)$ time algorithm for the inverse 1-maxian (or negative weight 1-median) problem on a tree with variable edge lengths. In 2008, Burkard et al. [5] solved the inverse Fermat-Weber problem with variable vertex weights in $O(n \log n)$ time for unit cost if the prespecified point that should become a 1-median does not coincide with a given point in the plane. Bonab, Burkard and Alizadeh [4] investigated the inverse 1-median problem with variable coordinates endowed with the rectilinear, the squared Euclidean or the Chebyshev norm. They showed that this problem is NP-hard if the rectilinear or Chebyshev norm is used, but it can be solved efficiently in polynomial time if the squared Euclidean norm is used. Yang and Zhang [14] proposed an $O(n^2 \log n)$ time algorithm for the inverse vertex center problem with variable edge lengths on tree networks where all the modified edge lengths remain positive. Alizadeh and Burkard [1] investigated the inverse absolute 1-center location problem on trees with variable edge lengths, and proposed an $O(n^2 r)$ time exact algorithm where r is the compressed depth of the underlying tree. Recently, Alizadeh, Burkard and Pferschy [2] used a set of suitably extended AVL-search trees and developed a combinatorial algorithm which solves the inverse 1-center location problem with edge length augmentation in $O(n \log n)$ time.

However, the costs incurred by modifying the weights are measured by l_1 norm in all the above inverse 1-median (1-center) problems. Hence, it is necessary to measure the costs by weighted l_∞ norm.

The paper is organized as follows. In Sect. 2, we present the model of inverse 1-median problem on trees. Then we propose two polynomial time algorithms to solve the problem under weighted l_∞ norm in Sect. 3. Conclusions and further research are given in Sect. 4.

In the following, for a function f defined on a set S, we define $f(S) := \sum_{s \in S} f(s)$.

2 The Inverse 1-median Problem on Trees

In this section, we obtain the model of inverse 1-median problem on trees.

We first analyze some properties of the 1-median problem on trees. The 1-median problem on trees has the interesting property that the solution is completely independent of the (positive) edge lengths and only depends on the weights of the vertices [12,13]. Let $W := w(V)$ be the sum of all vertex weights of the tree. Now, let v be an arbitrary vertex of degree k and let $v_1, v_2, \ldots, v_k$ be its immediate neighbors. If we root the tree at v, we get subtrees $T_1, T_2, \ldots, T_k$ which are rooted at $v_1, v_2, \ldots, v_k$, respectively. A consequence of the considerations of Hua et al. and Goldman is the following optimality criterion.

Lemma 1. *(Optimality criterion, [13,12]). A vertex v is a 1-median of the given tree with respect to w, if and only if the weights of all its subtrees $w(T_i)$ are not larger than $W/2$:*

$$\max_{1 \le i \le k} w(T_i) \le W/2. \tag{2.1}$$

Now let s^0 be a 1-median of the tree T with respect to w. If we root the tree at the given vertex s^*, we get subtrees $T_1, T_2, \ldots, T_k$, where k is the degree of vertex s^*. One of these subtrees contains s^0, say T_1. Let $W^* := w^*(V)$. Based on Lemma 1, we can formulate mathematically the inverse 1-median problem on trees as follows:

$$\min \|w^* - w\| \tag{2.2}$$

$$\text{s.t. } \max_{1 \le i \le k} w^*(T_i) \le W^*/2; \tag{2.3}$$

$$\underline{w} \le w^* \le \overline{w}. \tag{2.4}$$

Recall that s^* is not a 1-median with respect to w, then we know that $w(T_1) > W/2$ and $w(T_i) < W/2$ for all $i = 2, \ldots, k$ [6]. Let $T_{k+1} := \{s^*\}, W_{k+1} := w(s^*), W^*_{k+1} := w^*(s^*)$, $W_i = w(T_i)$ and $W^*_i = w^*(T_i), i = 1, \ldots, k$. In order to make s^* a 1-median with respect to w^*, we need to reduce the optimality gap $D := W_1 - W/2$ to 0.

Lemma 2. *[6]. Let $W^*_2 \le W^*/2, \ldots, W^*_k \le W^*/2$. The vertex s^* is a 1-median of the given tree with respect to w^*, if and only if the optimality gap $D^* := W^*_1 - W^*/2 = 0$, that is, $W^*_1 = W^*/2$.*

To decrease D by $\delta/2$, we either decrease the weight of a vertex in T_1 by δ, or increase the weight of a vertex in $T_i (i = 2, \ldots, k+1)$ by δ [6]. For any vertex v, let $p(v)$ and $q(v)$ be increment and decrement of vertex weight, respectively. Then we have

$$w^*(v) = w(v) + p(v) - q(v), \quad p(v), q(v) \ge 0;$$
$$p(v) = 0, w(v) - q(v) \ge \underline{w}(v), \ \forall v \in T_1;$$
$$q(v) = 0, \ w(v) + p(v) \le \overline{w}(v), \ \forall v \in T_i, i = 2, \ldots, k+1.$$

Furthermore, the constraint condition (2.3) means that $w^*(T_i) \leq W^*/2, i = 1, 2, \cdots, k$. That is,

$$w(T_i) + p(T_i) - q(T_i) \leq \frac{1}{2}(W + p(T) - q(T)), i = 1, 2, \cdots, k.$$

Furthermore, we have

$$\begin{cases} -q(T_1) - \frac{1}{2}\sum_{j=2}^{k+1} p(T_j) + \frac{1}{2}q(T_1) \leq \frac{1}{2}W - w(T_1); \\ p(T_i) - \frac{1}{2}\sum_{j=2}^{k+1} p(T_j) + \frac{1}{2}q(T_1) \leq \frac{1}{2}W - w(T_i),\ i = 2, \ldots, k+1. \end{cases}$$

Hence, we can transform the inverse 1-median problem into the following form.

$$\begin{aligned} &\min \|p - q\| \\ &\text{s.t. } q(T_1) + \sum_{j=2}^{k+1} p(T_j) \geq 2D; && (2.5) \\ &\quad p(T_i) - \frac{1}{2}\sum_{j=2}^{k+1} p(T_j) + \frac{1}{2}q(T_1) \leq \frac{1}{2}W - w(T_i),\ i = 2, \ldots, k+1; && (2.6) \\ &\quad 0 \leq q(v) \leq w(v) - \underline{w}(v), \forall v \in T_1; \\ &\quad 0 \leq p(v) \leq \overline{w}(v) - w(v), \forall v \in T_i, i = 2, \ldots, k+1. \end{aligned}$$

On the other hand, based on Lemma 2, we have

$$W_1^* = \frac{1}{2}W^* = \frac{1}{2}(W_1^* + \sum_{j=2}^{k+1} W_j^*).$$

It follows from $W_1^* = \sum_{j=2}^{k+1} W_j^*$ that

$$W_1 - q(T_1) = \sum_{j=2}^{k+1}(W_j + p(T_j)).$$

Therefore

$$q(T_1) + \sum_{j=2}^{k+1} p(T_j) = W_1 - \sum_{j=2}^{k+1} W_j = 2(W_1 - \frac{1}{2}\sum_{j=1}^{k+1} W_j) = 2(W_1 - \frac{1}{2}W) = 2D.$$

Obviously, the vectors p and q satisfying the above equality also satisfy the constraint condition (2.5). Furthermore, if we substitute $\sum_{j=2}^{k+1} p(T_j) = 2D - q(T_1)$ into the constraint condition (2.6), we get

$$p(T_i) - \frac{1}{2}(2D - q(T_1)) + \frac{1}{2}q(T_1) = p(T_i) - D + q(T_1) \leq \frac{1}{2}W - w(T_i),$$

which means

$$W_i^* = w(T_i) + p(T_i) \leq D + \frac{1}{2}W - q(T_1) = w(T_1) - q(T_1) = W_1^* = W^*/2.$$

So, there is an optimal solution (p, q) to the inverse 1-median problem, which also satisfies the problem below.

$$\min \|p - q\| \tag{2.7}$$

$$\text{s.t. } q(T_1) + \sum_{j=2}^{k+1} p(T_j) = 2D; \tag{2.8}$$

$$0 \leq q(v) \leq w(v) - \underline{w}(v), \forall v \in T_1; \tag{2.9}$$

$$0 \leq p(v) \leq \overline{w}(v) - w(v), \forall v \in T_i, i = 2, \ldots, k+1. \tag{2.10}$$

Theorem 3. *To solve the inverse 1-median problem (2.2)-(2.4), it is sufficient to solve the problem (2.7)-(2.10).*

3 The Problem under Weighted l_∞ Norm

Let $c^+(v)$ and $c^-(v)$ be the unit costs of v by increasing and decreasing the vertex weight of v, respectively. In this section, we consider the inverse 1-median problem on trees under weighted l_∞ norm, which can be formulated below.

$$\min \max\{\max_{v \in T_1} c^-(v)q(v), \max_{v \notin T_1} c^+(v)p(v)\} \tag{3.1}$$

$$\text{s.t. } \sum_{v \in T_1} q(v) + \sum_{v \notin T_1} p(v) = 2D; \tag{3.2}$$

$$0 \leq q(v) \leq w(v) - \underline{w}(v), \forall v \in T_1; \tag{3.3}$$

$$0 \leq p(v) \leq \overline{w}(v) - w(v), \forall v \notin T_1. \tag{3.4}$$

Let

$$\begin{cases} c(v) := c^-(v),\ u(v) := w(v) - \underline{w}(v),\ x(v) := q(v); & \text{if } v \in T_1; \\ c(v) := c^+(v),\ u(v) := \overline{w}(v) - w(v),\ x(v) := p(v); & \text{if } v \notin T_1 \end{cases} \tag{3.5}$$

Then the problem (3.1)-(3.4) can be simplified to the following form:

$$\min \max_{v \in V} c(v)x(v) \tag{3.6}$$

$$\text{s.t. } \sum_{v \in V} x(v) = 2D; \tag{3.7}$$

$$0 \leq x(v) \leq u(v), \forall v \in V. \tag{3.8}$$

Let x^* be an optimal solution of (3.6)-(3.8). Then the new vertex weights can be obtained by

$$w^*(v) := \begin{cases} w(v) - x^*(v), & \text{if } v \in T_1; \\ w(v) + x^*(v), & \text{if } v \notin T_1. \end{cases} \tag{3.9}$$

3.1 The Problem without Bound Constraints on Modifications

We first consider the unbounded case, that is, the problem (3.6)-(3.7).

To minimize the maximum cost, the costs for all the vertices should be equal. Let Q be the optimal objective value. Then we have

$$c(v)x(v) = Q, \forall v \in V,$$

and

$$x(v) = \frac{Q}{c(v)}, \forall v \in V. \tag{3.10}$$

It follows from

$$\sum_{v \in V} x(v) = \sum_{v \in V} \frac{Q}{c(v)} = 2D$$

that

$$Q = \frac{2D}{\sum_{v \in V} \frac{1}{c(v)}}. \tag{3.11}$$

Furthermore, x given by (3.10) is the only optimal solution of the problem.

Specially, when $c^-(v) = c^+(v) = 1$ for all vertices, we have $Q = \frac{2D}{n}$. In this case, an optimal solution can be given by

$$\begin{cases} x^*(v) = Q,\ w^*(v) = w(v) - Q,\ \forall v \in T_1; \\ x^*(v) = Q,\ w^*(v) = w(v) + Q,\ \forall v \notin T_1. \end{cases}$$

3.2 The Problem with Bound Constraints on Modifications

In this subsection, we present two algorithms to solve the problem (3.6)-(3.8) in the bounded case. Based on the approach of the unbounded case, we devise an $O(n^2)$ time algorithm. Based on the property of the optimal solution, we propose an improved algorithm using the binary search, which runs in $O(n \log n)$ time.

An $O(n^2)$ Time Algorithm. Notice that Q defined as (3.11) is a lower bound on the optimal objective value of the bounded case. In the unbounded case, we only need to distribute such a cost Q to each vertex. However, in the bounded case, we have to consider the impact on the upper bounds $u(v)$.

To minimize the maximum cost in the bounded case, we distribute the cost iteratively until the total modifications of weights reach to $2D$. In the first iteration, let $V^+ := V$, $\Delta := 2D$, $x := 0$ and

$$\overline{x}(v) := \min\{Q/c(v), u(v)\}, u(v) := u(v) - \overline{x}(v), \text{if } v \in V^+.$$

Let $\Delta := \Delta - \overline{x}(V^+)$ and $V^+ := \{v \in V | u(v) > 0\}$. If $\Delta > 0$ and $V^+ = \emptyset$, then $\overline{x}(v) = u(v) = 0, \forall v \in V^+$. In this case, all the upper bounds are met before the total modifications reach to $2D$, and hence the instance is infeasible. If $\Delta > 0$ and $V^+ \neq \emptyset$, then update

$$Q := \frac{\Delta}{\sum_{v \in V^+} 1/c(v)}, \tag{3.12}$$

$x := x + \overline{x}$, and repeat the above process until $\Delta = 0$, which means that $\overline{x}(v) = Q/c(v), \forall v \in V^+$. In this case, the total modifications reach to $2D$ in the most economical way, and an optimal solution is obtained.

Summarizing the above, we get the algorithm below. It is in fact a greedy type algorithm.

Algorithm 4. *(An $O(n^2)$ Time Algorithm.)*
Step 1. *Compute $W := w(V)$.*
Step 2. *Compute the weight of the subtrees rooted at s^*.*
If *all subtrees have a weight $\leq W/2$, then stop: the given weights are already optimal;*
else *let T_1 denote the subtree with largest weight W_1. Set $D := W_1 - W/2$. Calculate c and u by (3.5).*
Step 3. *Initialize $\Delta := 2D$, $x := 0$, $\overline{x} := 0$, $f_\infty := 0$ and $V^+ := V$.*
Step 4. ***While*** *$\Delta > 0$* ***do***
Let $u(v) := u(v) - \overline{x}(v)$ for $v \in V^+$. Let $V^+ := \{v \in V^+ | u(v) > 0\}$.
If *$V^+ = \emptyset$, then stop: the problem is infeasible;*
else *compute Q by (3.12). Update*

$$\overline{x}(v) := \min\{Q/c(v), u(v)\}, \qquad x(v) := x(v) + \overline{x}(v), \text{for } v \in V^+;$$
$$f_\infty := f_\infty + \max_{v \in V^+} \overline{x}(v)c(v); \text{ and } \Delta := \Delta - \overline{x}(V^+).$$

Step 5. *Output the optimal solution x, and the optimal objective value f_∞.*

To prove the optimality of Algorithm 4, we first present an important property of the optimal solution in the bounded case. Let $f(v) := c(v)u(v)$ for any vertex v, where $c(v)$ and $u(v)$ are defined as (3.5).

Theorem 5. *Suppose the problem (3.6)-(3.8) is feasible. Let f_∞ be the optimal objective value, $V_B := \{v \in V | f(v) < f_\infty\}$. Then we have*

$$f_\infty = \frac{2D - \sum_{v \in V_B} u(v)}{\sum_{v \notin V_B} \frac{1}{c(v)}}. \tag{3.13}$$

Furthermore,

$$x(v) := \begin{cases} u(v), & \text{if } v \in V_B \\ \frac{f_\infty}{c(v)}, & \text{if } v \notin V_B \end{cases} \tag{3.14}$$

is the only optimal solution of the problem.

Proof. Let f_∞ be the optimal objective value of the problem (3.6)-(3.8). Then we have $f_\infty = \max_{v \notin V_B} c(v)x(v)$. Furthermore, $\frac{f_\infty}{c(v)} \leq u(v)$ for any $v \notin V_B$. Therefore, f_∞ is also the optimal objective value of the unbounded problem defined below.

$$\min \quad \max_{v \notin V_B} c(v)x(v) \tag{3.15}$$
$$s.t. \sum_{v \notin V_B} x(v) = 2D - \sum_{v \in V_B} u(v) \tag{3.16}$$

It follows from (3.11) that the value of f_∞ is given by (3.13). Moreover, based on (3.10) we can obtain the only optimal solution x' of (3.15)-(3.16), where $x'(v) = \frac{f_\infty}{c(v)}, \forall v \notin V_B$.

It is obvious that $0 \leq x(v) \leq u(v)$. Furthermore,

$$\sum_{v \in V} x(v) = \sum_{v \in V_B} u(v) + \sum_{v \notin V_B} \frac{f_\infty}{c(v)} = \sum_{v \in V_B} u(v) + 2D - \sum_{v \in V_B} u(v) = 2D.$$

So x defined as (3.14) is a feasible solution whose objective value is f_∞, and hence is the only optimal solution. □

Now we prove the optimality of Algorithm 4. Let $x^k, \overline{x}^k, u^k, V_k^+, Q^k, f_\infty^k$ be the corresponding values in the k-th iteration. Suppose the While loop in Step 4 of Algorithm 4 is executed l times.

Theorem 6. *If the instance is feasible, then f_∞^l outputted by Algorithm 4 is the optimal objective value, and x^l satisfies the following formula*

$$x^l(v) = \begin{cases} u(v), & \text{if } v \notin V_l^+; \\ \frac{f_\infty^l}{c(v)}, & \text{if } v \in V_l^+. \end{cases} \tag{3.17}$$

and is the optimal solution.

Proof. If the instance is feasible, then the algorithm stops at the l-th iteration. First, note that in the k-th iteration,

$$\overline{x}^k(v)c(v) = \begin{cases} Q^k, & \text{if } v \in V_k^+ \text{ and } \overline{x}^k(v) = \frac{Q^k}{c(v)} \leq u^k(v); \\ u^k(v)c(v), & \text{if } v \in V_k^+ \text{ and } \overline{x}^k(v) = u^k(v) < \frac{Q^k}{c(v)}. \end{cases}$$

and hence $f_\infty^{k+1} := f_\infty^k + \max_{v \in V_k^+} \overline{x}^k(v)c(v) = f_\infty^k + Q^k$. If $v \in V_l^+$, then

$$x^l(v) = \frac{Q^l + Q^{l-1} + \ldots + Q^1}{c(v)} = \frac{f_\infty^l}{c(v)}.$$

If $v \notin V_l^+$, then there is an index $k \in \{1, 2, \ldots, l-1\}$, such that $v \in V_k^+ \backslash V_{k+1}^+$, and $\overline{x}^k(v) := \min\{Q^k/c(v), u^k(v)\} = u^k(v)$. Then

$$\begin{aligned} x^l(v) &= \overline{x}^k(v) + \overline{x}^{k-1}(v) + \ldots + \overline{x}^1(v) \\ &= u^k(v) + \overline{x}^{k-1}(v) + \ldots + \overline{x}^1(v) \\ &= u^{k-1}(v) + \overline{x}^{k-2}(v) + \ldots + \overline{x}^1(v) \\ &= \ldots = u^1(v) = u(v). \end{aligned}$$

Furthermore, if $v \notin V_l^+$, $f(v) < f_\infty^l$. Hence,

$$f_\infty^l = \frac{2D - \sum_{v \notin V_l^+} u(v)}{\sum_{v \in V_l^+} \frac{1}{c(v)}}. \tag{3.18}$$

Therefore, $0 \leq x^l(v) \leq u(v)$ for each vertex v, and x^l defined as (3.17) is a feasible solution with objective value f^l_∞. If f^l_∞ is not the optimal objective value, but f' is. Then we have $f' < f^l_\infty$, and the corresponding solution is

$$x'(v) := \begin{cases} u(v), & \text{if } v \in V'; \\ \frac{f'}{c(v)}, & \text{if } v \notin V'; \end{cases}$$

where $V' := \{v \in V | f(v) < f'\}$. Let $\overline{V}_l^+ := V \backslash V_l^+$. Then $V' \subseteq \overline{V}_l^+$, and

$$\begin{aligned} x'(V) &= \sum_{v \in V'} u(v) + \sum_{v \in \overline{V}_l^+ \backslash V'} \frac{f'}{c(v)} + \sum_{v \notin \overline{V}_l^+} \frac{f'}{c(v)} \\ &\leq \sum_{v \in \overline{V}_l^+} u(v) + \sum_{v \notin \overline{V}_l^+} \frac{f'}{c(v)} < \sum_{v \in \overline{V}_l^+} u(v) + \sum_{v \notin \overline{V}_l^+} \frac{f^l_\infty}{c(v)} = 2D, \end{aligned} \tag{3.19}$$

which means x' is not a feasible solution. Note that the first inequality "$\leq$" in (3.19) holds because $f' \leq f(v) < f^l_\infty$ for each $v \in \overline{V}_l^+ \backslash V'$. Thus, we obtain the conclusion that f^l_∞ is the optimal objective value, and x^l defined as (3.17) is the only optimal solution. □

Now we analyze the time complexity of the algorithm. It is obvious that $7|V_k^+|$ basic operations are needed in the k-th iteration of the While loop. Note that if the problem is feasible, then the algorithm stops with $\Delta = 0$, which involves the optimal solution. Therefore, in each iteration before termination, there is at least one vertex v reaching the upper bound $\overline{x}(v) = u(v)$, and hence the size of V_k^+ is reduced at least one. Hence, in the worst case, the total time is upper bounded by

$$\sum_{k=n}^{1} 7k = \frac{7n(n+1)}{2}.$$

As a conclusion, we have

Theorem 7. *Algorithm 4 can solve the inverse 1-median problem on trees under weighted l_∞ norm with bound constraint on modification in $O(n^2)$ time.*

An $O(n \log n)$ Time Algorithm. Sort all the costs $f(v)$ of vertices $v \in V$ in a strictly increasing order, i.e., $f(v_{i_1}) < f(v_{i_2}) < \cdots < f(v_{i_\tau})$. Obviously, f_∞ must belong to an interval $(f(v_{i_k}), f(v_{i_{k+1}})]$ for some index k. Based on Theorem 5, we propose an algorithm using the binary search to determine the optimal objective value first, and then give the optimal solution by (3.14).

Algorithm 8. *(An $O(n \log n)$ Time Algorithm.)*
Step 1 and 2. *Please see Algorithm 4.*
Step 3. *Sort all the costs $f(v)$ of vertices $v \in V$ in a strictly increasing order, i.e., $f(v_{i_1}) < f(v_{i_2}) < \cdots < f(v_{i_\tau})$. Put $a := 1$ and $b := \tau$.*
Step 4. If *$u(V) < 2D$, then stop: the instance is infeasible.*

Step 5. If $b - a = 1$*, then stop and output the optimal objective value* f_∞^a*, and the optimal solution* x *given by (3.14), where* f_∞ *and* V_B *are replaced with* f_∞^a *and* V_a*, respectively.*
Step 6. *Let* $k := \lfloor (a+b)/2 \rfloor$ *and* $V_k := \{v \in V | f(v) \leq f(v_{i_k})\}$*.* ***If***

$$f_\infty^k := \frac{2D - \sum_{v \in V_k} u(v)}{\sum_{v \notin V_k} \frac{1}{c(v)}} \leq f(v_{i_k}),$$

then put $b := k$*,* ***else*** *put* $a := k$*.* ***Return to Step 5.***

Now we prove the optimality of the algorithm.

Theorem 9. *If the instance is feasible, then* f_∞^a *outputted by the algorithm is the optimal objective value, and the optimal solution is*

$$x(v) := \begin{cases} u(v), & \text{if } v \in V_a; \\ \frac{f_\infty^a}{c(v)}, & \text{if } v \notin V_a. \end{cases} \tag{3.20}$$

Proof. The proof is omitted here due to limit of space. □

Now we analyze the time complexity of the algorithm. Sorting all the costs $f(v)$ in Step 3 can be done in $O(n \log n)$ operations. The binary search method can be done in $\lceil \log n \rceil$ iterations and $O(n)$ operations are needed in each iteration. As a conclusion, we have

Theorem 10. *Algorithm 8 can solve the inverse 1-median problem on trees under weighted* l_∞ *norm with bound constraint on modification in* $O(n \log n)$ *time.*

4 Conclusions and Further Research

In this paper, we first present the model of inverse 1-median problem on trees. Contrasting to the costs of modifications measured by l_1 norm in the corresponding references, we use weighted l_∞ norm to measure the costs. We propose two algorithms to solve the problem with bound constraints on modifications. Based on the approach of the unbounded case, we devise a greedy type algorithm, which runs in $O(n^2)$ time. Based on the property of the optimal solution, we propose an improved algorithm using the binary search, which runs in $O(n \log n)$ time.

In fact, by running the Matlab programs of these two algorithms with randomly generating lots of instances, the number of iterations of Algorithm 4 is as small as that of Algorithm 8. Note that if $|V_k^+| = |V_{k-1}^+| - 1$ in the k-th iteration, then there is a great reduction in Δ_k, while a small reduction in the denominator of Q_k, which means that Q_k is much smaller than Q_{k-1}. Hence the time complexity of Algorithm 4 may also be $O(n \log n)$, if an appropriate technique was used. The other reason, for presenting the $O(n^2)$ time algorithm when an $O(n \log n)$ time algorithm is available, is that the greedy type algorithm will be useful for some further related problems.

It is interesting that the inverse 1-median problem on trees has only one optimal solution under weighted l_∞ norm, while most inverse combinatorial optimization problems under weighted l_∞ norm have many optimal solutions. The reason is that the constraint condition (3.7) is just the equality constraint on the sum of all modifications of weights.

For further research, we can consider the inverse median problems and inverse center location problems under weighted Hamming distance, and weighted l_∞ norm, respectively.

Acknowledgement

Research are supported by NSFC (10626013,10801031), Science Foundation of Southeast University (9207011468), and Excellent Young Teacher Financial Assistance Scheme for teaching and research of Southeast University (4007011028).

References

1. Alizadeh, B., Burkard, R.E.: Combinatorial algorithms for inverse absolute and vertex 1-center location problems on trees. Technical Report, Graz University of Technology (2009), http://www.math.tugraz.at/fosp/pdfs/tugraz_0143.pdf
2. Alizadeh, B., Burkard, R.E., Pferschy, U.: Inverse 1-center location problems with edge length augmentation on trees. Computing 86(4), 331–343 (2009)
3. Balas, E., Zemel, E.: An algorithm for large zero-one knapsack problems. Operations Research 28, 1130–1154 (1980)
4. Bonab, F.B., Burkard, R.E., Alizadeh, B.: Inverse Median Location Problems with Variable Coordinates. Central European Journal of Operations Research (2009), doi:10.1007/s10100-009-0114-2
5. Burkard, R.E., Galavii, M., Gassner, E.: The inverse Fermat-Weber problem. Technical Report, Graz University of Technology (2008), http://www.math.tugraz.at/fosp/pdfs/tugraz_0107.pdf
6. Burkard, R.E., Pleschiutschnig, C., Zhang, J.: Inverse median problems. Discrete Optimization 1, 23–39 (2004)
7. Burkard, R.E., Pleschiutschnig, C., Zhang, J.Z.: The inverse 1- median problem on a cycle. Discrete Optimization 5, 242–253 (2008)
8. Cai, M., Yang, X., Zhang, J.: The complexity analysis of the inverse center location problem. J. Global Optim. 5, 213–218 (1999)
9. Galavii, M.: Inverse 1-Median Problems, Ph.D. Thesis, Institute of Optimization and Discrete Mathematics, Graz University of Technology, Graz, Austria (2008)
10. Gassner, E.: The inverse 1-maxian problem with edge length modiffication. Journal of Combinatorial Optimization 16, 50–67 (2008)
11. Gassner, E.: A game-theoretic approach for downgrading the 1-median in the plane with manhattan metric. Ann. Oper. Res. 172(1), 393–404 (2009)
12. Goldman, A.J.: Optimal center location in simple networks. Transportation Science 2, 77–91 (1962)
13. Hua, et al.: Applications of mathematical models to wheat harvesting. Acta Mathematica Sinica 11, 63–75 (1961) (in Chinese); English Translation in Chinese Math. 2, 77–91 (1962)
14. Yang, X., Zhang, J.: Inverse center location problem on a tree. Journal of Systems Science and Complexity 21, 651–664 (2008)

On the Approximability of the Vertex Cover and Related Problems

Qiaoming Han[1] and Abraham P. Punnen[2]

[1] School of Mathematics and Statistics, Zhejiang University of Finance & Economics, Hangzhou, Zhejiang 310018, China
qmhan@zufe.edu.cn

[2] Department of Mathematics, Simon Fraser University, 14th Floor Central City Tower, 13450 102nd Ave., Surrey, BC V3T5X3, Canada
apunnen@sfu.ca

Abstract. In this paper we show that the problem of identifying an edge (i, j) in a graph G such that there exists an optimal vertex cover S of G containing exactly one of the nodes i and j is NP-hard. Such an edge is called a weak edge. We then develop a polynomial time approximation algorithm for the vertex cover problem with performance guarantee $2 - \frac{1}{1+\sigma}$, where σ is an upper bound on a measure related to a weak edge of a graph. Further, we discuss a new relaxation of the vertex cover problem which is used in our approximation algorithm to obtain smaller values of σ.

1 Introduction

Let $G = (V, E)$ be an undirected graph on the vertex set $V = \{1, 2, \ldots, n\}$. A *vertex cover* (VC) of G is a subset S of V such that each edge of G has at least one endpoint in S. The *vertex cover problem* (VCP) is to compute a vertex cover of smallest cardinality in G. VCP is well known to be NP-hard on an arbitrary graph but solvable in polynomial time on a bipartite graph. A vertex cover S is said to be γ-optimal if $|S| \leq \gamma|S^0|$ where $\gamma \geq 1$ and S^0 is an optimal solution to the VCP.

It is well known that a 2-optimal vertex cover of a graph can be obtained in polynomial time by taking all the vertices of a maximal (not necessarily maximum) matching in the graph or by rounding up the LP relaxation solution of an integer programming formulation [18]. There has been considerable work (see e.g. survey paper [11]) on the problem over the past 30 years on finding a polynomial-time approximation algorithm with an improved performance guarantee. The current best known bound on the performance ratio of a polynomial time approximation algorithm for VCP is $2 - \Theta(\frac{1}{\sqrt{\log n}})$ [12]. It is also known that computing a γ-optimal solution in polynomial time for VCP is NP-Hard for any $1 \leq \gamma \leq 10\sqrt{5} - 21 \simeq 1.36$ [6]. In fact, no polynomial-time $(2-\epsilon)$-approximation algorithm is known for VCP for any constant $\epsilon > 0$ and existence of such an algorithm is one of the most outstanding open questions in approximation algorithms for combinatorial optimization problems. Under the assumption that the unique game conjecture [9,13,14] is true, a polynomial time $(2 - \epsilon)$-approximation algorithm with

B. Chen (Ed.): AAIM 2010, LNCS 6124, pp. 161–169, 2010.

constant $\epsilon > 0$ is not possible for VCP. For recent works on approximability of VCP, we refer to [1,4,5,6,7,8,10,12,15,16]. Recently, Asgeirsson and Stein [2,3] reported extensive experimental results using a heuristic algorithm which obtained no worse than $\frac{3}{2}$-optimal solutions for all the test problems they considered. Also, Han, Punnen and Ye [8] proposed a $(\frac{3}{2} + \xi)$-approximation algorithm for VCP, where ξ is an error parameter calculated by the algorithm.

A natural linear programming (LP) relaxation of VCP is

$$(LPR) \qquad \begin{array}{l} \min \sum_{i=1}^{n} x_i \\ s.t.\ x_i + x_j \geq 1, (i,j) \in E, \\ \quad\;\; x_i \geq 0, i = 1, 2, \cdots, n. \end{array} \tag{1}$$

It is well known that (e.g. [17]) any optimal basic feasible solution (BFS) $x^* = (x_1^*, x_2^*, \ldots, x_n^*)$ to the problem LPR, satisfies $x_i^* \in \{0, \frac{1}{2}, 1\}$. Let $S_{LP} = \{i \mid x_i^* = \frac{1}{2} \text{ or } x_i^* = 1\}$, then it is easy to see that S_{LP} is a 2-approximate solution to the VCP on graph G. Nemhauser and Trotter [18] have further proved that there exists an optimal VC on graph G, which agrees with x^* in its integer components.

An $(i,j) \in E$ is said to be a *weak edge* if there exists an optimal vertex cover V^0 of G such that $|V^0 \cap \{i,j\}| = 1$. Likewise, an $(i,j) \in E$ is said to be a *strong edge* if there exists an optimal vertex cover V^0 of G such that $|V^0 \cap \{i,j\}| = 2$. An edge (i,j) is *uniformly strong* if $|V^0 \cap \{i,j\}| = 2$ for any optimal vertex cover V^0. Note that it is possible for an edge to be both strong and weak. Also (i,j) is uniformly strong if and only if it is not a weak edge. In this paper, we show that the problems of identifying a weak edge and identifying a strong edge are NP-hard. We also present a polynomial time $(2 - \frac{1}{\sigma+1})$-approximation algorithm for VCP where σ is an appropriate graph theoretic measure (to be introduced in Section 3). Thus for all graphs for which σ is bounded above by a constant, we have a polynomial time $(2 - \epsilon)$-approximation algorithm for VCP. We give some examples of graphs satisfying the property that $\sigma = 0$. However, establishing tight bounds on σ, independent of graph structures and/or characterizing graphs for which σ is a constant remains an open question. VCP is trivial on a complete graph K_n since any collection of $n - 1$ nodes serves as an optimal solution. However, the LPR gives an objective function value of $\frac{n}{2}$ only on such graphs. We give a stronger relaxation for VCP and a complete linear programming description of VCP on complete graphs and wheels. This relaxation can also be used to find a reasonable expected guarantee for σ.

For a graph G, we sometimes use the notation $V(G)$ to represent its vertex set and $E(G)$ to represent its edge set.

2 Complexity of Weak and Strong Edge Problems

The *strong edge problem* can be stated as follows: "Given a graph, identify a strong edge of G or declare that no such edge exists."

Theorem 1. *The strong edge problem on a non-bipartite graph is NP-hard.*

Proof. Since G is not bipartite, then it must contain an odd cycle. For any odd cycle ω, any vertex cover must contain at least two adjacent nodes of ω and hence G must contain at least one strong edge. If such an edge (i, j) can be identified in polynomial time, then after removing the nodes i and j from G and applying the algorithm on $G - \{i, j\}$ and repeating the process we eventually reach a bipartite graph for which an optimal vertex cover $\hat{V}$ can be identified in polynomial time. Then $\hat{V}$ together with the nodes removed so far will form an optimal vertex cover of G. Thus if the strong edge problem can be solved in polynomial time, then the VCP can be solved in polynomial time.

It may be noted that the strong edge problem is solvable in polynomial time on a bipartite graph. The problem of identifying a weak edge is much more interesting. The *weak edge problem* can be stated as follows: "Given a graph G, identify a weak edge of G." It may be noted that unlike a strong edge, a weak edge exists for all graphs. We will now show that the weak edge problem is NP-hard. Before discussing the proof of this claim, we need to introduce some notations and definitions.

Let $x^* = (x_1^*, x_2^*, \ldots, x_n^*)$ be an optimal BFS of LPR, the linear programming relaxation of the VCP. Let $I_0 = \{i \ : \ x_i^* = 0\}$ and $I_1 = \{i \ : \ x_i^* = 1\}$. The graph $\bar{G} = G \setminus \{I_0 \cup I_1\}$ is called a $\{0, 1\}$-reduced graph of G. The process of computing $\bar{G}$ from G is called a *$\{0,1\}$-reduction.*

Lemma 1. *[18] If R is a vertex cover of $\bar{G}$ then $R \cup I_1$ is a vertex cover of G. If R is optimal for $\bar{G}$, then $R \cup I_1$ is an optimal vertex cover for G. If R is a γ-optimal vertex cover of $\bar{G}$, then $R \cup I_1$ is an γ-optimal vertex cover of G for any $\gamma \geq 1$.*

Let $e = (i, j)$ be an edge of G. Define $\Delta_{ij} = \{k \mid (i, k) \in E(G) \text{ and } (j, k) \in E(G)\}$, $D_{e,i} = \{s \in V(G) \mid (i, s) \in E(G), s \neq j, s \notin \Delta_{ij}\}$, and $D_{e,j} = \{t \in V(G) \mid (j, t) \in E(G), t \neq i, t \notin \Delta_{ij}\}$. Construct the new graph $G^{(i,j)}$ from G as follows. From graph G, delete Δ_{ij} and all the incident edges, connect each vertex $s \in D_{e,i}$ to each vertex $t \in D_{e,j}$ whenever such an edge is not already present, and delete vertices i and j with all the incident edges. The operation of constructing $G^{(i,j)}$ from G is called an *(i, j)-reduction.* When (i, j) is selected as a weak edge, then the corresponding (i, j)-reduction is called a *weak edge reduction.* The weak edge reduction is a modified version of the active edge reduction operation introduced in [8].

Lemma 2. *Let $e = (i, j)$ be a weak edge of G, $R \subseteq V(G^{(i,j)})$ and*

$$R^* = \begin{cases} R \cup \Delta_{ij} \cup \{j\}, & \text{if } D_{e,i} \subseteq R; \\ R \cup \Delta_{ij} \cup \{i\}, & \text{otherwise,} \end{cases}$$

1. *If R is a vertex cover of $G^{(i,j)}$, then R^* is a vertex cover of G.*
2. *If R is an optimal vertex cover of $G^{(i,j)}$, then R^* is an optimal vertex cover of G.*
3. *If R is a γ-optimal vertex cover in $G^{(i,j)}$, then R^* is a γ-optimal vertex cover in G for any $\gamma \geq 1$.*

Proof. Let R be a vertex cover of $G^{(i,j)}$. If $D_{e,i} \subseteq R$ then all arcs in G incident on i, except possibly (i,j) and (i,k) for $k \in \Delta_{ij}$, are covered by R. Then $R^* = R \cup \Delta_{ij} \cup \{j\}$ covers all arcs incident on j, including (i,j), (i,k) and (j,k) for $k \in \Delta_{ij}$, and hence R^* is a vertex cover in G. If at least one vertex of $D_{e,i}$ is not in R, then all vertices in $D_{e,j}$ must be in R by construction of $G^{(i,j)}$. Thus $R^* = R \cup \Delta_{ij} \cup \{i\}$ must be a vertex cover of G.

Suppose R is an optimal vertex cover of $G^{(i,j)}$. Since (i,j) is a weak edge, there exists an optimal vertex cover, say V^0, of G containing exactly one of the nodes i or j. Without loss of generality, let this node be i. For each node $k \in \Delta_{ij}$, (i,j,k) is a 3-cycle in G and hence $k \in V^0$ for all $k \in \Delta_{ij}$. Let $V^1 = V^0 - (\{i\} \cup \Delta_{ij})$, which is a vertex cover of $G^{(i,j)}$. Then $|R| = |V^1|$ for otherwise if $|R| < |V^1|$ we have $|R^*| < |V^0|$, a contradiction. Thus $|R^*| = |V^0|$ establishing optimality of R^*.

Suppose R is an γ-optimal vertex cover of $G^{(i,j)}$ and let $V^{(i,j)}$ be an optimal vertex cover in $G^{(i,j)}$. Thus

$$|R| \leq \gamma |V^{(i,j)}| \text{ where } \gamma \geq 1. \tag{2}$$

Let V^0 be an optimal vertex cover in G. Without loss of generality assume $i \in V^0$ and since (i,j) is weak, $j \notin V^0$. Let $V^1 = V^0 - (\{i\} \cup \Delta_{ij})$. Then $|V^1| = |V^{(i,j)}|$. Thus from (2), $|R| \leq \gamma |V^1|$. Thus

$$|R^*| \leq \gamma |V^1| + |\Delta_{ij}| + 1 \leq \gamma(|V^1| + |\Delta_{ij}| + 1) \leq \gamma |V^0|.$$

Thus R^* is γ-optimal in G.

Suppose that an oracle, say WEAK(G,i,j), is available which with input G outputs two nodes i and j such that (i,j) is a weak edge of G. It may be noted that WEAK(G,i,j) do not tell us which node amongst i and j is in an optimal vertex cover. It simply identifies the weak edge (i,j). Using the oracle WEAK(G,i,j), we develop an algorithm, called *weak edge reduction algorithm* or WER-algorithm to compute an optimal vertex cover of G.

The basic idea of the scheme is very simple. We apply $\{0,1\}$-reduction and weak edge reduction repeatedly until a null graph is reached, in which case the algorithm goes to a backtracking step. We record the vertices of the weak edge identified in each weak edge reduction step but do not determine which one to be included in the output vertex cover. In the backtrack step, taking guidance from Lemma 2, we choose exactly one of these two vertices to form part of the vertex cover we construct. In this step, the algorithm computes a vertex cover for G using all vertices in Δ_{ij} removed in the weak edge reduction steps, vertices with value 1 removed in the $\{0,1\}$-reduction steps, and the selected vertices in the backtrack step from the vertices corresponding to the weak edges recorded during the weak edge reduction steps. A formal description of the WER-algorithm is given below.

The WER-Algorithm

Step 1: {* Initialize *} $k = 1, G_k = G$.

Step 2: {* Reduction operations *} $\Delta_k = \emptyset$, $I_{k,1} = \emptyset$, $e_k = (i_k, j_k) = \emptyset$.

1. {* {0,1}-reduction *} Solve the LP relaxation problem LPR of VCP on the graph G_k. Let $x^k = \{x_i^k : i \in V(G_k)\}$ be the resulting optimal BFS, $I_{k,0} = \{i \mid x_i^k = 0\}$, $I_{k,1} = \{i \mid x_i^k = 1\}$, and $I_k = I_{k,0} \cup I_{k,1}$.
   If $V(G_k) \setminus I_k = \emptyset$ **goto** Step 3 **else** $G_k = G_k \setminus I_k$ **endif**
2. {* weak edge reduction *} Call WEAK(G_k, i, j) to identify the weak edge $e = (i, j)$. Let $G_{k+1} = G_k^{(i,j)}$, where $G_k^{(i,j)}$ is the graph obtained from G_k using the weak edge reduction operation. Compute Δ_{ij} for G_k as defined in the weak edge reduction. Let $\Delta_k = \Delta_{ij}, i_k = i, j_k = j, e_k = (i_k, j_k)$.
   If $G_{k+1} \neq \emptyset$ **then** $k = k + 1$ **goto** beginning of Step 2 **endif**

Step 3: L=k+1, $S_L = \emptyset$.

Step 4: {* Backtracking to construct a solution *}
Let $S_{L-1} = S_L \cup I_{L-1,1}$,
If $(i_{L-1}, j_{L-1}) \neq \emptyset$ **then** $S_{L-1} = S_{L-1} \cup \Delta_{L-1} \cup R^*$, where

$$R^* = \begin{cases} j_{L-1}, & \text{if } D_{e_{L-1}, i_{L-1}} \subseteq S_L; \\ i_{L-1}, & \text{otherwise,} \end{cases}$$

and $D_{e_{L-1}, i_{L-1}} = \{s : (i_{L-1}, s) \in G_{L-1}, s \neq j_{L-1}, s \notin \Delta_{L-1}\}$ **endif**
$L = L - 1$,
If $L \neq 1$ **then goto** beginning of step 4 **else** output S_1 and STOP **endif**

Using Lemma 1 and Lemma 2, it can be verified that the output S_1 of the WER-algorithm is an optimal vertex cover of G. It is easy to verify that the complexity of the algorithm is polynomial whenever the complexity of WEAK(G, i, j) is polynomial. Since VCP is NP-hard we established the following theorem:

Theorem 2. *The weak edge problem is NP-hard.*

3 An Approximation Algorithm for VCP

Let VCP(i, j) be the *restricted vertex cover problem* where feasible solutions are vertex covers of G using exactly one of the vertices from the set $\{i, j\}$ and looking for the smallest vertex cover satisfying this property. Note that a feasible solution to the restricted vertex cover problem is always exists, for instance, $V \backslash \{i\}$ or $V \backslash \{j\}$. More precisely, VCP(i, j) tries to identify a vertex cover V^* of G with smallest cardinality such that $|V^* \cap \{i, j\}| = 1$. Let δ and $\bar{\delta}(i, j)$ be the optimal objective function values of VCP and VCP(i, j) respectively. If (i, j) is indeed a weak edge of G, then $\delta = \bar{\delta}(i, j)$. Otherwise,

$$\bar{\delta}(i, j) = \delta + \sigma(i, j), \tag{3}$$

where $\sigma(i,j)$ is a non-negative integer. Further, using arguments similar to the proof of Lemma 2 it can be shown that

$$\zeta_{ij} + \Delta_{ij} + 1 = \bar{\delta}(i,j) = \delta + \sigma(i,j). \tag{4}$$

where ζ_{ij} is the optimal objective function value VCP on $G^{(i,j)}$.

Consider the optimization problem

$$\text{WEAK-OPT:} \qquad \begin{array}{l} \text{Minimize } \sigma(i,j) \\ \text{Subject to } (i,j) \in E(G) \end{array}$$

WEAK-OPT is precisely the weak edge problem in the optimization form and its optimal objective function value is always zero. However this problem is NP-hard by Theorem 2. We now show that an upper bound σ on the optimal objective function value of WEAK-OPT and a solution (i,j) with $\sigma(i,j) \leq \sigma$ can be used to obtain a $(2 - \frac{1}{1+\sigma})$-approximation algorithm for VCP. Let ALMOST-WEAK(G,i,j) be an oracle which with input G computes an approximate solution (i,j) to WEAK-OPT such that $\sigma(i,j) \leq \sigma$ for some σ. Consider the WER-algorithm with WEAK(G,i,j) replaced by ALMOST-WEAK(G,i,j). We call this the AWER-algorithm.

Let $G_k,\ k = 1,2,\ldots t$ be the sequence of graphs generated in Step 2(2) of the AWER-algorithm and (i_k, j_k) be the approximate solution to WEAK-OPT on $G_k,\ k = 1,2,\ldots,t$ identified by ALMOST-WEAK(G_k, i_k, j_k).

Theorem 3. *The AWER-algorithm identifies a vertex cover S_1 such that $|S_1| \leq (2 - \frac{1}{1+\sigma})|S^*|$ where S^* is an optimal solution to the VCP. Further, the complexity of the the algorithm is $O(n(\phi(n) + \psi(n)))$ where $n = |V(G)|$, $\phi(n)$ is the complexity of LPR and $\psi(n)$ is the complexity of ALMOST-WEAK(G,i,j).*

Proof. Without loss of generality, we assume that the LPR solution $x^1 = (x_1^1, x_2^1, \ldots, x_n^1)$ generated when Step 2(1) is executed for the first time satisfies $x_i^1 = \frac{1}{2}$ for all i. If this is not true, then we could replace G by a new graph $\bar{G} = G \setminus \{I_{1,1} \cup I_{1,0}\}$ and by Lemma 1, if $\bar{S}$ is a γ-optimal solution for VCP on $\bar{G}$ then $\bar{S} \cup I_{1,1}$ is a γ-optimal solution on G for any $\gamma \geq 1$. Thus, under this assumption we have

$$n \leq 2|S^*|. \tag{5}$$

Let t be the total number of iterations of Step 2 (2). For simplicity of notation, we denote $\sigma_k = \sigma(i_k, j_k)$ and $\bar{\delta}_k = \bar{\delta}(i_k, j_k)$. Note that δ_k and $\bar{\delta}_k$ are optimal objective function values of VCP and VCP(i_k, j_k), respectively, on the graph G_k. In view of equations (3) and (4) we have,

$$\bar{\delta}_k = \delta_k + \sigma_k, \qquad k = 1,2,\ldots,t \tag{6}$$

and

$$\delta_{k+1} + |\Delta_{i_k,j_k}| + |I_{k,1}| + 1 = \bar{\delta}_k, \qquad k = 1,2,\ldots,t. \tag{7}$$

From (6) and (7) we have

$$\delta_{k+1} - \delta_k = \sigma_k - |\Delta_{i_k,j_k}| - |I_{k,1}| - 1, \quad k = 1,2,\ldots,t. \tag{8}$$

Adding equations in (8) for $k = 1, 2, \ldots, t$ and using the fact that $\delta_{t+1} = |I_{t+1,1}|$, we have,

$$|S_1| = |S^*| + \sum_{k=1}^{t} \sigma_k, \tag{9}$$

where $|S^*| = \delta_1$, and by construction,

$$|S_1| = \Sigma_{k=1}^{t+1} I_{k,1} + \Sigma_{k=1}^{t} \Delta_k + t. \tag{10}$$

But,

$$|V(G)| = \Sigma_{k=1}^{t+1} I_k + \Sigma_{k=1}^{t} \Delta_k + 2t. \tag{11}$$

From (9), (10) and (11), we have

$$t = \frac{|V(G)| - \Sigma_{k=1}^{t+1} I_k - \Sigma_{k=1}^{t} \Delta_k}{2} \leq \frac{|V(G)| - |S^*| - \Sigma_{k=1}^{t}(\sigma_k - 1)}{2}. \tag{12}$$

From inequalities (5) and (12), we have

$$t \leq \frac{|S^*| - t(\bar{\sigma} - 1)}{2},$$

where $\bar{\sigma} = \frac{\Sigma_{k=1}^{t} \sigma_k}{t}$. Then we have

$$t \leq \frac{|S^*|}{\bar{\sigma} + 1}.$$

Thus,

$$\frac{|S_1|}{|S^*|} = \frac{|S^*| + \Sigma_{k=1}^{t} \sigma_k}{|S^*|} = \frac{|S^*| + t\bar{\sigma}}{|S^*|} \leq 1 + \frac{\bar{\sigma}}{\bar{\sigma} + 1} \leq 1 + \frac{\sigma}{\sigma + 1} = 2 - \frac{1}{1 + \sigma}.$$

The complexity of the algorithm can easily be verified.

The performance bound established in Theorem 3 is useful only if we can find an efficient way to implement our black-box oracle ALMOST-WEAK(G, i, j) that identifies a reasonable (i, j) in each iteration.

Any vertex cover must contain at least $s + 1$ vertices of an odd cycle of length $2s + 1$. This motivates the following *extended linear programming relaxation* (ELP) of the VCP, studied in [1,8].

$$(ELP) \quad \begin{aligned} \min & \sum_{i=1}^{n} x_i \\ s.t. \; & x_i + x_j \geq 1, (i, j) \in E, \\ & \sum_{i \in \omega_k} x_i \geq s_k + 1, \omega_k \in \Omega, \\ & x_i \geq 0, i = 1, 2, \ldots, n, \end{aligned} \tag{13}$$

where Ω denotes the set of all odd-cycles of G and $\omega_k \in \Omega$ contains $2s_k + 1$ vertices for some integer s_k. Note that although there may be an exponential number of odd-cycles in G, since the odd cycle inequalities has a polynomial-time separation scheme, ELP is polynomially solvable. Further, it is possible to compute an optimal BFS of ELP in polynomial time.

Let x^0 be an optimal basic feasible solution of ELP. An edge $(r, s) \in E$ is said to be an *active edge* with respect to x^0 if $x_i^0 + x_j^0 = 1$. There may or may not exist an active edge corresponding to an optimal BFS of the ELP as shown in [8]. For any arc (r, s), consider the *restricted ELP* (RELP(r, s)) as follows:

$$(RELP(r,s)) \quad \begin{aligned} \min \sum_{i=1}^{n} & x_i \\ s.t.\ & x_i + x_j \geq 1, (i,j) \in E \backslash \{(r,s)\}, \\ & x_r + x_s = 1, \\ & \sum_{i \in \omega_k} x_i \geq s_k + 1, \omega_k \in \Omega, \\ & x_i \geq 0, i = 1, 2, \ldots, n, \end{aligned} \tag{14}$$

Let $Z(r, s)$ be the optimal objective function value of RELP(r, s). Choose $(p, q) \in E(G)$ such that

$$Z(p, q) = \min\{Z(i, j) \ : \ (i, j) \in E(G)\}.$$

An optimal solution to RELP(p, q) is called *a RELP solution*. It may be noted that if an optimal solution x^* of the ELP contains an active edge, then x^* is also an RELP solution. Further $Z(p, q)$ is always a lower bound on the optimal objective function value of VCP.

The VCP on a complete graph is trivial since any collection of $(n - 1)$ nodes form an optimal vertex cover. However, for a complete graph, LPR yields an optimal objective function value of $\frac{n}{2}$ only and ELP yields an optimal objective function value of $\frac{2n}{3}$. Interestingly, the optimal objective function value of RELP on a complete graph is $n-1$, and the RELP solution is indeed an optimal vertex cover on a complete graph. In fact, it can be shown that for any $(i, j) \in E(G)$, an optimal BFS of the linear program RELP(i, j) gives an optimal vertex cover of G whenever G is a complete graph or a wheel.

Extending the notion of an active edge corresponding to an ELP solution [8], an edge $(i, j) \in E$ is said to be an *active edge* with respect to an RELP solution x^0 if $x_i^0 + x_j^0 = 1$. Unlike ELP, an RELP solution always contains an active edge. In AWER-algorithm, the output of ALMOST-WEAK(G, i, j) can be selected as an active edge with respect to RELP solution.

We believe that the value of $\sigma(i, j)$, i.e. the absolute difference between the optimal objective function value of VCP and the optimal objective function value of VCP(i,j), when (i, j) is an active edge corresponding to an RELP solution is a constant for a large class of graphs. Charactering such graphs is an open question. Nevertheless, our results provide new insight into the approximability of the vertex cover problem.

Acknowledgements. The authors thank the anonymous referees for their helpful comments and suggestions. Research of Qiaoming Han was supported in part by Chinese NNSF 10971187. Research of Abraham P. Punnen was partially supported by an NSERC discovery grant.

References

1. Arora, S., Bollobàs, B., Lovàsz, L.: Proving integrality gaps without knowing the linear program. In: Proc. IEEE FOCS, pp. 313–322 (2002)
2. Asgeirsson, E., Stein, C.: Vertex cover approximations on random graphs. In: Demetrescu, C. (ed.) WEA 2007. LNCS, vol. 4525, pp. 285–296. Springer, Heidelberg (2007)
3. Asgeirsson, E., Stein, C.: Vertex cover approximations: Experiments and observations. In: WEA, pp. 545–557 (2005)
4. Bar-Yehuda, R., Even, S.: A local-ratio theorem for approximating the weighted vertex cover problem. Annals of Discrete Mathematics 25, 27–45 (1985)
5. Charikar, M.: On semidefinite programming relaxations for graph coloring and vertex cover. In: Proc. 13th SODA, pp. 616–620 (2002)
6. Dinur, I., Safra, S.: The importance of being biased. In: Proc. 34th ACM Symposium on Theory of Computing, pp. 33–42 (2002)
7. Halperin, E.: Improved approximation algorithms for the vertex cover problem in graphs and hypergraphs. SIAM J. Comput. 31, 1608–1623 (2002)
8. Han, Q., Punnen, A.P., Ye, Y.: A polynomial time $\frac{3}{2}$-approximation algorithm for the vertex cover problem on a class of graphs. Operations Research Letters 37, 181–186 (2009)
9. Harb, B.: The unique games conjecture and some of its implications on inapproximability (May 2005) (manuscript)
10. Håstad, J.: Some optimal inapproximability results. JACM 48, 798–859 (2001)
11. Hochbaum, D.S.: Approximating covering and packing problems: set cover, independent set, and related problems. In: Hochbaum, D.S. (ed.) Approximation Algorithms for NP-Hard Problems, pp. 94–143. PWS Publishing Company (1997)
12. Karakostas, G.: A better approximation ratio for the vertex cover problem. In: Caires, L., Italiano, G.F., Monteiro, L., Palamidessi, C., Yung, M. (eds.) ICALP 2005. LNCS, vol. 3580, pp. 1043–1050. Springer, Heidelberg (2005)
13. Khot, S.: On the power of unique 2-Prover 1-Round games. In: Proceedings of 34th ACM Symposium on Theory of Computing (STOC), pp. 767–775 (2002)
14. Khot, S., Regev, O.: Vertex cover might be hard to approximate to within $2 - \epsilon$. In: Complexity (2003)
15. Kleinberg, J., Goemans, M.: The Lovász theta function and a semidefinite programming relaxation of vertex cover. SIAM J. Discrete Math. 11, 196–204 (1998)
16. Monien, B., Speckenmeyer, E.: Ramsey numbers and an approximation algorithm for the vertex cover problem. Acta Informatica 22, 115–123 (1985)
17. Nemhauser, G.L., Trotter Jr., L.E.: Properties of vertex packing and independence system polyhedra. Mathematical Programming 6, 48–61 (1974)
18. Nemhauser, G.L., Trotter Jr., L.E.: Vertex packings: Structural properties and algorithms. Mathematical Programming 8, 232–248 (1975)

Feasibility Testing for Dial-a-Ride Problems*

Dag Haugland[1] and Sin C. Ho[2]

[1] Department of Informatics, University of Bergen, Bergen, Norway
dag.haugland@ii.uib.no
[2] CORAL, Department of Business Studies, Aarhus School of Business, Aarhus University, Århus, Denmark
sinch@asb.dk

Abstract. Hunsaker and Savelsbergh have proposed an algorithm for testing feasibility of a route in the solution to the dial-a-ride problem. The constraints that are checked are load capacity constraints, time windows, ride time bounds and wait time bounds. The algorithm has linear running time. By virtue of a simple example, we show in this work that their algorithm is incorrect. We also prove that by increasing the time complexity by only a logarithmic factor, a correct algorithm is obtained.

Keywords: dial-a-ride problem, feasibility, algorithm.

1 Introduction

In the *Dial-a-Ride Problem* (DARP), a set of users specify transportation requests between given origins and destinations. Users may provide a time window on their desired departure or arrival time, or on both. A fleet of vehicles based at a common depot is available to operate the routes. Each vehicle can carry a load bounded by the vehicle load capacity. The time each user spends in the vehicle is bounded by a threshold, and there is also a bound on the waiting time at each location. The DARP consists of constructing a set of feasible minimum cost routes.

A common application of the DARP arises in door-to-door transportation of the elderly and the disabled [1,2,3,4]. For surveys of models and algorithms for the DARP, the reader is referred to [5], [6] and [7].

The purpose of this work is to develop a fast algorithm for checking feasibility of any given route. In a paper published in *Operations Research Letters*, Hunsaker and Savelsbergh [8] proposed one such algorithm with linear running time. Unfortunately, their algorithm is not correct, and does therefore not prove that feasibility can be verified in linear time.

The remainder of this work is organized as follows: In Sect. 2, we define the feasibility checking problem in precise terms, and provide the necessary notations and assumptions. In Sect. 3, we briefly review the algorithm in [8], and give a small example showing that it may fail to give the correct conclusion. An

* This work was supported by *the Norwegian Research Council*, *Gassco* and *Statoil* under contract 175967/S30.

B. Chen (Ed.): AAIM 2010, LNCS 6124, pp. 170–179, 2010.

alternative algorithm together with a correctness proof and running time analysis are provided in Sect. 4, while Sect. 5 concludes the paper.

2 Problem Definition

A transportation request consists of a pickup and a delivery location, and is referred to as a *package*. Assume now that a partial solution to a DARP instance is suggested. By that we mean that all packages have been assigned to vehicles, and for each vehicle, a route is suggested. A route is simply an order in which the locations corresponding to the packages are visited, with the addition of initial and final locations of the vehicle as head and tail, respectively. The remaining problem is to check whether the constraints on load capacity, time windows, ride time and waiting time can be satisfied by this selection of routes.

Since there is no constraint across the set of routes, the problem is decomposed into a set of feasibility testing problems, each of which corresponds to a unique route (vehicle). Consequently, we will henceforth consider an arbitrary vehicle in the fleet, along with the set of packages assigned to the vehicle. That is, our attention is directed to the *single-vehicle* variant of DARP, but all our results apply to the general version of the problem.

We include a dummy package consisting of the initial and final vehicle locations, and for simplicity, we assume that all parameters and constraints defined for other packages are also defined for the dummy package. The initial and final locations are typically identical physical sites, referred to as the depot.

We define the following input data:

- I = set of packages assigned to the vehicle, including the dummy package.
- $n = |I| - 1$ = number of packages without counting the dummy package.
- (i^+, i^-) = pickup and delivery location of package $i \in I$.
- $J = (0, 1, \ldots, 2n+1)$ = route = ordered set consisting of all pickup and delivery locations including the initial (0) and the final $(2n+1)$ locations. For notational simplicity, we thus let each location be identified by an integer corresponding to its position in the route.
- $N = 2n + 1$.
- $J^+ \subseteq J$ = pickup locations including location 0.
- $J^- \subseteq J$ = delivery locations including location $2n+1$.
- $[e_j, \ell_j]$ = time window of location $j \in J$.
- t_j = travel time from location j to location $j+1$.
- $a_i = a_{i^+} = a_{i^-}$ = upper bound on the ride time for package i.
- w_j = upper bound on the wait time at location j.
- $d_i = d_{i^+} = -d_{i^-} \geq 0$ = demand, package i (quantity transported from i^+ to i^-).
- Q = load capacity of the vehicle.

As decision variables, we let D_j denote the departure time for location $j = 0, \ldots, N$. Following [8], we also define $A_j = D_{j-1} + t_{j-1}$ as the arrival time for location $j = 1, \ldots, N$, and let $A_0 = e_0$.

In this work, we exclude infeasible instances violating any of the following conditions, all of which can be checked in linear time.

Assumption 1. *For all* $i \in I$, $i^+ < i^-$ *and* $\sum_{j=i^+}^{i^- -1} t_j \leq a_i$. *For all* $j \in J$, $0 \leq e_j \leq \ell_j$, $t_j \geq 0$, $w_j \geq 0$, *and* $\sum_{k=0}^{j} d_k \leq Q$.

By the last inequality, the load capacity constraint is assumed to be satisfied by the route, and will not be considered further. The problem of checking feasibility of the route is now reduced to decide whether the following inequalities are consistent:

$$e_j \leq D_j \leq \ell_j, \qquad j \in J, \tag{1}$$

$$D_{i^-} - D_{i^+} \leq a_i, \qquad i \in I, \tag{2}$$

$$0 \leq D_j - D_{j-1} - t_{j-1} \leq w_j, \; j \in J \setminus \{0\} \; . \tag{3}$$

The focus of this work is to develop an algorithm that solves the above decision problem, and outputs one feasible assignment to D for input instances where the inequalities are consistent.

3 The Algorithm of Hunsaker and Savelsbergh

Three passes constitute the algorithm in [8] for testing feasibility of a given route. First, all arrival and departure times are left-adjusted by taking the precedence constraints (first inequality of (3)) and the lower time window bounds (first inequality of (1)) into consideration. In the second pass, the departure times are right-adjusted in order to satisfy ride time bounds (2). Finally, the third pass right-adjusts arrival and departure times in order to satisfy precedence relations violated in the second pass, and checks whether any upper time window bounds (second inequality of (1)) or ride time bounds hence are violated.

In Algorithm 1, we give a concise description of how the algorithm works in instances where the bounds on the waiting time are sufficiently large to be neglected. This simplification is made in order to exclude algorithmic details irrelevant to the conclusion of our work. It is straightforward to verify that each pass of Algorithm 1 has linear running time.

3.1 An Instance Where the Algorithm Fails

Let $n = 2$, and let the packages be denoted $(0^+, 0^-)$, $(1^+, 1^-)$, and $(2^+, 2^-)$. Consider the route $J = (0^+, 1^+, 2^+, 1^-, 2^-, 0^-)$, where the following data are defined:

- $t_j = 1 \; \forall j \neq 0^+, 2^-$ (travel times on all legs except those to/from the depot are 1),
- $t_{0^+} = t_{2^-} = 0$ (travel times on the legs to/from the depot are 0),
- $w_j = \infty \; \forall j \in J$ (bounds on waiting time can be neglected),
- $a_1 = a_2 = 2$, $a_0 = \infty$ (maximum ride time is 2 for packages $(1^+, 1^-)$ and $(2^+, 2^-)$).
- $[e_{0^+}, \ell_{0^+}] = [e_{0^-}, \ell_{0^-}] = [0, 6]$, $[e_{1^+}, \ell_{1^+}] = [0, 6]$, $[e_{2^+}, \ell_{2^+}] = [2, 6]$, $[e_{1^-}, \ell_{1^-}] = [4, 6]$, $[e_{2^-}, \ell_{2^-}] = [6, 6]$.

Algorithm 1. `Hunsaker&Savelsbergh`

First pass:
$D_0 \leftarrow e_0$
for $j \leftarrow 1, \ldots, N$ **do**
  $A_j \leftarrow D_{j-1} + t_{j-1}$, $D_j \leftarrow \max\{A_j, e_j\}$
  if $A_j > \ell_j$ **then**
    Print "Infeasible route", and stop
Second pass:
for $j \leftarrow N-1, \ldots, 0$ **do**
  $i \leftarrow$ the package to which location j belongs
  if $j = i^+$ **then**
    if $D_{i^-} - D_{i^+} > a_i$ **then**
      $D_{i^+} \leftarrow D_{i^-} - a_i$
    if $D_j > \ell_j$ **then**
      Print "Infeasible route", and stop
Third pass:
for $j \leftarrow 1, \ldots, N$ **do**
  $i \leftarrow$ the package to which location j belongs
  $A_j \leftarrow D_{j-1} + t_{j-1}$, $D_j \leftarrow \max\{A_j, D_j\}$
  if $D_j > \ell_j$ **then**
    Print "Infeasible route", and stop
  if $j = i^-$ and $D_{i^-} - D_{i^+} > a_i$ **then**
    (*) Print "Infeasible route", and stop
Print "Feasible route"

The first pass of the algorithm then assigns the following values to the arrival and departure times:

- $D_{0^+} = 0$
- $A_{1^+} = 0$, $D_{1^+} = 0$
- $A_{2^+} = 1$, $D_{2^+} = 2$
- $A_{1^-} = 3$, $D_{1^-} = 4$
- $A_{2^-} = 5$, $D_{2^-} = 6$
- $A_{0^-} = 6$

After the second pass, we get:

- $D_{0^+} = 0$
- $A_{1^+} = 0$, $D_{1^+} = 2$
- $A_{2^+} = 1$, $D_{2^+} = 4$
- $A_{1^-} = 3$, $D_{1^-} = 4$
- $A_{2^-} = 5$, $D_{2^-} = 6$
- $A_{0^-} = 6$

In the third pass, we get:

- $D_{0^+} = 0$
- $A_{1^+} = 0$, $D_{1^+} = 2$

- $A_{2^+} = 3$, $D_{2^+} = 4$
- $A_{1^-} = 5$, $D_{1^-} = 5$

and in the statement labeled (*), it is concluded that the route is infeasible. However, the schedule

- $D_{0^+} = 3$
- $A_{1^+} = 3$, $D_{1^+} = 3$
- $A_{2^+} = 4$, $D_{2^+} = 4$
- $A_{1^-} = 5$, $D_{1^-} = 5$
- $A_{2^-} = 6$, $D_{2^-} = 6$
- $A_{0^-} = 6$

is feasible.

The small example above demonstrates the shortcoming of Algorithm 1 in instances where the ride times are critical and the packages are interleaved. That is, the pickup location 2^+ is visited between the locations of package 1, while the delivery location 2^- is visited after 1^- $(1^+, 2^+, 1^-, 2^-)$. Since both ride time bounds are violated in the first pass, D_{1^+} and D_{2^+} are both increased in the second pass. In the third pass, the departure times of later visited locations must be increased as well, since otherwise the precedence constraints would be violated. However, this also involves D_{1^-}, which hence is increased by the same amount as D_{1^+}, and the ride time bound remains violated. The algorithm neglects the fact that running the two last passes a second time would increase the value of D_{1^+} to 3 without delaying the departure from later visited locations.

To tackle interleaved packages adequately, the principle applied in Algorithm 1 requires a loop over the two last passes. Unless the number of iterations of the loop can be bounded by a constant, which seems unlikely, such a modification leads to an algorithm with superlinear running time.

In general, Algorithm 1 gives the correct conclusion if the given route is infeasible. However, the algorithm does not conclude correctly for all feasible routes, as it may fail to recognize some of them.

4 An Alternative Algorithm

Define the *net ride times* between locations j and h $(0 \leq j \leq h \leq N)$ $T_{jh} = \sum_{r=j}^{h-1} t_r$. For any pickup location $j \in J^+$, let j^- denote the corresponding delivery location, and for any delivery location $j \in J^-$, let j^+ be the corresponding pickup location.

The idea of our algorithm is to consider first the relaxed version of (1)-(3) where $D_j \leq \ell_j$ is removed for all $j \in J$. Below, we demonstrate that (1)-(3) are consistent if and only if the vector of earliest departure times in the relaxed problem satisfies the upper time window bounds.

To compute this relaxed solution, we determine for every location $j \in J$ whether its departure can be earlier if the precedence constraint $D_j \geq D_{j-1} + t_{j-1}$ is removed. If this is not possible, location j is said to be *critical*. Lemma

4 below provides a simple formula for minimization of relaxed departure times at critical locations. At non-critical locations, the precedence constraints are binding, and the departure times are simply found by adding travel time to the departure time of their predecessors. To identify critical locations, Lemma 5 below gives an inductive formula.

In formal terms, let the feasible region of the relaxed problem be defined as

$$\bar{X} = \{D \in \mathbb{R}^{N+1} : D_j \geq e_j \ (j \in J),$$
$$0 \leq D_j - D_{j-1} - t_{j-1} \leq w_j \ (j = 1, \dots, N), \ D_{i^-} - D_{i^+} \leq a_i \ (i \in I)\},$$

and let $X = \{D \in \bar{X} : D_j \leq \ell_j (j \in J)\}$ denote the set of departure times satisfying (1)-(3).

Lemma 1. $\bar{X} \neq \emptyset$.

Proof. Choose e.g. $D_0 = \max\{e_j : j \in J\}$ and $D_j = D_{j-1} + t_{j-1} \ \forall j > 0$. Then, $D_{i^-} - D_{i^+} \leq a_i \ \forall i \in I$, and thereby also $D \in \bar{X}$, follow directly from Assumption 1. □

Observation 1. *Let $a_1, a_2, b_1, b_2, c_1, c_2$ be real numbers such that $c_1 \leq a_1 - a_2 \leq c_2$ and $c_1 \leq b_1 - b_2 \leq c_2$. Then $c_1 \leq \min\{a_1, b_1\} - \min\{a_2, b_2\} \leq c_2$.*

Proof. The lower bound on $a_1 - a_2$ and $b_1 - b_2$ yields $\min\{a_1, b_1\} - \min\{a_2, b_2\} \geq \min\{c_1 + a_2, c_1 + b_2\} - \min\{a_2, b_2\} = c_1$. The second inequality is proved analogously. □

By Lemma 1 and the lower time window bounds, we have that $D^{\min} = (D_0^{\min}, \dots, D_N^{\min})$, where $D_j^{\min} = \min\{D_j : D \in \bar{X}\}$, exists.

Lemma 2. $D^{\min} \in \bar{X}$.

Proof. Let $D' \in \arg\min\left\{\sum_{j \in J} D_j : D \in \bar{X}\right\}$. To prove that $D'_j = D_j^{\min} \ \forall j \in J$, observe that $D'_j \geq D_j^{\min}$, and assume $D'_k > D_k^{\min}$ for some $k \in J$. Since $D_k^{\min}$ is a feasible value of D_k, there exists some $D^k \in \bar{X}$ such that $D_k^k = D_k^{\min}$. Define $\hat{D} \in \mathbb{R}^{N+1}$ by $\hat{D}_j = \min\{D'_j, D_j^k\}$ $(j \in J)$, which yields $\sum_{j \in J} \hat{D}_j < \sum_{j \in J} D'_j$. Obviously, $\hat{D} \geq e$, and Observation 1 implies that $\hat{D}$ also satisfies (2)-(3). Hence, $\hat{D} \in \bar{X}$, contradicting the definition of D'. □

Define $E \in \mathbb{R}^{N+1}$ by $E_0 = D_0^{\min}$ and for all $h = 1, \dots, N$,
$E_h = \min\{D_h : D \in \bar{X}^h\}$, where $\bar{X}^h$ is defined by the same inequalities as $\bar{X}$, but with $D_h - D_{h-1} - t_{h-1} \geq 0$ removed.

That is, E_h is the earliest departure from location h if we eliminate all upper time window bounds and the precedence constraint between locations $h-1$ and h. The set of critical locations is defined as $J' = \{j \in J : E_j = D_j^{\min}\} = \{j_1, j_2, \dots, j_K\}$, where $K = |J'|$. We assume $j_1 > j_2 > \dots > j_K = 0$, since our algorithm below computes the critical locations in an order of decreasing indices. We let $E_{N+1} = e_{N+1} > D_N^{\min}$ and $w_{N+1} > e_{N+1} - D_N^{\min}$ be arbitrarily chosen, and define $j_0 = N + 1$.

Note that if $D_j^{\min} > D_{j-1}^{\min} + t_{j-1}$ then $j \in J'$, but the converse is not necessarily true. It follows that

$$D_j^{\min} = D_{j_k}^{\min} + T_{j_k j} \ \forall j \in B_k, \tag{4}$$

where $B_k = \{j_k, j_k + 1, \ldots, j_{k-1} - 1\}$, and knowledge to J' and E is thus sufficient for assessing $D^{\min}$.

Lemma 3. *Let $j \in B_k$. Then the following inequality is satisfied, and it holds with equality if $j = j_k$:*

$$E_j \geq \begin{cases} \max\left\{e_j, D_{j^-}^{\min} - a_j, D_{j+1}^{\min} - w_{j+1} - t_j\right\}, & j \in J^+, j+1 \notin B_k, \\ \max\left\{e_j, D_{j+1}^{\min} - w_{j+1} - t_j\right\}, & j \in J^-, j+1 \notin B_k, \\ \max\left\{e_j, D_{j^-}^{\min} - a_j\right\}, & j \in J^+, j+1 \in B_k, j^- \notin B_k, \\ e_j, & \textit{otherwise.} \end{cases} \tag{5}$$

Proof. The lower bound on E_j is obvious. In the first two cases, it is also obvious that it holds with equality for $j = j_k$.

To prove equality for $j = j_k$ in the third case, assume that $E_{j_k} > \max\left\{e_{j_k}, D_{j_k^-}^{\min} - a_{j_k}\right\}$. Since $j_k \in J'$, we have that $D^{\min} \in \arg\min\left\{D_{j_k} : D \in \bar{X}^{j_k}\right\}$ and $E_j < D_j^{\min}$ for all $j \in B_k \setminus \{j_k\}$. Starting from the solution $D^{\min}$, we construct a new solution D' in $\bar{X}^{j_k}$ by moving departures from all locations in B_k forward in time as much as any available slack δ permits. This slack is found by minimizing the slack over all $j \in B_k$:

$$\delta' = \min\left\{D_j^{\min} - E_j : j \in B_k \setminus \{j_k\}\right\},$$
$$\delta = \min\left\{E_{j_k} - \max\left\{e_{j_k}, D_{j_k^-}^{\min} - a_{j_k}\right\}, \delta'\right\},$$

and the new solution becomes

$$D'_j = D_j^{\min} - \delta, \ \ j \in B_k,$$
$$D'_j = D_j^{\min}, \quad j \in J \setminus B_k \ .$$

We then have $D' \in \bar{X}^{j_k}$. By the assumption, $\delta > 0$, leading to the contradiction that $D'_{j_k} < D_{j_k}^{\min} = E_{j_k}$.

The last case is proved by repeating the above arguments with $\max\left\{e_{j_k}, D_{j_k^-}^{\min} - a_{j_k}\right\}$ replaced by e_{j_k}. □

To compute E_j for all $j \in J'$, we define a sequence $E^0, E^1, \ldots, E^K \in \mathbb{R}^{N+2}$ of lower bounds by $E^0 = e$, and for $k = 1, \ldots, K$, $j \in J$:

$$E_j^k = \begin{cases} E_j^{k-1}, & j \geq j_{k-1}, \\ \max\left\{e_j, E_{j_m}^{k-1} + T_{j_m j^-} - a_j, \right. & \\ \left. E_{j_{k-1}}^{k-1} - w_{j_{k-1}} - t_j\right\}, & j = j_{k-1} - 1 \in J^+, j^- \in B_m, m < k, \\ \max\left\{e_j, E_{j_{k-1}}^{k-1} - w_{j_{k-1}} - t_j\right\}, & j = j_{k-1} - 1 \in J^-, \\ \max\left\{e_j, E_{j_m}^{k-1} + T_{j_m j^-} - a_j\right\}, & j_{k-1} - 1 > j \in J^+, j^- \in B_m, m < k, \\ e_j, & \text{otherwise.} \end{cases} \quad (6)$$

Lemma 4. *For all $k = 0, 1, \ldots, K$ and all $j \in J$, we have $E_j \geq E_j^k$ and $E_{j_k} = E_{j_k}^k$.*

Proof. The proof is by induction in k. For $k = 0$, the result is trivial. Assume it holds for $0, 1, \ldots, k-1$. It follows from the induction hypothesis and (4) that $E_{j_{k-1}}^{k-1}$ and $E_{j_m}^{k-1} + T_{j_m j^-}$ $(m < k)$ equal $D_{j_{k-1}}^{\min}$ and $D_{j^-}^{\min}$, respectively. Hence, for $j \geq j_k$, the right hand sides of (5) and (6) coincide, and the result follows from Lemma 3. □

Lemma 5.

$$j_k = \max \arg\max \left\{E_j^{k-1} + T_{jN} : j = 0, \ldots, j_{k-1} - 1\right\}, \forall k = 1, \ldots, K. \quad (7)$$

Proof. Denote the right hand side of (7) by r_k, and assume $j_k < r_k$. Then $r_k \notin J'$. By utilizing Lemma 4 for $j = r_k$ and $j = j_k$, we get $E_{r_k} < D_{r_k}^{\min} = D_{j_k}^{\min} + T_{j_k r_k} = E_{j_k}^k + T_{j_k N} - T_{r_k N} \leq E_{r_k}^k + T_{r_k N} - T_{r_k N} \leq E_{r_k}$, which is a contradiction.

Assume $j_k > r_k$. Then $D_{j_k}^{\min} \geq D_{r_k}^{\min} + T_{r_k j_k} \geq E_{r_k}^k + T_{r_k j_k} = E_{r_k}^k + T_{r_k N} - T_{j_k N} > E_{j_k}^k + T_{j_k N} - T_{j_k N} = E_{j_k}$, which contradicts $j_k \in J'$. □

Equation (6) and Lemma 5 give simple formulae for the computation of $D_j^{\min}$ and J'. This is exploited in Algorithm 2. We conclude that the route is feasible if and only if $D_j^{\min} \leq \ell_j$ for all $j \in J$. By the second condition of Assumption 1, we can disregard the ride time constraint of package i if there is some $k = 1, \ldots, K$ such that $i^+, i^- \in B_k$.

Proposition 1. *Algorithm 2 returns $(\texttt{false}, D^{\min})$ if $X = \emptyset$, and $(\texttt{true}, D^{\min})$, otherwise.*

Proof. It follows from (4) and Lemmata 3-5 that the computed value of D is $D^{\min}$. Since $X = \left\{D \in \bar{X} : D \leq \ell\right\}$ and $D \geq D^{\min}$ $\forall D \in \bar{X}$, it follows from Lemma 2 that $X \neq \emptyset$ if and only if $D^{\min} \leq \ell$, which completes the proof. □

Proposition 2. *The time complexity of Algorithm 2 is $\mathcal{O}(n \log n)$.*

Algorithm 2. `FeasibilityCheck`

```
k ← 0
E ← e
j_k ← N + 1
feas ← true
while j_k > 0 do
    k ← k + 1
    j_k ← max arg max {E_j^{k-1} + T_{jN} : j = 0, ..., j_{k-1} - 1}
    for j ← j_k, ..., j_{k-1} - 1 do
        D_j ← E_{j_k} + T_{j_k j}
        if D_j > ℓ_j then
            feas ← false
        if j ∈ J^- and j^+ < j_k then
            E_{j^+} ← max{E_{j^+}, D_j - a_j}
    E_{j_k - 1} ← max{E_{j_k - 1}, D_{j_k} - w_{j_k} - t_{j_k - 1}}
return (feas, D)
```

Proof. In each iteration of the while-loop, let locations $j = 0, \ldots, j_{k-1} - 1$ be represented by a heap with primary and secondary keys $E_j + T_{jN}$ and j, respectively. Each location is pushed on the heap prior to execution of the loop. Location j is removed from the heap when the value of j_k is assigned a value no larger than j. Each location is hence removed exactly once. While in the heap, the primary key of the location is updated at most twice. Since the while-loop is executed at most $N+1$ times, only $\mathcal{O}(n)$ retrievals of the location with maximum key value are required. The total number of heap operations (counting insertions, retrievals and key updates), each of which has time complexity $\mathcal{O}(\log n)$, is thus linear.

The proof is complete by observing that the algorithm needs access to net ride times between only $\mathcal{O}(n)$ pairs of locations, and that each of these is computed in constant time once T_{jN} is computed for all $j \in J$. □

It is straightforward to verify that when applied to the instance studied in Sect. 3.1, Algorithm 2 identifies the critical locations $j_1 = 2^-$, $j_2 = 2^+$, $j_3 = 1^+$ and $j_4 = 0^+$. The algorithm concludes that the route is feasible, and suggests the schedule $D_{0^+} = 0$, $D_{1^+} = 3$, $D_{2^+} = 4$, $D_{1^-} = 5$, $D_{2^-} = 6$, and $D_{0^-} = 6$.

5 Conclusions

In this paper, we have studied the problem of verifying feasibility of a suggested solution to the dial-a-ride problem. We have shown that an algorithm with linear running time published by Hunsaker and Savelsbergh may give incorrect answer when the suggested route is feasible. Hence, we have proposed a new algorithm, and shown that feasibility of a route can be checked in linearithmic time. Whether feasibility of a solution to the DARP can be verified in linear time still remains an open question.

References

1. Madsen, O.B., Ravn, H.F., Rygaard, J.M.: A heuristic algorithm for a dial-a-ride problem with time windows, multiple capacities and multiple objectives. Annals of Operations Research 60(1-4), 193–208 (1995)
2. Ioachim, I., Desrosiers, J., Dumas, Y., Solomon, M.M., Villeneuve, D.: A request clustering algorithm for door-to-door handicapped transportation. Transportation Science 29(1), 63–78 (1995)
3. Borndörfer, R., Grötschel, M., Klostermeier, F., Küttner, C.: Telebus Berlin: Vehicle scheduling in a dial-a-ride system. Technical Report SC 97-23, Konrad-Zuse-Zentrum für Informationstechnik Berlin (1997)
4. Toth, P., Vigo, D.: Heuristic algorithms for the handicapped persons transportation problem. Transportation Science 31(1), 60–71 (1997)
5. Cordeau, J.-F., Laporte, G.: The dial-a-ride problem: models and algorithms. Annals of Operations Research 153(1), 29–46 (2007)
6. Cordeau, J.-F., Laporte, G., Potvin, J.Y., Savelsbergh, M.W.P.: Transportation on demand. In: Barnhart, C., Laporte, G. (eds.) Transportation, Handbook in OR & MS, vol. 14, pp. 429–466. Elsevier, Amsterdam (2007)
7. Parragh, S.N., Doerner, K.F., Hartl, R.F.: A survey on pickup and delivery problems. Part II: Transportation between pickup and delivery locations. Journal für Betriebswirtschaft 58(2), 81–117 (2008)
8. Hunsaker, B., Savelsbergh, M.: Efficient feasibility testing for dial-a-ride problems. Operations Research Letters 30(3), 169–173 (2002)

Indexing Similar DNA Sequences

Songbo Huang[1], T.W. Lam[1], W.K. Sung[2], S.L. Tam[1], and S.M. Yiu[1]

[1] Department of Computer Science, The University of Hong Kong, Hong Kong
{sbhuang,twlam,sltam,smyiu}@cs.hku.hk
[2] Department of Computer Science, National University of Singapore, Singapore
ksung@comp.nus.edu.sg

Abstract. To study the genetic variations of a species, one basic operation is to search for occurrences of patterns in a large number of very similar genomic sequences. To build an indexing data structure on the concatenation of all sequences may require a lot of memory. In this paper, we propose a new scheme to index highly similar sequences by taking advantage of the similarity among the sequences. To store r sequences with k common segments, our index requires only $O(n + N \log N)$ bits of memory, where n is the total length of the common segments and N is the total length of the distinct regions in all texts. The total length of all sequences is $rn + N$, and any scheme to store these sequences requires $\Omega(n + N)$ bits. Searching for a pattern P of length m takes $O(m + m \log N + m \log(rk) psc(P) + occ \log n)$, where $psc(P)$ is the number of prefixes of P that appear as a suffix of some common segments and occ is the number of occurrences of P in all sequences. In practice, $rk \leq N$, and $psc(P)$ is usually a small constant. We have implemented our solution[1] and evaluated our solution using real DNA sequences. The experiments show that the memory requirement of our solution is much less than that required by BWT built on the concatenation of all sequences. When compared to the other existing solution (RLCSA), we use less memory with faster searching time.

1 Introduction

The study of genetic variations of a species often involves mining very similar genomic sequences. For example, when studying the association of SNPs (single nucleotide polymorphism) with a certain disease [3,1] in which the differences on a few characters in the genomes cause the disease, the same regions of individual genomes from different normal people and patients are extracted and compared. These different sequences are almost identical except on those SNPs. The length of each sequence can be from several million to several hundred million, and the number of copies can be up to a few hundreds.

When studying these similar sequences, a basic operation is to search the occurrences of different patterns. This seems to be straightforward as one can consider a given set of similar sequences as a single long sequence and exploit classical text indexes like suffix trees or even better, compressed indexes like BWT

[1] The software is available at http://i.cs.hku.hk/~sbhuang/SimDNA/

B. Chen (Ed.): AAIM 2010, LNCS 6124, pp. 180–190, 2010.

(Burrows-Wheeler Transform) to perform very fast pattern searching [9, 10]. However, these indexes would demand much more memory than ordinary computers can support. Roughly speaking, a suffix tree requires more than 10 bytes per nucleotide, and BWT requires 0.5 to 1 byte per nucleotide (other compressed indexes like CSA [6] and FM-index [4, 5] have slightly higher memory requirement). Consider a case involving 250 sequences each of 200 million nucleotides. To index all these sequences, even BWT would require about 40 Gigabytes, far exceeding the capacity of a workstation. The problem of these naive solutions lies on that they do not take advantage of the high similarity of these sequences. In this paper we propose a new scheme to index highly similar sequences. It takes advantage of the similarity to obtain a very compact index, while allowing very efficient searching for any given patterns.

We first consider the following model of similarity of the input sequences. We assume that the positions in a sequence at which the symbols are different from other sequences are more or less the same for all sequences. One example is the SNP locations in a set of genes. Details are given as follows.

Model 1: Consider r sequences $T_0, T_1, \ldots, T_{r-1}$, not necessarily of the same length. Assume that they have k segments $C_0, C_1, \ldots, C_{k-1}$ in common. That is, each T_i is equal to $R_{i,0}C_0R_{i,1}C_1 \ldots C_{k-1}R_{i,k}$, where $R_{i,j}$ $(0 \leq j \leq k)$ is a segment varies according to T_i. Note that some $R_{i,j}$ can be empty and, for any $i' \neq i$, $R_{i,j}$ and $R_{i',j}$ may not have the same length.

Below we use n to denote the total length of all C_j's, and use N to denote the total length of all $R_{i,j}$'s over all T_i's. Note that the total length of all T_i's is exactly $rn + N$. Since the sequences are highly similar, the length of each T_i is dominated by n, and $N << rn$. Furthermore, N is usually larger than rk. The problem is to design a space-efficient index to store the sequences while allowing efficient search for any given pattern.

Our contributions: We develop a solution to solve the above problem by exploiting BWT and the suffix array data structures. Our solution requires $O(n + N \log N)$ bits of memory. To search a pattern P of length m, our solution takes $O(m + m \log N + m(\log rk)psc(P) + occ \log n)$ time, where $psc(P)$ is the number of prefixes of P that appear as a suffix of some common segments, and occ is the total number of occurrences of P in all sequences. In practice, $psc(P)$ is usually a small constant. It can also be shown that $psc(P)$ is upper bounded by $O(\log n)$ for random sequences. We implemented our solution. To store 250 versions of a sequence about 200M long, our solution only requires 2.7G memory. The memory requirement is only less than 7% of the memory required to store all 250 sequences using BWT. To search a pattern of length 500 in this collection of sequences, it takes less than 0.5ms, thus our solution is practical.

We also extend our solution to another model of similarity of the input sequences. In this model, every pair of sequences have a few positions with different nucleotides, but such positions vary from sequence to sequence. A typical example is a set of genes from closely related species. In this case, we arbitrarily take a sequence as a reference sequence, and the model is defined as follows.

Model 2: Consider r sequences $T_0, T_1, \ldots, T_{r-1}$, all of the same length n. Each T_i ($i \neq 0$) differs from T_0 in x_i positions, where $x_i << n$. For $i \neq i'$, the positions at which T_i is different from T_0 may not be the same as those of $T_{i'}$.

With minor modification, our solution can also be applied to handle Model 2 with similar space and time complexity. [12] provides a solution (referred as RLCSA) for Model 2. Their core idea is to make use of the *run-length encoding* [11] to further compress BWT by considering the maximal segments in BWT in which all symbols are the same (called *runs*). The key observation is that if we concatenate all r sequences as a long sequence T, then construct BWT for T, the expected value for the number of runs X_T in T is bounded by $X + O(s \log L)$, where X is the number of runs of BWT for T_0, s is the sum of all x_i's, and L is the total length of all sequences. Based on this bound, using our notation, their space complexity is $O((n + N \log(rn + N)) \log[(rn + N)/(n + N \log(rn + N))])$, which is slightly worse than our solution. In practice, X_T can be small. The searching time complexities of both our and their solutions depend on a factor which is related to the input data. We compare the two solutions based on real experiments in Section 4.

2 Preliminaries

We give a brief review of two indexing data structures, namely, suffix array and Burrows-Wheeler Transform (BWT) [2]. In the paper, we only consider DNA sequences which are strings of 4 symbols, {A, C, G, T}, only.

Suffix array: Given a text $T[0..n-1]$, we define the suffix array of T, denoted $SA[0..n-1]$, as follows. $SA[i] = j$ if the suffix $T[j..n-1]$ is lexicographically the i-th smallest suffix among all suffixes of T (and we say that the *rank* of the suffix $T[j..n-1]$ is i). In other words, SA stores the starting positions of all suffixes of T in lexicographical order. For any pattern P, suppose P appears in T. We define the *SA range* of P with respect to T as $[s, e]$ such that s and e are respectively the rank of the lexicographically-smallest and largest suffix of T that contains P as a prefix.

To find all occurrences of a pattern P in T, we can first compute the SA range of P (using $O(m \log n)$ time [7]), afterwards the occurrences of P can be retrieved from the suffix array directly one by one in constant time. To store the suffix array of a text with n characters, we need to store n positions (more precisely, $n \log n$ bits of memory) in addition to the text. Suffix array can also be defined on a set of strings $D_0, D_1, \ldots, D_{q-1}$ as follows. $SA[i] = j$ if the string D_j is lexicographically the i-th smallest among all given strings. The rank of D_j is defined to be i. For any pattern P, suppose P appears as a prefix of some D_i. The SA range of P is defined to be $[s, e]$ where s and e are respectively the rank of the lexicographically-smallest and largest D_i that has P as a prefix.

Burrows-Wheeler Transform (BWT): Given a text $T[0..n-1]$, the BWT data structure, $BWT[0..n-1]$, is defined as $BWT[i] = T[j-1]$ where $j = SA[i]$ for $SA[i] \neq 0$, otherwise, set $BWT[i] = \$$, where \$ is a special character not in the alphabet Σ and assumed to be lexicographically smaller than all other characters. That is, $BWT[i]$ stores the character immediately before the i-th smallest suffix. BWT requires only the same amount of memory as for storing the text. Using BWT and some auxiliary functions, we can compute the SA range of a given pattern of length m in a backward manner (*backward search*) in $O(m)$ time [8,9].

To retrieve the positions of an SA range, we only store part of the suffix array, called *sampled suffix array*. Intuitively, we store one SA value for every α entries for some constant α. More precisely, we store the $SA[i]$ value for $i = k\alpha$ for $0 \leq k \leq \lceil \frac{n}{\alpha} \rceil$, i.e., we store the $SA[i]$ value if the rank of the suffix $T[SA[i]..n-1]$ is a multiple of α. Retrieving the value for $SA[i]$ where i is not a multiple of α can be done by searching repeatedly the BWT data structure [8].

3 Our Solution

For Model 1, recall that the input is a set of r DNA sequences $T_0, T_1, \ldots, T_{r-1}$, each T_i can be partitioned into $R_{i,0}C_0R_{i,1}C_1 \ldots C_{k-1}R_{i,k}$, where C_j $(0 \leq j \leq k-1)$ is a common segment that all sequences agree, and $R_{i,j}$ $(0 \leq j \leq k)$ is called an R segment, which may be different for different T_i's. Let n be the total length of all common segments and N be the total length of all R segments.

Indexing data structure: Let $C = C_0\$C_1\$\ldots\$C_{k-1}\$$, where \$ is a new symbol and is lexicographically smaller than all other symbols. We store C as an array of characters, together with another array $Start_C$ of the starting positions of each C_i in C which can be used to recover the text for any common segment (see Figure 1 for an example, (b) shows the common segments, C, C^R together with $Start_C$). We construct a BWT index for C^R. Given a string P, we can search the BWT index using P^R to determine the SA range of P^R, which captures all the occurrences of P^R in C^R (that is, all occurrences of P in C).

Each $T_i = R_{i,0}C_0R_{i,1}C_1 \ldots C_{k-1}R_{i,k}$. We consider every suffix d of T_i that starts inside an R segment and refer it as a *differentiating suffix*. Note that N is the total number of *differentiating suffixes* for all T_i's. We sort the *differentiating suffixes* of all T_i's and construct a suffix array $SAR[0..N-1]$ such that $SAR[i]$ stores a reference to i-th lexicographically smallest differentiating suffix.

Before defining what is a reference, we need to show how each T_i and its R segments are represented. For each T_i, we only store its R segments. The segments $R_{0,0}, \ldots, R_{0,k}, R_{1,0}, \ldots, R_{1,k}, \ldots, R_{r-1,0}, \ldots, R_{r-1,k}$ are stored sequentially in a character array $T[0..N-1]$. We assign a segment number to each $R_{i,j}$, which is its order in T. Precisely, the segment number of $R_{i,j}$ is $(k+1)i + j$. For example, $R_{0,0}$ is segment 0, $R_{0,1}$ segment 1, and $R_{1,0}$ segment $k+1$ (see Figure 1(c) for an example of all R segments). Note that a segment number w can be used to identify to which T_i this segment belongs, namely, $i = \lfloor w/(k+1) \rfloor$. We also construct an array $Start_T[0..r(k+1)-1]$ such that $Start_T[j]$ stores the starting position of segment j in T. We are now ready to define a reference to a differentiating suffix. It is essentially a pair of integers, (segment number w, offset o), where $T[Start_T[w] + o]$ stores the first character of the differentiating

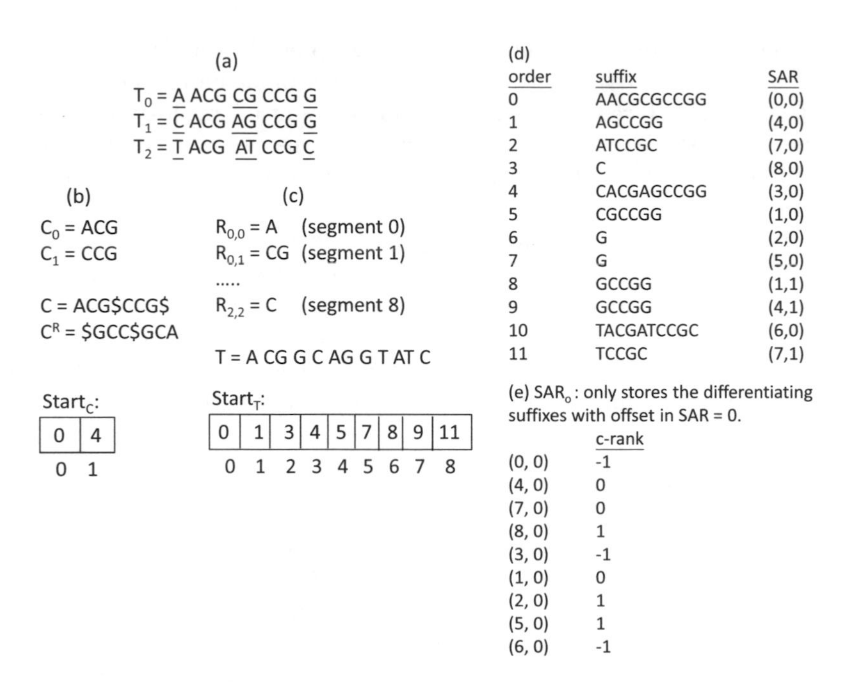

Fig. 1. An example for the indexing data structure. (a) shows the input sequences, the underlined segments are the R segments. (b) shows the common segments and content of $Start_C$. (c) shows the R segments and content of $Start_T$. (d) shows all differentiating suffixes in lexicographical order and also SAR, note that each entry in SAR stores the pair (segment number, offset). (e) shows the example for SAR_o.

suffix. Given such a reference, we can recover every character of the suffix by referencing the corresponding segments from T and C. Figure 1(d) shows an example of all differentiating suffixes in lexicographical order and the contents of SAR for the same example.

From SAR, we construct a subarray SAR_o that contains only those entries with offset zero. In other words, SAR_o includes the differentiating suffixes starting from the first character of an R segment. To save space, we only need to store a segment number w in each entry of SAR_o as the offest is always zero. In addition, let w correspond to the segment $R_{i,j}$, we store a number c-rank that is the rank of the suffix $\$C^R_{j-1}\$C^R_{j-2}\cdots\$C^R_0$ among all suffixes of C^R (if $j = 0$, we set c-rank $= -1$). The latter is useful to determine if a pattern crosses the boundary between a common segment and an R segment of some T_i. See Figure 1(e) for an example for SAR_o.

Space complexity: Given r similar sequences $T_0, T_1, \ldots, T_{r-1}$, each can be partitioned into k common segments and $(k+1)$ R segments. For C^R, we have a BWT index and the array $Start_C$ of k entries. For T, we have the suffix arrays SAR and SAR_o and the array $Start_T$ for the starting positions of all R segments. The whole data structure requires $O(n + N \log rk + rk(\log n + \log N))$ bits.

Searching algorithm: Given a pattern P of length m, if it occurs in any of the text T_i, there are three cases.
(1) P is completely inside a common segment.
(2) P is a prefix of a differentiating suffix of T_i (i.e., P starts in an R region).
(3) P can be partitioned into non-empty substrings P_1P_2 where P_1 is a suffix of a common segment C_j of T_i and P_2 is a prefix of the following differentiating suffix in T_i.

For Case 1, we search the BWT index of C^R for P^R. This takes $O(m + occ_1)$ time, where occ_1 is the number of Case 1 occurrences of P. For Case 2, we search SAR for P. This takes $O(m \log N + occ_2)$ time, where occ_2 is the number of Case 2 occurrences of P.

Case 3 is the only non-trivial case, detailed as follows. Consider a particular partition (P_1, P_2) of P. First, we search the BWT index of C^R for the presence of $(P_1\$)^R$. Let LR_1 be the resulting SA range. If LR_1 is non-empty, P_1 appears as a suffix of some common segment. We then search the suffix array SAR_o for P_2 and let LR_2 be the resulting SA range. For each x in LR_2, suppose $SAR[x] = (w, \text{c-rank})$. Let $i = \lfloor w/(k+1) \rfloor$ and $t = w \mod (k+1)$. By definition, P_2 has an occurrence in T_i, starting from the first character of $R_{i,t}$. Thus, if there is a corresponding occurrence of P_1 at the end of the common segment C_{t-1}, we find a valid occurrence of P. We can make use of the c-rank stored in SAR_o to perform the above checking in constant time. By definition, the c-rank value stored in $SAR[x]$ is the rank of the suffix $\$C^R_{t-1}\$C^R_{t-2}\ldots\$C^R_0$ with respect to all suffixes of C^R. Thus, if c-rank in within LR_1, P_1 must appear as a suffix of C_{t-1}. This is summarized in the following lemma.

Lemma 1. *Suppose P is partitioned into P_1P_2 as described above. For any x in LR_2, let $SAR[x] = (w, \text{c-rank})$. Let $i = \lfloor w/(k+1) \rfloor$ and $t = w \mod (k+1)$. Then, if c-rank $\in LR_1$, P has an occurrence in T_i, precisely, the concatenation of the last $|P_1|$ characters of C_{t-1} and the first $|P_2|$ characters of the differentiating suffix $R_{i,t}C_tR_{i,t+1}\ldots R_{i,k}$.*

Note that C^R, $Start_C$, T, and $Start_T$ are required for searching the suffix arrays SAR and SAR_o. We put together the above ideas in Algorithm 1, which forms the basics of the searching algorithm of Case 3. Note that BWT supports backward searching. With the BWT index of C^R, we can compute the SA ranges for $P[0], P[0..1]^R, P[0..2]^R, \ldots P[0..m-1]^R$ incrementally using $O(m)$ time; of course, we can terminate the search as soon as an empty SA range is found. Furthermore, from the SA range of $P[0..i]^R$, we can compute the SA range of $(P[0..i]\$)^R$ in $O(1)$ time using the auxiliary functions for BWT.

Time complexity: Below we denote $psc(P)$ to be the number of prefixes of P which are suffixes of some common segments. In an iteration where $(P[0..j]\$)^R$ is found to have a non-empty LR_1 (i.e., $P[0..j]$ is a suffix of some common segment), the corresponding LR_2 may contain up to rk candidates and the verification may take $O(rk)$ time. The overall time required by Case 3 searching is $O(m + psc(P)(m\log(rk) + rk) + occ_3)$ time where occ_3 is the number of Case 3 occurrences of P.

Algorithm 1. *Case_3_Search*$(P[0..m-1])$

Require: $|P| > 0$
Ensure: All Case 3 occurrences of P in $T_0, \ldots, T_{r-1}$

1: **for** $j = 0$ to $m - 2$ **do**
2: Using BWT of C^R, find SA range LR_1 of $(P[0..j]\$)^R$.
3: **if** LR_1 is non-empty **then**
4: Find SA range LR_2 of $P[j+1..m-1]$ using SAR_o.
5: **for** each $x \in LR_2$ **do**
6: Let $SAR_o[x] = (w, \text{c-rank})$.
7: **if** c-rank $\in LR_1$ **then**
8: Report the corresponding occurrence of P.
9: **end if**
10: **end for**
11: **end if**
12: **end for**

The problem of Algorithm 1 is that the time required in each iteration depends on the size of LR_2, yet it is possible that no entries in LR_2 could form a valid occurrence of P. To speed up the checking whether the c-ranks of the entries captured by LR_2 fall in LR_1, we construct a 2-dimensional range search index [13] of the c-ranks stored in SAR_o. Note that the value of each c-rank is in the range $[1..n]$. Then given LR_1 and LR_2, we can find the existence of any c-ranks specified by LR_2 fall in the range LR_1 in $O(\log n)$ time, and each occurrence can be retrieved in $O(\log n)$ time. The time complexity of Case 3 becomes $O(m + psc(P)(m\log(rk)) + occ_3 \log n)$ time. The range search index requires $O(rk \log n)$ bits. The overall result is summarized in the following theorem.

Theorem 1. *Given r similar sequences $T_0, T_1, \ldots, T_{r-1}$, each can be partitioned into k common segments and $(k+1)$ R segments. Let n be the total length of the common segments, and let N be the total length of the R segments. We can build an indexing data structure using $O(n + N \log rk + rk(\log n + \log N))$ bits such that locating the occurrences of a pattern P of length m in the sequences can be done in $O(m + m\log N + psc(P)(m\log(rk)) + occ \log n)$ time, where occ is the total number of occurrences of P.*

Extension to Model 2: Recall that in Model 2, we are given a set of r DNA sequences $T_0, T_1, \ldots, T_{r-1}$, all with the *same length* n, each T_i $(i \neq 0)$ differs from T_0 in x_i positions. Let $N = \sum x_i$. We describe the version without using range search index. Modifying it to use range search index is straightforward.

We redefine a differentiating suffix as a suffix in T_i such that this suffix starts at one of the x_i positions. There are N differentiating suffixes. Similar to Model 1, we store T_0^R as an array of characters and construct a BWT index for T_0^R. Note that we do not have an array similar to $Start_C$. Then, construct a suffix array SAR for all differentiating suffixes. We can define an R segment as a maximal region of characters in T_i which differ from the corresponding characters in T_0 and label the R segments from 0 to $s-1$ where s is the total number of R segments. We store $T = R_0R_1 \ldots R_{s-1}$ as an array of characters. To recover each differentiating suffix, we need an array $Start_T$ to store the starting position of each segment in T and also the starting position of this segment in its original sequence T_i. So, SAR will store a pair of integers (segment number w, offset o) in order to reference a differentiating suffix.

From SAR, we construct SAR_o that contains only those entries with offset zero. We construct an array B' as follows. Let $d = T_i[j..n-1]$ be a differentiating suffix of SAR_o and rank of d in SAR_o is x. Then, consider $T_0[0..j-1]$ and let rank of $T_0[0..j-1]^R$ with respect to T_0^R be y. Then, we set $B'[x] = y$.

The searching algorithm is very similar. Given a pattern P, search P in SAR to locate all occurrences of P which are completely inside a differentiating suffix. Search P^R in T_0^R using BWT to locate all occurrences of P in T_0. The additional checking we need here is for each occurrence of P in T_0, we check if it also occurs in the same position in each of T_i for all $i > 0$. The last case is to partition P into $P_1 = P[0..j]$ ($j < m-2$) and $P_2 = P[j+1..m-1]$. Then, search P_1^R using BWT. Let the SA range returned be LR_1. Search P_2 using SAR_o and let the SA range returned be LR_2. For each x in LR_2, check if $B'[x] \in LR_1$. If yes, an occurrence of P is found. The space complexity is $O(n + N \log N + s(\log n + \log N))$ bits while the time complexity is $O(m + m \log N + psc'(P)(m \log s + s) + occ)$ where $psc'(P)$ is the number of prefix of P that occurs in T_0.

4 Evaluation

We first compare the memory consumption of our solution[2] based on Model 1 with the following. Concatenate all sequences to a long sequence and apply BWT directly on the long resulting sequence. We generate the texts as follow. We use two chromosomes (Chromosome Y of length 25M and Chromosome 1 of length 217M) and download the positions of SNPs for these chromosomes from NCBI[3]. For Chromosome Y, 0.1% positions are SNPs. For Chromosome 1, 0.5% positions are SNPs. Most of these SNPs are not consecutive, i.e., most of the R segments are of length 1. The number of R segments $(k+1)$ are very similar to the number of SNPs in both cases (23,077 and 980,618 for Chromosome Y and 1 respectively). We construct four test cases. Tests 1 and 2 use Chromosome Y as Tests 3 and 4 use Chromosome 1. For Tests 1 and 3, we generate 50 texts while for Tests 2 and 4, we generate 250 texts. For each text, for each position of SNPs, we randomly generate a nucleotide. For the patterns, we randomly select them from the texts with length varying from 50 to 500. For each length, we repeat the experiment 100 times and obtain the average searching time. All experiments were conducted in a personal computer with 8G memory and a dual core 2.66GHz CPU.

For the BWT data structure for C^R, we use 0.75bytes per character which store 1/8 sampled SA for all our experiments. Table 1 shows the memory

[2] There are a few implementation tricks we used to speed up the searching process. Details will be given in the full paper.

[3] ftp://ftp.ncbi.nih.gov/snp/organisms/human_9606/chr_rpts

consumed by our data structure and the memory required by using BWT to store all texts. We can see that the amount of memory required by our solution is about 2-9% of that required by using BWT. In fact, the smaller the amount of SNPs, the more memory our scheme can save. For the searching performance, Table 2 shows the searching time for different pattern lengths in both Test 2 and Test 4. The searching time for Test 1 is similar to Test 2 and that for Test 3 is similar to Test 4. Note that the searching time for Test 2 is longer than that for Test 4. The reason is because of the higher percentage of SNPs in Test 4. The common segments are shorter and the searching will stop earlier (refer to Step 3 of Algorithm 1). But then even for the slower case, searching a pattern of length 500 can still be done in less than 0.5ms which is reasonably fast. This shows that our solution is practical.

Table 1. Memory consumption of our solution

Chromosome	Text length	No. of Texts (r)	No. of SNPs	No. of common segments (k)	Total length of R segments (N)	Memory Ours	Memory BWT
(Test 1) Y	25M	50	0.0247M (0.1%)	0.0237M	1.2M	35M	958M
(Test 2) Y	25M	250	0.0247M (0.1%)	0.0237M	5.9M	61M	4.8G
(Test 3) 1	217M	50	1.06M (0.5%)	0.98M	51M	716M	8.3G
(Test 4) 1	217M	250	1.06M (0.5%)	0.98M	253M	2.7G	41.6G

Table 2. Searching performance (in milliseconds) for Test 2 (250 copies of Chromosome Y (25M)) and Test 4 (250 copies of Chromosome 1 (217M))

	Ave. Searching Time ($\times 10^{-3}$ seconds)									
Pattern length	50	100	150	200	250	300	350	400	450	500
Test 2	0.083	0.127	0.170	0.211	0.252	0.295	0.334	0.372	0.410	0.450
Test 4	0.079	0.115	0.145	0.169	0.192	0.204	0.219	0.231	0.241	0.251

Table 3. Comparison of our solution with RLCSA (RLCSA v.1 requires less memory with longer searching time; RLCSA v.2 can search faster but requires more memory. The last column shows the ratio of our average searching time over theirs based on RLCSA v.2).

No. of	Memory Comparison in M					Searching Time
Texts	Ours (A)	RLCSA v.1 (B)	(A)/(B)	RLCSA v.2 (C)	(A)/(C)	(Ours/RLCSA v.2)
25	102	113	90.3%	156	65.4%	13.2%
50	152	197	77.2%	277	54.9%	15.7%
75	203	277	73.3%	396	51.3%	17.6%
100	253	362	69.9%	514	49.2%	21.5%
125	304	445	68.3%	623	48.8%	23.2%

We also compare our solution with RLCSA for Model 2. We use simulated data with text length 25M, mutation rate 1%. We compare their performance using different number of texts. For the patterns, we follow [12] and use shorter patterns with length 10 to 50. For each pattern length, we randomly retrieve

1000 patterns from the texts and take the average searching time. RLCSA has different settings, to compare memory consumption, we use the option which uses the smallest amount of memory (with slower searching time). When comparing the searching time, we use their option which can search faster but use more memory. The results are shown in Table 3. The results show that we use less memory and can search faster than RLCSA.

5 Discussion and Conclusions

Recall that the searching time of our solution depends on $psc(P)$, the number of prefixes of P that is the suffix of some common segment C_i. If the sequences are random texts, we can show that $psc(P)$ is upper bounded by $O(\log n)$ as follows. Let $P[0..c^*]$ be the longest prefix in P such that it is a suffix of some common segment, say $C_i[s-c^*+1..s]$. It is obvious that $psc(P) = psc(C_i[s-c^*+1..s])$. Then, for any P, $psc(P) \leq max_{S' \in \Delta} psc(S')$ where Δ is the set of all suffixes over all common segments. Let $Y = max_{S' \in \Delta} psc(S')$. To bound Y, we consider a generalized suffix tree G for all common segments. For any $S' \in \Delta$, let u be the node representing S' in G, $psc(S') \leq$ node depth of u which is upper bounded by $O(\log n)$ for random texts [14].

We can further reduce the space complexity of our solution to $O(n + N + rk\log(rk))$ bits, closer to the lower bound of $O(n+N)$ bits with a slightly increase in searching time. Details will be shown in the full paper. For the evaluation of our solution, we also tried different mutation rates, the performance is consistent with the ones shown in the paper. We are now investigating how to extend the scheme for approximate pattern matching.

Acknowledgements. The project is partially supported by the Seed Funding Programme for Basic Research of the University of Hong Kong (200811159089). We would also like to thank the authors of [12] for providing us the programs of RLCSA.

References

1. Briniza, D., He, J., Zelikovsky, A.: Combinatorial search methods for multi-SNP disease association. In: EMBS, pp. 5802–5805 (2006)
2. Burrows, M., Wheeler, D.J.: A block-sorting lossless data compression algorithm. Technical Report 124, Digital Equipment Corporation, California (1994)
3. Emahazion, T., Feuk, L., Jobs, M., Sawyer, S.L., Fredman, D., Clair, D.S., Prince, J.A., Brookes, A.J.: SNP association studies in Alzheimer's disease highlight problems for complex disease analysis. Trends in Genetics 17(7), 407–413 (2001)
4. Ferragina, P., Manzini, G.: Opportunistic data structures with applications. In: FOCS, pp. 390–398 (2000)
5. Ferragina, P., Manzini, G.: An experimental study of an opportunistic index. In: SODA, pp. 269–278 (2001)
6. Grossi, R., Vitter, J.S.: Compressed suffix arrays and suffix trees with applications to text indexing and string matching. In: STOC, pp. 397–406 (2000)

7. Gusfield, D.: Algorithms on strings, trees, and sequences. Cambridge University Press, Cambridge (1997)
8. Kao, M.-Y. (ed.): Encyclopedia of Algorithms. Springer, Heidelberg (2008)
9. Lam, T.W., Sung, W.K., Tam, S.L., Wong, C.K., Yiu, S.M.: Compressed indexing and local alignment of DNA. Bioinformatics 24(6), 791–797 (2008)
10. Lippert, R.A.: Space-efficient whole genome comparisons with Burrows-Wheeler transforms. Journal of Computational Biology 12(4), 407–415 (2005)
11. Mäkinen, V., Navarro, G.: Succinct suffix arrays based on run-length encoding. Nordic Journal of Computing 12(1), 40–66 (2005)
12. Mäkinen, V., Navarro, G., Sirén, J., Välimäki, N.: Storage and retrieval of individual genomes. In: Batzoglou, S. (ed.) RECOMB 2009. LNCS, vol. 5541, pp. 121–137. Springer, Heidelberg (2009)
13. Nekrich, Y.: Orthogonal range searching in linear and almost-linear space. Computational Geometry: Theory and Applications 42(4), 342–351 (2009)
14. Szpankowski, W.: Probabilistic analysis of generalized suffix trees. In: CPM, pp. 1–14 (1992)

Online Scheduling on Two Uniform Machines to Minimize the Makespan with a Periodic Availability Constraint

Ming Liu[1], Chengbin Chu[1,2], Yinfeng Xu[1], and Lu Wang[3]

[1] School of Management, Xi'an Jiaotong University, Xi'an, Shaanxi Province, 710049, P.R. China
minyivg@gmail.com

[2] Laboratoire Génie Industriel, Ecole Centrale Paris, Grande Voie des Vignes, 92295 Châtenay-Malabry Cedex, France

[3] Shanghai Vocational School of CAAC, Shanghai, 200232, P.R. China

Abstract. We consider the problem of online scheduling on 2 uniform machines where one machine is periodically unavailable. The problem is online in the sense that when a job presents, we have to assign it to one of the 2 uniform machines before the next one is seen. Preemption is not allowed. The objective is to minimize makespan. Assume that the speed of the periodically unavailable machine is normalized to 1, while the speed of the other one is s. Given a constant number $\alpha > 0$, we also suppose that $T_u = \alpha T_a$, where T_u and T_a are the length of each unavailable time period and the length of the time interval between two consecutive unavailable time periods, respectively. In the case where $s \geq 1$, we show a lower bound of the competitive ratio $1 + \frac{1}{s}$ and prove that LS algorithm is optimal. We also show that for the problem $P2, M1PU|online, T_u = \alpha T_a|C_{max}$, LS algorithm proposed in [7] is optimal with a competitive ratio 2. After that, we give some lower bounds of competitive ratio in the case $0 < s < 1$. At last, we study a special case $P2, M1PU|online, T_u = \alpha T_a, non-increasing\ sequence|C_{max}$, where non-increasing sequence means that jobs arrive in a non-increasing order of their processing times. We show that LS algorithm is optimal with a competitive ratio $\frac{3}{2}$.

Keyword: Online scheduling; Makespan; Competitive analysis; Uniform machines; Periodic availability constraint.

1 Introduction

In the class scheduling, one of the basic assumptions made in deterministic scheduling is that all the useful information of the problem instance was known in advance. However, in practice, this assumption is usually not possible. Online scheduling becomes more and more concerned. In the literature of online scheduling, two online models have been widely researched [2]. The first one assumes that there are no release dates and that the jobs arrive in a list (one by

B. Chen (Ed.): AAIM 2010, LNCS 6124, pp. 191–198, 2010.

one or called over list). The online algorithm has to schedule (or assign) the first job in this list before it sees the next job in the list. The other model assumes that jobs arrive over time. There exists a release time (or called release date) with regard to each job. At each time when the machine is idle, the algorithm decides which one of the available jobs is scheduled, if any. In this paper, we consider the first model where jobs arrive in a list.

Online algorithm is developed to cope with online (scheduling) problems. For a certain online scheduling problem, we would like to find the optimal (or called best possible) algorithm. This algorithm has the best possible performance. In order to compare the performance of online algorithms, we need a tool to measure the performance of each algorithm. In the literature, competitive analysis [1] is such a tool to measure the performance of an online algorithm. Specifically, for any input job sequence I, let $C_{ON}(I)$ denote the makespan of the schedule produced by the online algorithm $\mathcal{A}_{ON}$ and $C_{OPT}(I)$ denote the makespan of an optimal schedule. We say that $\mathcal{A}_{ON}$ is ρ-competitive if

$$C_{ON}(I) \leq \rho C_{OPT}(I) + v$$

where v is a constant number. We also say that ρ is the competitive ratio of $\mathcal{A}_{ON}$.

If we have some online algorithms and their competitive ratios, do we need to further design new algorithms which may perform better with respect to their competitive ratios? In other words, we need a method to judge whether an algorithm is optimal or not. A lower bound (on competitive ratios of all online algorithms for the problem) serves such a purpose. For an online minimization problem, a lower bound means that there exists no online algorithm with a competitive ratio smaller than this bound. If an online algorithm's competitive ratio achieves this lower bound, this algorithm is called optimal and the corresponding lower bound is called tight.

For an online problem, a general idea is to first give a lower bound then prove the competitive ratio of an online algorithm. Then we can justify whether or not the proposed algorithm is optimal.

In the classical scheduling problem, we assume that machines are available simultaneously at all times. However, this availability assumption may not be true in practice [3]. In the real industry settings, machines may be unavailable because of preventive maintenances, periodical repairs and tool changes.

This paper studies the *online* version of this problem with 2 uniform machines with a periodic availability constraint. We are given a sequence of independent jobs which arrive in a list. We have to assign a job to one of 2 uniform machines before the next job shows up. When all information is available at one time before scheduling, the problem is called *offline*. In the offline settings, Lee [5] investigated some parallel machine scheduling problems where at least one machine is always available and each of the other machines has at most one unavailable period. He gave some results of competitive ratios for different objectives, such as minimization of total completion time and makespan. Liao et al. [4] considered a special case of one of the scheduling problems studied in [5]. They partitioned the problem into four subproblems, each of which was optimally solved. In the online settings, Tan et al. [6] considered the online scheduling on two identical machines with machine availability constraint to minimize makespan. They assumed that the unavailable periods of two machines do not overlap and proposed an optimal online algorithm with a competitive ratio $\frac{5}{2}$. Xu et al. [7] showed that for the problem of online scheduling on two identical machines where one machine is periodically unavailable with the objective of minimizing makespan, the competitive ratio of LS algorithm is 2.

We use machine-1 to denote the machine which is periodically unavailable and machine-s to denote the other one. Without loss of generality, the speeds of machine-1 and machine-s are normalized to 1 and s, respectively. Given a constant number $\alpha > 0$, we also suppose that $T_u = \alpha T_a$, where T_u and T_a are the length of each unavailable time period and the length of the time interval between two consecutive unavailable time periods, respectively. Our problem can be denoted by $Q2, M1PU|online, T_u = \alpha T_a|C_{max}$, where $M1PU$ denotes that machine-1 is periodically unavailable.

The remainder of this paper is organized as follows. In section 2, we present some notations and problem definition. In Section 3, we study the problem in the condition $s \geq 1$. We show a lower bound of $1 + \frac{1}{s}$ and prove that LS algorithm has a matching competitive ratio of $1 + \frac{1}{s}$. We also show that for the problem $P2, M1PU|online, T_u = \alpha T_a|C_{max}$, LS algorithm proposed in [7] is optimal with a competitive ratio 2. In section 4, we give some lower bounds of competitive ratio in the case $0 < s < 1$. In section 5, we study a special case $P2, M1PU|online, T_u = \alpha T_a, non-increasing\ sequence|C_{max}$ where non-increasing sequence means that jobs arrive in a non-increasing sequence processing time. We prove that LS algorithm is optimal with a competitive ratio $\frac{3}{2}$.

2 Notations and Problem Definition

We first give some notations:

- machine-1: the machine which is periodically unavailable and the speed of which is 1.
- machine-s: the machine is always available and the speed of which is s.
- p_j: the processing time of job J_j.
- T_a: the length of the time interval between two consecutive unavailable time periods.
- T_u: the length of each unavailable time period.

In this paper, we always assume that machine-1 begins with a available time period. On machine-1, we suppose $T_u = \alpha T_a$. We are given 2 uniform machines where machine-1 is periodically unavailable. The speeds of machine-1 and machine-s are 1 and s, respectively. A sequence of jobs $\sigma = \{J_1, J_2, ..., J_n\}$ which arrive online have to be scheduled irrevocably on one of the machines at the time of their arrivals. The new job shows up only after the current job is scheduled. We use p_j to denote the processing time of job J_j. p_i is not known until the previous job J_{j-1} has been scheduled, except job J_1. The schedule can be seen as a partition of job sequence σ into two subsets, denoted by S_1 and S_2, where S_1 and S_2 consist of jobs assigned to machine-1 and machine-s, respectively. Let L_1 and L_2 denote the completion times of the last jobs on machine-1 and machine-s, respectively. The makespan of the schedule is $max\{L_1, L_2\}$. The online problem can be written as:

Given σ, find S_1 and S_2 to minimize $max\{L_1, L_2\}$.

Let C_{ON} and C_{OPT} denote the makespan of online algorithm and offline optimal algorithm (for short, offline algorithm), respectively.

3 A Matching Lower and Upper Bound in the Case Where $s \geq 1$

In this section, we first show a lower bound on competitive ratios for the problem. Then we present an optimal online algorithm: *Greedy* algorithm.

3.1 A Lower Bound

Theorem 1. *In the case where $s \geq 1$, for the problem of scheduling 2 uniform machines with a periodic availability constraint, no online algorithm exists, whose competitive ratio is less than $1 + \frac{1}{s}$.*

Proof. Assume $T_a = 1$, so $T_u = \alpha$. Let ϵ be a sufficiently small positive number. We give a job sequence which consists of at most 4 jobs to show that the competitive ratios of all online algorithms cannot be less than $1 + \frac{1}{s}$. We begin with job J_1 with $p_1 = \epsilon$.

Case 1: J_1 is assigned to machine-1.
J_2 with $p_2 = 1$ arrives. If the online algorithm assigns J_2 to machine-1, then no jobs arrive. The optimal algorithm can schedule J_1 and J_2 on machine-1 and machine-s, respectively. Therefore, $C_{ON} \geq 2 + \alpha$, $C_{OPT} = \frac{1}{s}$ and $\frac{C_{ON}}{C_{OPT}} \geq (2+\alpha)s > 1+\frac{1}{s}$. Otherwise, namely if the online algorithm assigns J_2 to machine-s, then J_3 with $p_3 = s$ arrives. Thus, $C_{ON} \geq \frac{1+s}{s} = 1+\frac{1}{s}$. The optimal algorithm can assign J_1 to machine-s, J_2 to machine-1 and J_3 to machine-s. Therefore, $C_{OPT} = \frac{\epsilon+s}{s}$ and

$$\frac{C_{ON}}{C_{OPT}} \geq \frac{1+\frac{1}{s}}{\frac{\epsilon+s}{s}} = 1 + \frac{1}{s}, \quad \epsilon \to 0.$$

Case 2: J_1 is assigned to machine-s.
J_2 with $p_2 = \frac{\epsilon}{s}$ arrives. If the online algorithm schedules J_2 on machine-s, then no jobs come in the future. The optimal algorithm can schedule J_1 and J_2 on machine-s and machine-1, respectively. Therefore, $C_{ON} \geq \frac{\epsilon+\frac{\epsilon}{s}}{s} = \frac{\epsilon}{s}(1+\frac{1}{s})$, $C_{OPT} = \frac{\epsilon}{s}$ and $\frac{C_{ON}}{C_{OPT}} \geq 1 + \frac{1}{s}$. Otherwise, i.e., if the online algorithm assigns J_2 to machine-1, consider J_3 with $p_3 = 1$ and J_4 with $p_4 = s$. Then, we have

$C_{ON} \geq \frac{\epsilon+1+s}{s}$. In an optimal schedule, J_1, J_2 and J_4 are assigned to machine-s, while J_3 is assigned to machine-1. Therefore, $C_{OPT} = \frac{\epsilon+\frac{\epsilon}{s}+s}{s}$ and

$$\frac{C_{ON}}{C_{OPT}} \geq \frac{\epsilon+1+s}{\epsilon+\frac{\epsilon}{s}+s} = 1+\frac{1}{s}, \quad \epsilon \to 0.$$

Therefore, the theorem holds.

Corollary 1. *For the problem of online scheduling on two identical machines with one machine periodically unavailable, there is no online algorithm with a competitive ratio less than 2.*

3.2 An Optimal Online Algorithm

In this subsection we prove that LS algorithm is optimal with a competitive ratio $1+\frac{1}{s}$.

LS **algorithm** runs as follows:
Assign the job to the machine on which it can be finished as early as possible. If there is a choice, schedule the job on machine-s.

Theorem 2. *In the case where $s \geq 1$, for the problem of scheduling 2 uniform machines with a periodic availability constraint, LS algorithm is optimal with a competitive ratio $1+\frac{1}{s}$.*

Proof. For any job instance I, let $P(I) = \sum_{J_i \in I} p_i$ denote the total processing time of all jobs in I. We have $C_{ON} \leq \frac{P(I)}{s}$ and $C_{OPT} \geq \frac{P(I)}{1+s}$. Therefore,

$$\frac{C_{ON}}{C_{OPT}} \leq \frac{\frac{P(I)}{s}}{\frac{P(I)}{1+s}} = 1+\frac{1}{s}.$$

By Theorem 1, we know a lower bound of competitive ratio is $1+\frac{1}{s}$. Thus, the theorem follows.

Theorem 3. *(Dehua Xu [7]) For $P2, M1PU|online, T_u = \alpha T_a|C_{max}$, the competitive ratio of the LS algorithm is 2.*

Corollary 2. *For the problem $P2, M1PU|online, T_u = \alpha T_a|C_{max}$, LS algorithm proposed in [7] is optimal with a competitive ratio 2.*

For any instance which contains n jobs, the time complexity of LS algorithm is $O(n)$ since when a job is available LS algorithm has only one comparison.

An example:
Let $s = 2$ and then $1+\frac{1}{s} = 1.5$. Set $\alpha = 1$ and $T_a = T_u = 8$. We give a job sequence $\{J_1, J_2, J_3, J_4, J_5, J_6\}$ with $p_1 = 2, p_2 = 3, p_3 = 4, p_4 = 4, p_5 = 3, p_6 = 8$. For this instance, the optimal algorithm can assign J_6 to machine-1, and the others to machine-s. Therefore, $C_{OPT} = 8$. LS algorithm schedules J_1, J_2, J_4, J_5, J_6 on machine-s and J_3 on machine-1. Thus, $C_{ON} = 10$ and $\frac{C_{ON}}{C_{OPT}} = 1.25 < 1.5$.

4 A Lower Bound in the Case Where $0 < s < 1$

In this section, we show a lower bound lower bound in the case where $0 < s < 1$.

Theorem 4. *In the case $0 < s < 1$, for the problem of scheduling 2 uniform machines with a periodic availability constraint, no online algorithm exists, whose competitive ratio is less than: (1) 2 in the case where $\frac{1}{2} \leq s < 1$; (2) $\frac{1}{s}$ in the case where $s < \frac{1}{2}$.*

Proof. Assume $T_a = 1$, so $T_u = \alpha$. Let ϵ be a sufficiently small positive number.

(1) $\frac{1}{2} \leq s < 1$.
We give a job sequence which consists of at most 4 jobs to show that the competitive ratios of all online algorithms cannot be less than 2. We first give job J_1 with $p_1 = \epsilon$.

Case 1: J_1 is assigned to machine-1.
J_2 with $p_2 = 1$ arrives. If the online algorithm assigns J_2 to machine-1, then no job arrives. We have $C_{ON} \geq 2 + \alpha$. The optimal algorithm can schedule J_1 and J_2 on machine-s and machine-1, respectively. Therefore, $C_{OPT} = 1$ and $\frac{C_{ON}}{C_{OPT}} \geq 2 + \alpha > 2$. Otherwise, i.e., if the online algorithm assigns J_2 to machine-s, then J_3 with $p_3 = 1$ arrives. Thus, $C_{ON} \geq \frac{1+1}{s} = \frac{2}{s}$. The optimal algorithm can assign J_1 to machine-s, J_2 to machine-1 and J_3 to machine-s. Therefore,$C_{OPT} = \frac{\epsilon+1}{s}$. We have

$$\frac{C_{ON}}{C_{OPT}} \geq \frac{\frac{2}{s}}{\frac{\epsilon+1}{s}} = 2, \quad \epsilon \to 0.$$

Case 2: J_1 is assigned to machine-s.

J_2 with $p_2 = \frac{\epsilon}{s}$ arrives. If the online algorithm schedules J_2 on machine-s, then no job comes in the future. The optimal algorithm can schedule J_1 and J_2 on machine-s and machine-1, respectively. Therefore, $C_{ON} \geq \frac{\epsilon+\frac{\epsilon}{s}}{s} = \frac{\epsilon}{s}(1 + \frac{1}{s})$, $C_{OPT} = \frac{\epsilon}{s}$ and $\frac{C_{ON}}{C_{OPT}} \geq 1 + \frac{1}{s} > 2$. Otherwise, i.e., the online algorithm assigns J_2 to machine-1, consider J_3 with $p_3 = 1$ and J_4 with $p_4 = 1$. Then, we have $C_{ON} \geq \frac{\epsilon+2}{s}$. In an optimal schedule, J_1, J_2 and J_4 are assigned to machine-s, while J_3 is assigned to machine-1. Therefore, $C_{OPT} = \frac{\epsilon+\frac{\epsilon}{s}+1}{s}$ and

$$\frac{C_{ON}}{C_{OPT}} \geq \frac{\epsilon + 2}{\epsilon + \frac{\epsilon}{s} + 1} = 2, \quad \epsilon \to 0.$$

(2) $0 < s < \frac{1}{2}$.
We give a job sequence which consists at most 2 jobs to show that the competitive ratios of all online algorithms cannot be less than $\frac{1}{s}$. We first give job J_1 with $p_1 = \epsilon$.

Case 1: J_1 is assigned to machine-1.
J_2 with $p_2 = 1$ arrives. Therefore, $C_{ON} \geq \frac{1}{s}$. The optimal algorithm can schedule J_1 and J_2 on machine-s and machine-1, respectively. Thus, we have $C_{OPT} = 1$ and $\frac{C_{ON}}{C_{OPT}} = \frac{1}{s}$.
Case 2: J_1 is assigned to machine-s.
No job comes in the future. We have $C_{ON} \geq \frac{\epsilon}{s}$. The optimal algorithm schedules J_1 on machine-1. Thus, $C_{OPT} = \epsilon$ and $\frac{C_{ON}}{C_{OPT}} \geq \frac{1}{s}$.
The theorem follows.

5 A Special Case

In this section, we discuss a special case where $s = 1$ and jobs arrive in a non-increasing sequence. Non-increasing sequence means that jobs arrive in a non-increasing sequence (or order) of their processing times. We denote this problem by $P2, M1PU|online, non - increasing\ sequence|C_{max}$.

Theorem 5. *For the problem $P2, M1PU|online, non - increasing\ sequence|$ C_{max}, there is no online algorithm with a competitive ratio less than $\frac{3}{2}$.*

Proof. $s = 1$. We also use machine-s to denote the always available machine. Assume $T_a = 1$, so $T_u = \alpha$. Let ϵ be a sufficiently small positive number. We give a job sequence which consists of at most 4 jobs to show that the competitive ratios of all online algorithms cannot be less than $\frac{3}{2}$. We begin with job J_1 with $p_1 = \frac{1}{2} + \epsilon$.
Case 1: J_1 is assigned to machine-1.
J_2 with $p_2 = \frac{1}{2}$ arrives. If the online algorithm assigns J_2 to machine-1, then no jobs arrive. Note that J_2 is scheduled in the next available period on machine-1. The optimal algorithm can schedule J_1 and J_2 on machine-1 and machine-s, respectively. Therefore, $C_{ON} \geq 1 + \alpha + \frac{1}{2} = \frac{3}{2} + \alpha$, $C_{OPT} = \frac{1}{2} + \epsilon$ and $\frac{C_{ON}}{C_{OPT}} \geq \frac{\frac{3}{2}+\alpha}{\frac{1}{2}+\epsilon} > \frac{3}{2}$. Otherwise, i.e., if the online algorithm assigns J_2 to machine-s, then J_3 and J_4 with $p_3 = p_4 = \frac{1}{2}$ arrives. Thus, $C_{ON} \geq \frac{1}{2} + \frac{1}{2} + \frac{1}{2} = \frac{3}{2}$. The optimal algorithm can assign J_1 and J_2 to machine-s, J_3 and J_4 to machine-1. Therefore,$C_{OPT} = \frac{1}{2} + \epsilon + \frac{1}{2} = 1 + \epsilon$ and

$$\frac{C_{ON}}{C_{OPT}} \geq \frac{\frac{3}{2}}{1+\epsilon} = \frac{3}{2}, \quad \epsilon \to 0.$$

Case 2: J_1 is assigned to machine-s.
J_2 with $p_2 = \frac{1}{2}+\epsilon$ arrives. If the online algorithm schedules J_2 on machine-s, then no job comes in the future. The optimal algorithm can schedule J_1 and J_2 on machine-1 and machine-s, respectively. Therefore, $C_{ON} \geq 1 + 2\epsilon$, $C_{OPT} = \frac{1}{2} + \epsilon$ and $\frac{C_{ON}}{C_{OPT}} \geq \frac{1+2\epsilon}{\frac{1}{2}+\epsilon} = 2 > \frac{3}{2}$. Otherwise, i.e., the online algorithm assigns J_2 to machine-1. We consider J_3 and J_4 with $p_3 = p_4 = \frac{1}{2}$. Then, we have $C_{ON} \geq \frac{1}{2} + \epsilon + \frac{1}{2} + \frac{1}{2} = \frac{3}{2} + \epsilon$. In an optimal schedule, J_1, J_2 are assigned to machine-s, and J_3, J_4 are assigned to machine-1. Therefore, $C_{OPT} = \frac{1}{2} + \epsilon + \frac{1}{2} + \epsilon = 1 + 2\epsilon$ and

$$\frac{C_{ON}}{C_{OPT}} \geq \frac{\frac{3}{2}+\epsilon}{1+2\epsilon} = \frac{3}{2}, \quad \epsilon \to 0.$$

Therefore, the theorem holds.

In order to prove LS algorithm is optimal, we must restate some results in the literature.

LPT **algorithm** (Dehua Xu [7]):
Re-order all the job in non-increasing order of their processing times, i.e., $p_1 \geq p_2 \geq ... \geq p_n$; for $i = 1, ..., n$, assign J_i to the machine on which it can be finished as early as possible.

Theorem 6. *(Dehua Xu [7]) For* $P2, M1PU||C_{max}$*, the worst-case ratio of the LPT algorithm is* $\frac{3}{2}$*.*

By Theorem 6, we have the following result.

Corollary 3. *For the problem* $P2, M1PU|online, non-increasing\ sequence|C_{max}$*, LS algorithm is optimal with a competitive ratio* $\frac{3}{2}$*.*

Acknowledgements

This work is partially supported by NSF of China under Grants 70525004 and 70702030.

References

1. Borodin, A., El-Yaniv, R.: Online Computation and Competitive Analysis. Cambridge University Press, Cambridge (1998)
2. Pruhs, K., Sgall, J., Torng, E.: Online scheduling. In: Leung, J.Y.-T. (ed.) Handbook of Scheduling: Algorithms, Models, and Performance Analysis (2004)
3. Lee, C.Y., Lei, L., Piendo, M.: Current trends in deterministic scheduling. Ann. Oper. Res. 70, 1–41 (1997)
4. Liao, C.J., Shyur, D.L., Lin, C.H.: Makespan minimization for two parallel machines with an availability constraint. European Journal of Operational Research 160, 445–456 (2005)
5. Lee, C.Y.: Machine scheduling with an availability constraint. Jounral of Global Optimization 9, 395–416 (1996)
6. Tan, Z., He, Y.: Optimal online algorithm for scheduling on two identical machines with machine availability constraints. Information Processing Letters 83, 323–329 (2002)
7. Xu, D., Cheng, Z., Yin, Y., Li, H.: Makespan minimization for two parallel machines scheduling with a periodic availability constraint. Computers and Operations Research (2008), doi:10.1016/j.cor.2008.05.001

A New Smoothing Newton Method for Symmetric Cone Complementarity Problems*

Lixia Liu and Sanyang Liu

Department of Applied Mathematics, Xidian University, Xi'an, 710071, China
liulixia@mail.xidian.edu.cn

Abstract. Based on a new smoothing function, a smoothing Newton-type method is proposed for the solution of symmetric cone complementarity problems (SCCP). The proposed algorithm solves only one linear system of equations and performs only one line search at each iteration. Moreover, it does neither have restrictions on its starting point nor need additional computation which keep the iteration sequence staying in the given neighborhood. Finally, the global and Q-quadratical convergence is shown. Numerical results suggest that the method is effective.

Keywords: Symmetric cone; Complementarity; Smoothing Newton method; Global convergence; Q-quadratical convergence.

The symmetric cone complementarity problem (SCCP) is stated as follows: finding $x \in \mathcal{J}$ satisfying

$$x \in \mathcal{K}, F(x) \in \mathcal{K}, \langle x, F(x) \rangle = 0, \tag{1}$$

where $\mathcal{K} = \{x^2 | x \in \mathcal{J}\}$ is the symmetric cone in $\mathcal{J}$, $A = (\mathcal{J}, \circ, \langle \cdot, \cdot \rangle)$ is a Euclidean Jordan algebra (see Section 2 for definition) with $\mathcal{J}$ being a finite-dimensional vector space over the real field $\mathbb{R}$ endowed with the inner product $\langle \cdot, \cdot \rangle$, and " $\circ$ " denoting the Jordan product. Let $F : \mathcal{J} \to \mathcal{J}$ is a continuous function.

SCCP have wide applications in engineering, management science and other fields. Furthermore, they provide a unified framework for various complementarity problems, such as Semidefinite Complementarity Problems (SDCP), Second-order Cone Complementarity Problems (SOCCP), and Nonlinear Complementarity Problems (NCP). So they have attracted more and more attentions [2, 9, 14, 13, 7, 5, 6] recently. Various methods have been developed to solve them, such as interior-point algorithm [13], regularized smoothing Newton method [7], and smoothing Newton algorithm [5, 6]. Among them, the smoothing Newton methods are sometime superior to the class of interior-point methods since they do not require strict complementarity of the solution [11].

In this paper, a new smoothing function of the well known minimum function is given and then a smoothing Newton method for SCCP based on the new

* The project is supported by the NSF of China (NO. 60974082) and the Fundamental Research Funds for the Central Universities (JY10000970009).

B. Chen (Ed.): AAIM 2010, LNCS 6124, pp. 199–208, 2010.

smoothing function is proposed by modifying and extending the Qi-Sun-Zhou (QSZ) algorithm in [11]. It is shown that our method has the following good properties:

(i) The algorithm can start from an arbitrary initial point;

(ii) the method solves only one linear system of equations and performs only one line search at each iteration;

(iii) if an accumulation point of the iteration sequence satisfies a nonsingularity assumption, the whole iteration sequence converges to the accumulation point globally and locally quadratically without strict complementarity.

The paper is organized as follows. In the next Section, we list out some definitions and properties of the Euclidean Jordan algebra without proofs, give a new smoothing function of minimum function. In Section 3 we propose a smoothing Newton method for SCCP based on the new smoothing function. In Section 4, we analyze the global convergence and local quadratic convergence properties of our algorithm . Some preliminary numerical results are reported in Section 5 and some conclusions are given in Section 6.

1 Preliminaries

In this section, we review some preliminaries that will be used throughout this paper.

1.1 Euclidean Jordan Algebras

We first give a brief description to Euclidean Jordan algebras, which is a basic tool extensively used in this paper. Our presentation is concise and without proofs. For more details, see [1].

A Euclidean Jordan algebra is a triple $(\mathcal{J}, \circ, \langle\cdot,\cdot\rangle)$, where $(\mathcal{J}, \langle\cdot,\cdot\rangle)$ is a finite dimensional inner product space over R and $(x, y) \mapsto x \circ y : \mathcal{J} \times \mathcal{J} \to \mathcal{J}$ is a bilinear mapping satisfying the following conditions:

(i) $x \circ y = y \circ x$ for all $x, y \in \mathcal{J}$;

(ii)$x \circ (x^2 \circ y) = x^2 \circ (x \circ y)$ for all $x, y \in \mathcal{J}$,where $x^2 = x \circ x$, and

(iii)$\langle x \circ y, z\rangle = \langle y, x \circ z\rangle$ for $x, y, z \in \mathcal{J}$.

We call $x \circ y$ the Jordan product of x and y. In addition, thoughout the paper we assume that there is an unit element e such that $x \circ e = e \circ x = x$ for all $x \in \mathcal{J}$, which is called the *identity element* in $\mathcal{J}$. Although a Jordan algebra does not necessarily have an identity element.

The set of squares

$$\mathcal{K} := \{x^2 : x \in \mathcal{J}\}.$$

is called a *symmetric cone* [1]. That is, $\mathcal{K}$ is a self-dual closed convex cone with nonempty interior, and for any two elements $x, y \in \mathcal{J}$, there exists an invertible linear transformation $\Gamma : \mathcal{J} \to \mathcal{J}$ such that $\Gamma(\mathcal{J}) = (\mathcal{J})$ and $\Gamma(x) = y$.

An element $c \in \mathcal{J}$ is called *idempotent* if $c \circ c = c$. Idempotents c and c' are *orthogonal* if and only if $c \circ c' = 0$. An idempotent c is *primitive* if c cannot be written as a sum of two idempotents. We denote the maximum possible number of primitive orthogonal idempotents by r, which is called the *rank* of $\mathcal{J}$. In general, the rank of $\mathcal{J}$ is different from the dimension of $\mathcal{J}$. A set of idempotents $\{c_1, \cdots, c_r\}$ is called a *Jordan frame* if they are orthogonal to each other and $c_1 + \cdots + c_r = e$. For any element $x \in \mathcal{J}$, we have the following important spectral decomposition theorem.

Theorem 1. *[1] Let J be a Euclidean Jordan algebra with rank r. Then for any $x \in \mathcal{J}$, there exist a Jordan frame $\{c_1, c_2, \cdots, c_r\}$ and real numbers $\lambda_1(x), \lambda_2(x), \cdots, \lambda_r(x)$, arranged in decreasing order $\lambda_1(x) \geq \lambda_2(x) \geq \cdots \geq \lambda_r(x)$ such that*

$$x = \lambda_1(x)c_1 + \lambda_2(x)c_2 + \cdots + \lambda_r(x)c_r. \tag{2}$$

The numbers $\lambda_1(x), \lambda_2(x), \cdots, \lambda_r(x)$ are called the *eigenvalues* of x, which are uniquely determined by x and are continuous functions of x. The *trace* of x is defined by $\sum_{i=1}^{r} \lambda_i(x)$, denoted as $tr(x)$, which is a linear function of x. The *determinant* of x is defined by $\Pi_{i=1}^{r}\lambda_i(x)$, denoted as $det(x)$.

In a Jordan algebra $\mathcal{J}$, for an $x \in \mathcal{J}$, we define the corresponding *Lyapunov transformation* $L_x : \mathcal{J} \to \mathcal{J}$ by

$$L_x(y) = x \circ y.$$

We say that elements x and y *operator commute* if L_x and L_y commute, i.e.$L_xL_y = L_yL_x$. It is well known that x and y operator commute if and only if x and y have their spectral decompositions with respect to a common Jordan frame.

Lemma 1. *[8] Suppose that $c \in int\ \mathcal{K}$, then the inverse operator L_c^{-1} of linear mapping L_c exists. Moreover, L_c^{-1} is bounded and continuous on $\mathcal{J}$.*

We define the inner product $\langle \cdot, \cdot \rangle$ by $\langle x, y \rangle := tr(x \circ y)$ for any $x, y \in \mathcal{J}$. Thus, we define norm on $\mathcal{J}$ by

$$\|x\| := \sqrt{\langle x, x \rangle} = \sqrt{tr(x^2)} = \sqrt{\sum_{i=1}^{r} \lambda_i(x)^2}, \quad x \in \mathcal{J}.$$

And

$$[x]_+ = \sum_{i=1}^{r}[\lambda_i]_+c_i, \quad [x]_- = \sum_{i=1}^{r}[\lambda_i]_-c_i, \quad and \quad |x| = \sum_{i=1}^{r}|\lambda_i|c_i.$$

In particular, if $x \succeq 0$, then $\lambda_i \geq 0 (i = 1, 2, \cdots, r)$. When $x \succeq 0$, we define the (unique) sqrare root of x by $\sqrt{x} := \sum_{i=1}^{r} \sqrt{\lambda_i}c_i$.

1.2 A Smoothing Function

Definition 1. *[14] For a nondifferentiable function $h : R^n \to R^m$, we consider a function $h_\mu : R^n \to R^m$ with a parameter $\mu > 0$ that has the following properties: (i) h_μ is differentiable for any $\mu > 0$, (ii) $\lim_{\mu \downarrow 0} h_\mu(x) = h(x)$ for any $x \in R^n$. Such a function h_μ is called a smoothing function of h.*

In this subsection, we discuss a new smoothing function for SCCP. In [14], it has been shown that the minimum function $\phi_{min} : \mathcal{J} \times \mathcal{J} \to \mathcal{J}$ defined by

$$\phi_{min}(x, s) = x + s - \sqrt{(x-s)^2} \tag{3}$$

satisfies

$$\phi_{min}(x, s) = 0 \Leftrightarrow x \in \mathcal{K}, s \in \mathcal{K}, x \circ s = 0. \tag{4}$$

It is well-known that ϕ_{min} is nonsmooth. In the case of $\mathcal{K} = R^n_+$, the symmetric perturbed technique was originally proposed in [3,4]. By smoothing the symmetric perturbed function of ϕ_{min}, we now obtain the new vector-valued function $\phi : R_{++} \times \mathcal{J} \times \mathcal{J} \to \mathcal{J}$, defined by

$$\phi(\mu, x, s) = (\cos\mu + \sin\mu)(x + s) - \sqrt{(\cos\mu - \sin\mu)^2(x-s)^2 + 2\mu^2 e}. \tag{5}$$

In the following, we show that the function ϕ given in (5) is a smoothing function of ϕ_{min}.

Theorem 2. *(i) ϕ is globally Lipschitz continuous and strongly semismooth everywhere. Moreover, ϕ is continuously differentiable at any $(\mu, x, s) \in R_{++} \times \mathcal{J} \times \mathcal{J}$ with its Jacobian*

$$D\phi(\mu, x, s) = \begin{pmatrix} \phi'_\mu \\ \phi'_x \\ \phi'_s \end{pmatrix} = \begin{pmatrix} (\cos\mu - \sin\mu)(x+s) + L_\omega^{-1}[\cos 2\mu(x-s)^2 - 2\mu e] \\ (\cos\mu + \sin\mu)I - (\cos\mu - \sin\mu)^2 L_\omega^{-1} L_{(x-s)} \\ (\cos\mu + \sin\mu)I + (\cos\mu - \sin\mu)^2 L_\omega^{-1} L_{(x-s)} \end{pmatrix} \tag{6}$$

where

$$\omega := \omega(\mu, x, s) = \sqrt{(\cos\mu - \sin\mu)^2(x-s)^2 + 2\mu^2 e}. \tag{7}$$

(ii) $\lim_{\mu \downarrow 0} \phi(\mu, x, s) = \phi_{min}(x, s)$ for any $(x, s) \in \mathcal{J} \times \mathcal{J}$. That is $\phi(\mu, x, s)$ is a smoothing function of $\phi_{min}(x, s)$.

Proof. By Proposition 3.4 in [12], it is easy to know that ϕ is globally Lipschitz continuous, strongly semismooth (for its definition, please refer to [10]) everywhere, and continuously differentiable at any $(\mu, x, s) \in R_{++} \times \mathcal{J} \times \mathcal{J}$. Now we prove (6). For any $(\mu, x, s) \in R_{++} \times \mathcal{J} \times \mathcal{J}$, it is easy to see $\omega \in \text{int}\mathcal{K}$ by (7). So L_ω is invertible by Lemma 1. From (7), we know

$$\omega^2 = (\cos\mu - \sin\mu)^2(x-s)^2 + 2\mu^2 e.$$

By finding the derivative on both sides of the above relation, where the chain rule for differentiation is used, we obtain

$$D\omega(\mu, x, s) = \begin{pmatrix} -L_\omega^{-1}[\cos 2\mu(x-s)^2 - 2\mu e]) \\ (\cos\mu - \sin\mu)^2 L_\omega^{-1} L_{(x-s)} \\ -(\cos\mu - \sin\mu)^2 L_\omega^{-1} L_{(x-s)} \end{pmatrix}. \tag{8}$$

Then, the desired Jacobian formula is obtained.

Next, we show (ii). By Theorem 1, we denote $x+s = \sum_{i=1}^{r} \alpha_i u_i$ and $(x-s)^2 = \sum_{i=1}^{r} \beta_i v_i$, then ϕ_{min} and ϕ can be expressed as $\phi_{min} = \sum_{i=1}^{r} \alpha_i u_i - \sum_{i=1}^{r} \sqrt{\beta_i} v_i$. Moreover, there is

$$\begin{aligned} \phi(\mu, x, s) &= \sum_{i=1}^{r}((\cos\mu + \sin\mu)\alpha_i)u_i - \sum_{i=1}^{r} \sqrt{(\beta_i(\cos\mu - \sin\mu)^2 + 2\mu^2)} v_i \\ &:= \sum_{i=1}^{r} \alpha_i(\mu) u_i - \sum_{i=1}^{r} \sqrt{\beta_i(\mu)} v_i. \end{aligned} \tag{9}$$

It is easy to see that $\lim\limits_{\mu\downarrow 0} \alpha_i(\mu) = \alpha_i$ and $\lim\limits_{\mu\downarrow 0} \beta_i(\mu) = \beta_i$, so $\phi(\mu, x, s)$ is a smoothing function of $\phi_{min}(x, s)$.

2 Algorithm for SCCP

Our smoothing Newton method aims to reformulate the SCCP as a nonlinear system of equations

$$H(\mu, x, y) := \begin{pmatrix} \mu \\ F(x) - y \\ \phi(\mu, x, y) \end{pmatrix} = 0, \tag{10}$$

and then apply Newton's method to the system. According to (4), we know (x^*, y^*) is the solution of SCCP if and only if $(0, x^*, y^*)$ is a root of the system $H(z) = 0$. So let us talk about some properties of $H(z)$. For any $z = (\mu, x, y) \in R \times \mathcal{J} \times \mathcal{J}$, let

$$\phi(z) := \phi(\mu, x, y), \quad \phi_0(z) := \phi(0, x, y)$$

and

$$\theta(z) = \|H(z)\|^2 = \mu^2 + \|F(x) - y\|^2 + \|\phi(z)\|^2.$$

Let $\gamma \in (0, 1)$ and denote $\beta : R \times \mathcal{J} \times \mathcal{J} \to R$ by

$$\beta(z_k) := \gamma \min\{1, \theta(z_k)\}. \tag{11}$$

The smoothing Newton method is defined as follows.

Algorithm 1. ***Step 0.*** *Choose $\delta, \sigma \in (0, 1)$. Let $z_0 := (\mu_0, x_0, y_0) \in R_{++} \times \mathcal{J} \times \mathcal{J}$ be an arbitrary point and $\bar{z} = (\mu_0, 0, 0)$. Choose $\gamma \in (0, 1)$ such that $\gamma\mu_0 < 1$. Set $k := 0$.*

Step 1. *If $H(z_k) = 0$, stop. Otherwise, let $\beta_k := \beta(z_k)$.*

Step 2. *Compute $\Delta z_k := (\Delta\mu_k, \Delta x_k, \Delta y_k) \in R \times \mathcal{J} \times \mathcal{J}$ by*

$$H(z_k) + DH(z_k)\Delta z_k = \beta_k \bar{z}. \tag{12}$$

Step 3. *Let l_k be the smallest nonnegative integer l such that*

$$\theta(z_k + \delta^l \Delta z_k) \le [1 - 2\sigma(1 - \gamma\mu_0)\delta^l]\theta(z_k). \tag{13}$$

Set $\lambda_k := \delta^{l_k}$ and $z_{k+1} := z_k + \lambda_k \Delta z_k$.

Step 4. *Set $k := k + 1$ and go to Step 1.*

For analyzing our algorithm, we study some properties of the function $H(z)$ defined by (10). Moreover, we derive the computable formula for the Jacobian of the function $H(z)$ and give the condition for the Jacobian to be invertible. Firstly, we give a lemma which will be used in the following.

Lemma 2. *[6] If $a\in\mathcal{K}, b\in\mathcal{K}$, and $a\circ b=\alpha e$, where $\alpha>0$, then $\langle (L_b^{-1}L_a)(w), w\rangle \geq 0$ for any $w\in\mathcal{J}$, and $\langle (L_b^{-1}L_a)(w), w\rangle = 0$ implies that $w=0$.*

Theorem 3. *Let $z=(\mu,x,y)\in R\times\mathcal{J}\times\mathcal{J}$, F is a continuously differentiable monotone function, and H be defined by (10), then the following results hold.*
(i) $H(z)$ is strongly semismooth everywhere in $R\times\mathcal{J}\times\mathcal{J}$ and continuously differentiable at any $z=(\mu,x,y)\in R_{++}\times\mathcal{J}\times\mathcal{J}$ with its Jacobian

$$DH(\mu,x,y):=\begin{pmatrix}1 & 0 & 0\\ 0 & DF(x) & -I\\ \phi'_\mu & \phi'_x & \phi'_y\end{pmatrix}, \tag{14}$$

where the expressions of ϕ'_μ, ϕ'_x and ϕ'_y are in (6).
(ii)DH is invertible for any $z\in R_{++}\times\mathcal{J}\times\mathcal{J}$.

Proof. It is obvious that (i) is true by Theorem 2. Now let we prove (ii). Fix any $\mu>0$ and let $\triangle z:=(\triangle\mu,\triangle x,\triangle y)\in R\times\mathcal{J}\times\mathcal{J}$ be a vector in the null space of $DH(z)$, it suffices to show that $\triangle z=0$. According to (14), we have

$$\begin{aligned}\triangle\mu=0,\\ DF(x)\triangle x-\triangle y=0,\\ (\cos\mu+\sin\mu)(\triangle x+\triangle y)-L_\omega^{-1}[(\cos\mu-\sin\mu)^2(x-y)(\triangle x-\triangle y)]+\phi'_\mu\triangle\mu=0.\end{aligned} \tag{15}$$

By the second equation in (15) and the monotonicity of F, we have

$$\langle\triangle x,\triangle y\rangle=\langle x, DF(x)\triangle x\rangle\geq 0. \tag{16}$$

In addition, by the other two equations in (15), we have

$$L_\omega[(\cos\mu+\sin\mu)(\triangle x+\triangle y)]-(\cos\mu-\sin\mu)^2(x-y)(\triangle x-\triangle y)=0,$$

which implies

$$\begin{aligned}&[\omega-(\cos\mu-\sin\mu)(x-y)](\triangle x\cos\mu+\triangle y\sin\mu)\\ &+[\omega+(\cos\mu-\sin\mu)(x-y)](\triangle x\sin\mu+\triangle y\cos\mu)=0,\end{aligned}$$

i.e.,

$$L_{\omega-(\cos\mu-\sin\mu)(x-y)}(\triangle x\cos\mu+\triangle y\sin\mu)+L_{\omega+(\cos\mu-\sin\mu)(x-y)}(\triangle x\sin\mu+\triangle y\cos\mu)=0.$$

Therefore, we have

$$\sin\mu\triangle x+\cos\mu\triangle y=-L^{-1}_{\omega+(\cos\mu-\sin\mu)(x-y)}L_{\omega-(\cos\mu-\sin\mu)(x-y)}(\triangle x\cos\mu+\triangle y\sin\mu), \tag{17}$$

and

$$\langle L^{-1}_{\omega+(\cos\mu-\sin\mu)(x-y)} L_{\omega-(\cos\mu-\sin\mu)(x-y)}(\triangle x\cos\mu + \triangle y\sin\mu), \triangle x\cos\mu + \triangle y\sin\mu\rangle$$
$$= -\langle \triangle x\sin\mu + \triangle y\cos\mu, \triangle x\cos\mu + \triangle y\sin\mu\rangle$$
$$= -\sin\mu\cos\mu(\|\triangle x\|^2 + \|\triangle y\|^2) - (\sin^2\mu + \cos^2\mu)\langle \triangle x, \triangle y\rangle \leq 0 \tag{18}$$

where the last inequality comes from (16). By the definition of ω, it easy to know that $\omega - (\cos\mu - \sin\mu)(x-y) \in \mathcal{K}$, $\omega + (\cos\mu - \sin\mu)(x-y) \in \mathcal{K}$ and

$$[\omega - (\cos\mu - \sin\mu)(x-y)] \circ [\omega + (\cos\mu - \sin\mu)(x-y)] = 4\mu^2 e. \tag{19}$$

Then we have $\triangle x\cos\mu + \triangle y\sin\mu = 0$ by Lemma 2. This together with (17) implies that $\triangle x\sin\mu + \triangle y\cos\mu = 0$. Furthermore, the above two equalitics imply that $\triangle x = 0$ and $\triangle y = 0$. Thus the null space of $H(z)$ consists of only the origin, and hence $DH(z)$ is invertible.

Theorem 4. *Suppose that F is a continuously differentiable monotone function, then $\mu_k \in R_{++}$ for all $k > 1$ and the Algorithm 1 is well-defined.*

Proof. Since F is a continuously differentiable monotone function, it follows from Theorem 3 that $DH(z_k)$ is nonsingular for any $\mu_k > 0$. While, by the equation (12), we have

$$\mu_{k+1} = (1 - \lambda_k)\mu_k + \lambda_k\beta_k\mu_0 > 0, \tag{20}$$

since $\mu_0 > 0$, then $\mu_k > 0$ for all $k > 1$. Hence Step 2 is well-defined at the k-th iteration. Now we show that Step 3 is well-defined. For all $\alpha \in (0, 1]$, define

$$h(\alpha) := \theta(z_k + \alpha\Delta z_k) - \theta(z_k) - \alpha(\theta'(z_k))^T\Delta z_k. \tag{21}$$

By the similar analysis of (20), we know $\mu_k + \alpha\Delta\mu_k > 0$. Associated with Theorem 3, we know that $H(\cdot)$ is continuously differentiable around z_k, so is $\theta(z_k)$. Then it is easy to know

$$\|h(\alpha)\| = o(\alpha) \tag{22}$$

and

$$\begin{aligned} \theta(z_k + \alpha\Delta z_k) &= \theta(z_k) + \alpha\theta'(z_k)^T\Delta z_k + h(\alpha) \\ &= (1 - 2\alpha)\theta(z_k) + 2\alpha H(z_k)^T\beta_k\bar{z}_k + h(\alpha) \\ &\leq (1 - 2\alpha)\theta(z_k) + 2\alpha\gamma\mu_0\theta(z_k) + o(\alpha) \\ &= [1 - 2\alpha(1 - \gamma\mu_0)]\theta(z_k) + o(\alpha). \end{aligned} \tag{23}$$

the first equality comes from (12) and $\theta'(z_k) = 2H(z_k)^T DH(z_k)$, the first inequality followed by the Hölder inequality and the definition of β_k. So the inequality (23) implies that there exists a constant $\bar{\alpha} \in (0, 1]$ such that

$$\theta(z_k + \delta^l\Delta z_k) \leq [1 - 2\sigma(1 - \gamma\mu_0)\alpha]\theta(z_k)$$

holds for any $\alpha \in (0, \bar{\alpha}]$. This demonstrates that Step 3 is well-defined at the k-th iteration.

3 Global and Local Convergence

Lemma 3. *Suppose that F is a continuously monotone function, and that H is defined by (3.2). Then $H(\mu, x, y)$ is coercive in any $(\mu, x, y) \in R_+ \times \mathcal{J} \times \mathcal{J}$, i.e., $\lim_{\|(\mu,x,y)\| \to \infty} \|H(\mu, x, y)\| = +\infty$.*

The proof can be found in [6].

Lemma 4. *Suppose that F is a continuously differentiable monotone function and that $\{z_k\}$ is the iteration sequence generated by Algorithm 1. Then $z_k \in \Omega$ for any $k \geq 0$, where*

$$\Omega = \{z = (\mu, x, y) \in R \times \mathcal{J} \times \mathcal{J} : \mu \geq \beta(z)\mu_0\}.$$

Proof. By the definition of β, we know $z_0 \in \Omega$. Suppose that $z_k \in \Omega$, it is sufficient to prove that $z_{k+1} \in \Omega$. We know that

$$\mu_{k+1} = (1 - \lambda_k)\mu_k + \lambda\beta_k\mu_0 \geq (1 - \lambda_k)\beta_k\mu_0 + \lambda\beta_k\mu_0 = \beta_k\mu_0,$$

the first inequality comes from $\mu_k \in \Omega$. In order to prove $\mu_{k+1} \in \Omega$, it just need to show that $\beta_k \geq \beta_{k+1}$. By the Step 3, we know $\theta(z_{k+1}) < \theta(z_k)$. Then $\beta_k \geq \beta_{k+1}$ comes from the definition of β_k.

Theorem 5. *Suppose that F is a continuously differentiable monotone function and the solution set of SCCP is nonempty and bounded and $z^* := (\mu^*, x^*, y^*)$ is an accumulation point of the sequence $\{z_k\}$ generated by Algorithm 1. Then,*
(i) z^* is a solution of $H(z) = 0$;
(ii) if all $V \in \partial H(z^*)$ are nonsingular, then,
(a)$\lambda_k = 1$ for all $\{z_k\}$ sufficiently close to z^*;
(b)the whole sequence $\{z_k\}$ converges to z^* Q-quadratically if ∇F is Lipschitz continuous, i.e.

$$\|z_{k+1} - z^*\| = O(\|z_k - z^*\|^2) \quad \textit{and} \quad \mu_{k+1} = O(\mu_k^2).$$

Proof. Using Lemma 4, we can obtain that (i) holds in a similar way as in Theorem 3.2 of [15]. Using (i) and Theorem 3, we can prove (ii) similarly as in Theorem 8 of [11]. For brevity, we omit the details here.

4 Numerical Results

In this section, we have conducted some numerical experiments to evaluate the efficiency of Algorithm 1. All experiments were done at a PC with 3.06GHz CPU and 0.99G memory. The operating system was Windows XP and the implementations were done in MATLAB 7.0.1. In our experiments, we test the SOCCP, specially case of SCCP, of the following form: Finding $x \in R^n$ such that

$$\langle F(x), x\rangle = 0, x \in \mathcal{K}, F(x) = Mx + q \in \mathcal{K},$$

where $\mathcal{K}$ is a second order cone, i.e. $\mathcal{K} = \{(x_0, x_1) \in R^1 \times R^{n-1} : x_0 \geq \|x_1\|\}$, and $M \in R^n \times R^n, q \in R^n$ were generated by the following procedure. Elements of q were chosen randomly from the interval [-1, 1] and M is a symmetric sparse metric (the density of M is the nonzero density)whose eigenvalues are chosen random from the interval [0, 1] so that M is positive semidefinite.

The parameters used in this test were as follows: $\gamma = 0.01\min\{1, 1/\|H(z_0)\|\}$, $\eta = \gamma\mu_0, \sigma = 0.001, \delta = 0.9$, and the stopping criterion was set as $\theta(z) \leq 10^{-6}$. The CPU time is in seconds. The testing problems are generated by using the method mentioned above. The starting points (μ_0, x_0, y_0) of the algorithm in the experiments are all chosen randomly.

In the first experiment, we generated 10 test problems with various problem sizes for each nonzero density 1% and the results were showed in Table 1. In this Table and the following Table 2, **Iteration, cpu** are the averages of 10 trials. In the second experiments, we generated 10 test problems with n = 1000 for each nonzero density 0.5%, 5%, 10%, 20%, 50% and 80%, the results were summarized in Table 2. The above results indicate that Algorithm 1 performs very well. We also observed similar results for other examples.

Table 1. Numerical results for the affine SOCCP of various problem size (n)

n	Iteration	cpu(s)	n	Iteration	cpu(s)
100	5.1	0.0438	600	5.9	4.2783
200	5.6	0.2436	800	6.0	8.7530
400	6.0	1.5406	1000	6.0	15.6734

Table 2. Numerical results for the affine SOCCP with different degrees of sparsity Dens (%).

Dens(%)	Iteration	cpu(s)	Dens(%)	Iteration	cpu(s)
0.5	6.0	15.6248	20	6.0	15.9154
5	6.0	15.7202	50	6.3	16.6999
10	6.1	15.8798	80	6.0	15.7002

5 Conclusions

In this paper, the symmetric cone complementarity (SCCP) problem was discussed in detail. We give a smoothing Newton method for SCCP based on a new smoothing function of the minimum function and prove that the given algorithm is globally and locally Q-quadratically convergent. Since ϕ'_x in (6) is not necessarily a diagonal matrix which is true in the case of NCP, we can not have the nonsingular of DH under the assumption of F being a P_0 function. It is yet unknown whether the assumption on F is been weakened, on which we will keep an eye in the future.

Acknowledgements. The authors are grateful to the associate editor and the referees for their valuable comments on the paper.We thank some of the pioneers of the smoothing Newton method field: J. Sun, D. Sun, L. Qi, Z. Huang and L. Zhang.

References

1. Faraut, J., Koranyi, A.: Analysis on Symmetric Cones. Clarendon Press, Oxford (1994)
2. Han, D.R.: On the coerciveness of some merit functions for complementarity problems over symmetric cones. J. Math. Anal. Appl. 336, 727–737 (2007)
3. Huang, Z.H., Han, J.Y., Xu, D.C., et al.: The non-interior continuation methods for solving the function nonlinear complementarity problem. Sci. in China (Series A) 44, 1107–1114 (2001)
4. Huang, Z.H., Han, J., Chen, Z.: Predictor-Corrector Smoothing Newton Method, Based on a New Smoothing Function, for Solving the Nonlinear Complementarity Problem with a Function. J. Optim. Theory Appl. 117, 39–68 (2003)
5. Huang, Z.H., Liu, X.H.: Extension of smoothing Newton algorithms to solve linear programming over symmetric cones. Technique Report, Department of Mathematics, School of Science, Tianjin University, China (2007)
6. Huang, Z.H., Tie, N.: Smoothing algorithms for complementarity problems over symmetric cones. Comput. Optim. Appl. 45, 557–579 (2010)
7. Kong, L.C., Sun, J., Xiu, N.H.: A regularized smoothing Newton method for symmetric cone complementarity problems. SIAM J. on Optim. 9, 1028–1047 (2008)
8. Liu, Y.J., Zhang, L.W., Liu, M.J.: Extension of smoothing functions to symmetric cone complementarity problems. Appl. Math. A Journal of Chinese Universities B 22, 245–252 (2007)
9. Liu, Y.J., Zhang, L.W., Wang, Y.H.: Some properties of a class of merit functions for symmetric cone complementarity problems. Asia-Pacific J. Oper. Res. 23, 473–495 (2006)
10. Mifflin, R.: Semismooth and semiconvex functions in constrained optimization. SIAM J. Control Optim. 15, 957–972 (1977)
11. Qi, L., Sun, D., Zhou, G.: A new look at smoothing Newton methods for nonlinear complementarity problems and box constrained variational inequalities. Math. Program. 87, 1–35 (2000)
12. Sun, D., Sun, J.: Löwner operator and spectral functions in Euclidean Jordan algebras. Math. Oper. Res. 33, 421–445 (2008)
13. Schmieta, S., Alizadeh, F.: Extension of primal-dual interior-point algorithms to symmetric cones. Math. Program. 96, 409–438 (2003)
14. Tao, J., Gowda, M.S.: Some P-properties for nonlinear transformations on Euclidean Jordan algebras. Mathe. Oper. Res. 30, 985–1004 (2005)
15. Zhang, L., Gao, Z.: Superlinear/quadratic one-step smoothing Newton method for P0-NCP without strict complementarity. Math. Meth. Oper. Res. 56, 231–241 (2002)

Approximation Algorithms for Scheduling with a Variable Machine Maintenance

Wenchang Luo[1,2], Lin Chen[1], and Guochuan Zhang[3]

[1] Department of Mathematics, Zhejiang University, Hangzhou, 310027, China
[2] Faculty of Science, Ningbo University, Ningbo, 315211, China
luowenchang@nbu.edu.cn
[3] College of Computer Science, Zhejiang University, Hangzhou, 310027, China
zgc@zju.edu.cn

Abstract. In this paper, we investigate the problem of scheduling weighted jobs on a single machine with a maintenance whose starting time is prior to a given deadline and whose duration is a nondecreasing function of the starting time. We are asked not only to schedule the jobs but also the maintenance such that the total weighted job completion time is minimum. The problem is shown to be weakly NP-hard. In the case that the duration of the maintenance is a concave (and nondecreasing) function of its starting time, we provide two approximation algorithms with approximation ratio of 2 and at most $1 + \sqrt{2}/2 + \epsilon$, respectively.

Keywords: Approximation algorithms; scheduling with maintenance; total weighted completion time.

1 Introduction

Scheduling problems with preventive maintenance on machines have received considerable attention in the last two decades. In such models the machines are not always available. One is asked to schedule the given jobs by considering the unavailable period of the machines. There are two parameters to define a maintenance: its starting time and its duration. Relying on them we have different problems. The model that both the parameters are fixed in advance has been extensively studied [2,5,6,7,11,15]. The model assuming that the starting time for maintenance can be chosen but the duration is fixed was also well studied [9,12]. We refer interested readers to the survey papers by Schmidt [14] and Ma et al. [10] for more details.

Very recently a seminal paper by Kubzin and Strusevich [8] dealt with the more general scenario in which both parameters are variables. More precisely the starting time of the maintenance is given within a time window and the duration is a nondecreasing function of the starting time. In other words, the duration will not be shorter if the maintenance starts later, and we have to arrange the maintenance by a given time point. They considered both flow shop

B. Chen (Ed.): AAIM 2010, LNCS 6124, pp. 209–219, 2010.

and open shop for the two-machine case, in which each machine has a maintenance. The goal is to minimize the makespan. They showed that the open shop problem is polynomially solvable. By contrast, the flow shop problem is weakly NP-hard even for the case that the duration of maintenance is a linear function of the starting time. They further derived an FPTAS (Fully Polynomial Time Approximation Scheme) and a fast 3/2-approximation algorithm. Mosheiov and Sidney [13] studied the problem on a single machine with the following objective functions: makespan, flow time, maximum lateness, total earliness, tardiness and due-date cost, and number of tardy jobs. All of these problems were shown to be polynomially solvable. In this paper, motivated by the previous work of Kubzin and Strusevich [8] and Mosheiov and Sidney [13], we consider the objective minimizing the total weighted completion time on a single machine. The problem is formally presented below.

Problem statement. Given a set $J = \{J_1, J_2, \cdots, J_n\}$ of jobs, each job J_i has a processing time p_i and a weight w_i. There is a maintenance that must be performed prior to a given deadline s_d. If the maintenance starts at time $s \leq s_d$, its duration is defined by $f(s)$ which is a nondecreasing function of time s. The goal is to well arrange the maintenance and schedule the set of jobs such that the total weighted completion time $\sum_i w_i c_i$ is minimized. We extend the 3-field notation $\alpha|\beta|\gamma$ by Graham et al. [4] and denote our problem as $1|VM|\sum_i w_i c_i$, where VM stands for the variable maintenance. Let $P = \sum_i p_i$. Without loss of generality, we can assume that $s_d < P$, since otherwise our problem becomes trivial (all jobs can be completed before the maintenance starts). Analogously, by $1|M|\sum_i w_i c_i$ we can denote the special case of our problem, in which the duration of the maintenance is fixed.

Our model has many applications in real-world manufacturing where maintenance includes cleaning, recharging, refilling, or partial replacement of tools or parts that have been subject to essential wear. Delaying the maintenance may results in a longer duration. On the other hand, our problem can also be viewed as a two-agent competitive scheduling problem [1]. We may assume that the jobs belong to agent A while agent B is responsible for performing the maintenance. The objective is to minimize the total weighted completion time of agent A subject to the starting time of agent B performing maintenance prior to a given deadline, provided that the processing time of the maintenance is a variable.

Related work. The problem minimizing the total weighted completion time on a single machine with a fixed maintenance period was considered in [5] and [6]. Kacem [5] presented a 2-approximation algorithm and the bound is tight. Kellerer et al. [6] gave a $(2+\epsilon)$-approximation algorithm by employing an FPTAS for the knapsack problem. For the case that jobs are resumable Wang et al. [15] designed an approximation algorithm with a tight bound of two. Megow and Verschae [11] further improved the result by giving a $((1+\sqrt{5})/2+\epsilon)$-approximation algorithm. Finally, Kellerer and Strusevich [7] showed that both of the above two problems admit an FPTAS. Mosheiov and Sarig [12] dealt with the problem $1|M|\sum_i w_i c_i$, in which the duration of the maintenance is fixed while the starting time of the maintenance has a deadline. They showed that a WSPT-based heuristic algorithm may work arbitrarily bad, and presented a pseudo-polynomial time algorithm via dynamic programming. Interestingly, it was pointed out in [7] that this problem is actually equivalent to the problem in [15]. Lee and Chen [9] investigated parallel machine scheduling in which there is a maintenance on each machine.

Our results. We first point out that our problem is weakly NP-hard by designing a pseudo-polynomial time algorithm for any function $f(s)$ (not necessary to be nondecreasing). Then we turn to approximation algorithms, assuming that $f(s)$ is nondecreasing and concave. There are two critical points to be handled. One is to guess a time interval when the maintenance starts in an optimal schedule, while another is to select those jobs which are scheduled before the maintenance. To this end we propose a number of candidate schedules, among which we select the best one. It is proved that the resulting schedule has an approximation ratio of two. Finally we improve the algorithm by employing an FPTAS for the scheduling problem with a common deadline and achieve a better bound of $1 + \sqrt{2}/2 + \epsilon$. The remainder of the paper is organized as follows. In Section 2, we simply state the complexity result via dynamic programming. Sections 3 and 4 present the two approximation algorithms. Some concluding remarks are given in Section 5.

2 The Complexity

Kellerer and Strusevich [7] have pointed out that NP-hardness of the problem $1|M|\sum_i w_i c_i$. It is clear that our problem $1|VM|\sum_i w_i c_i$ is NP-hard. In the following we introduce a pseudo-polynomial time algorithm based on dynamic programming. Assume all input data of the problem are integers. Recall that $s_d < P = \sum p_i$. There are at most P possible starting time points for the maintenance. After we select a time point for the maintenance we only need to partition the jobs into two subsets S_1 and S_2. S_1 is scheduled before the maintenance while S_2 after the maintenance. It is worthy to note that in an optimal schedule the jobs before (after) the maintenance must obey a WSPT (weighted shortest processing time) order. Thus, as we get S_1 and S_2 the schedule is clear.

First we sort the jobs in a WSPT order, i.e., $w_1/p_1 \geq w_2/p_2 \geq \cdots \geq w_n/p_n$. Let $F(i, v, s)$ denote the minimum total weighted completion time of jobs $J_1, J_2, \cdots, J_i$, where s is the starting time of the maintenance and v is the total processing time of jobs scheduled before s. Let $A_i = \sum_{j=1}^{i} p_j$.

For $s = 0, 1, \cdots, s_d$ and $v \leq s$, let $F(0, v, s) = 0$. The recursive equation for $i = 1, 2, \cdots, n$, $s = 0, 1, \cdots, s_d$, and $v \leq s$, is defined as

$$F(i, v, s) = \min\{F(i-1, v-p_i, s) + w_i v, F(i-1, v, s) + w_i(s + f(s) + A_i - v)\}.$$

Moreover, we define $F(i, v, s) = \infty$ if $v < 0$.

During the recursion, as a new job is added, it always has the smallest (weight)/(processing time) ratio. No matter it is assigned to S_1 or S_2, it is always the last job in the class. Computing the best schedule becomes much simpler. It is easy to verify the correctness of the dynamic program. The optimal value is determined by minimizing $F(n, v, s)$ over $s = 0, 1, \cdots, s_d; v \leq s$. The running time of the procedure is bounded by $O(ns_d^2)$ (or $O(nP^2)$, which is pseudo-polynomial. Therefore, we have the following theorem.

Theorem 1. *The problem* $1|VM|\sum_i w_i c_i$ *is weakly NP-hard.*

3 A 2-Approximation Algorithm

In this section we propose an approximation algorithm with an approximation ratio of 2 for problem $1|VM|\sum_i w_i c_i$ and also show that the bound is tight, assuming that the function $f(s)$ is nondecreasing and concave. Throughout the paper we always suppose that the jobs are already sorted in a WSPT order, i.e., $w_1/p_1 \geq w_2/p_2 \geq \cdots \geq w_n/p_n$. Furthermore, as in the last section, we use S_1 and S_2 to denote respectively the set of jobs scheduled before the maintenance and the set of jobs scheduled after the maintenance, when a schedule is specified.

Before presenting the algorithm we investigate an optimal schedule π^* and give two lower bounds. Let s^* be the starting time of the maintenance in the schedule π^*, and let Z^* denote the total weighted completion time of π^*. We determine i and θ such that $\sum_{j=1}^{i-1} p_j + \theta p_i = s^*$, $0 \leq \theta < 1$. Then we split J_i into two jobs J_i' and J_i'', where J_i' has a weight of θw_i and a processing time of θp_i, and J_i'' has a weight of $(1-\theta)w_i$ and a processing time of $(1-\theta)p_i$. We get a new job set $J' = \{J \setminus \{J_i\}\} \cup \{J_i'\} \cup \{J_i''\}$. Moreover, we denote by Z_1 the optimal objective value for scheduling J'. There are two observations:

1. $Z_1 \leq Z^*$. To see this, we simply replace job J_i by J_i' and J_i'' in the schedule π^* maintaining the assignment for the other jobs, that results in a feasible schedule for J'. Note that the total weighted completion time of J_i' and J_i'' is not larger than the weighted completion time of J_i, while the costs incurred by the other jobs remain unchanged. Thus the inequality holds.
2. The schedule assigning $J_1, J_2, \cdots, J_{i-1}, J_i'$ before time s^* and $J_i'', J_{i+1}, \cdots, J_n$ after the maintenance (both in a WSPT order) is an optimal schedule for J', which is easily proved as in [15]. The total weighted completion time Z_1 of this schedule can be easily computed as

$$\begin{aligned} Z_1 = &\sum_{j=1}^{i-1} w_j(\sum_{k=1}^{j} p_k) + \theta w_i(\sum_{k=1}^{i-1} p_k + \theta p_i) \\ &+(1-\theta)w_i(\sum_{k=1}^{i} p_k + f(\sum_{k=1}^{i-1} p_k + \theta p_i)) \\ &+ \sum_{j=i+1}^{n} w_j(\sum_{k=1}^{j} p_k + f(\sum_{k=1}^{i-1} p_k + \theta p_i)). \end{aligned} \tag{1}$$

The first term is the total weighted completion time of jobs $J_1, J_2, \cdots, J_{i-1}$, the second and the third terms are contributions from J_i' and J_i'', and the fourth term is owing to the jobs $J_{i+1}, \cdots, J_n$.

Hence, we have the first lower bound for Z^*.

Lemma 1. *$Z^* \geq Z_1$, where Z_1 is determined in (1).*

Note that Z_1 may not correspond to a feasible schedule for J since the job J_i is split. Now we slightly modify the schedule without splitting J_i. There are two possibilities: schedule J_i either right after or right before the maintenance. Ignore the maintenance deadline and denote the objective values by Z_1^A and Z_1^B, respectively. Then,

$$Z_1^A = \sum_{j=1}^{i-1} w_j(\sum_{k=1}^{j} p_k) + \sum_{j=i}^{n} w_j(\sum_{k=1}^{j} p_k + f(\sum_{k=1}^{i-1} p_k))$$

and

$$Z_1^B = \sum_{j=1}^{i} w_j(\sum_{k=1}^{j} p_k) + \sum_{j=i+1}^{n} w_j(\sum_{k=1}^{j} p_k + f(\sum_{k=1}^{i} p_k)).$$

Lemma 2. $Z^* \geq Z_1 \geq \frac{3}{4} \min\{Z_1^A, Z_1^B\}$

Proof. Since $f(s)$ is concave, we have

$$f(\sum_{k=1}^{i-1} p_k + \theta p_i) \geq \theta f(\sum_{k=1}^{i} p_k) + (1-\theta) f(\sum_{k=1}^{i-1} p_k).$$

It's easy to verify that

$$Z_1 - \theta Z_1^B > (1-\theta) \sum_{j=1}^{i-1} w_j(\sum_{k=1}^{j} p_k) + (1-\theta) w_i(\sum_{k=1}^{i} p_k) + (1-\theta) w_i f(\sum_{k=1}^{i-1} p_k)$$
$$+ (1-\theta) \sum_{j=i+1}^{n} w_j(\sum_{k=1}^{j} p_k + f(\sum_{k=1}^{i-1} p_k)) + (\theta^2 - \theta) w_i p_i.$$

Then

$$Z_1 - \theta Z_1^B \geq (1-\theta) Z_1^A + (\theta^2 - \theta) Z_1^B,$$

$$Z_1 \geq (1 + \theta^2 - \theta) \min\{Z_1^A, Z_1^B\}.$$

Note that $1 + \theta^2 - \theta \geq 3/4$, since $0 \leq \theta < 1$. Thus

$$Z^* \geq Z_1 \geq \frac{3}{4} \min\{Z_1^A, Z_1^B\}.$$

This lemma shows the better one of the above two schedules for J has an objective value close to Z_1. It is clear that both of the two schedules are feasible if

$$\sum_{k=1}^{i} p_k \leq s_d. \tag{2}$$

In this case the optimal cost can be very well bounded. However, we do not know how much s^* is and there is no way to exactly determine the index i. We can first determine i_1 such that $\sum_{j=1}^{i_1-1} p_j \leq s_d < \sum_{j=1}^{i_1} p_j$. Divide $[0, s_d]$ into a number of disjoint intervals $[0, p_1], (p_1, p_1 + p_2], \cdots, (\sum_{j=1}^{i_1-1} p_j, s_d]$.

Clearly s^* must fall into one of the above intervals. If s^* is out of the last interval $(\sum_{j=1}^{i_1-1} p_j, s_d]$, then $1 \leq i \leq i_1 - 1$, and the inequality (2) holds. We only need to try the $i_1 - 1$ intervals and get an upper bound of 4/3 by Lemma 2. In the following we assume that $s^* \in (\sum_{j=1}^{i_1-1} p_j, s_d]$. Denote by Z_{i_1} the objective value of the schedule π_{i_1}, in which $S_1 = \{J_1, J_2 \cdots, J_{i_1-1}\}$ (the set of jobs scheduled before the maintenance) and $S_2 = \{J_{i_1}, \cdots, J_n\}$ (the set of jobs scheduled after the maintenance), i.e., the maintenance starts immediately after job J_{i_1-1} is completed. The following lemma analyzes this case.

Lemma 3. *If $Z_{i_1} > 2Z^*$, then job J_{i_1} must be scheduled before the maintenance in any optimal solution.*

Proof. Assume that J_{i_1} is processed after the maintenance in an optimal solution. By Lemma 1, we have

$$Z_{i_1} > 2\sum_{j=1}^{i_1} w_j(\sum_{k=1}^{j} p_k) + 2(\theta^2 - \theta)w_{i_1}p_{i_1} + 2(1-\theta)w_{i_1}f(\sum_{k=1}^{i_1-1} p_k + \theta p_{i_1})$$
$$+ 2\sum_{j=i_1+1}^{n} w_j(\sum_{k=1}^{j} p_k + f(\sum_{k=1}^{i_1-1} p_k + \theta p_{i_1})).$$

Thus

$$w_{i_1}f(\sum_{k=1}^{i_1-1} p_k) > \sum_{j=1}^{i_1-1} w_j(\sum_{k=1}^{j} p_k) + \sum_{j=i_1+1}^{n} w_j(\sum_{k=1}^{j} p_k + f(\sum_{k=1}^{i_1-1} p_k)),$$

$$2Z^* > \sum_{j=1}^{i_1} w_j(\sum_{k=1}^{j} p_k) + w_{i_1}f(\sum_{k=1}^{i_1-1} p_k) + \sum_{j=i_1+1}^{n} w_j(\sum_{k=1}^{j} p_k + f(\sum_{k=1}^{i_1-1} p_k)).$$

Thus, $2Z^* > Z_{i_1}$, which is an contradiction.

Now we have to consider that some jobs with smaller (weight)/(processing time) ratio are assigned to S_1, and give an improved lower bound for an optimal schedule π^*. Again, assume that s^* is the starting time of the maintenance in

π^*. Recall that the jobs are already in a WSPT order. Let $J^l = \{J_{i_1}, \cdots, J_{i_l}\}, l = 1, 2, \cdots, m$, be a set of jobs that must be scheduled before s^*, i.e., $J^l \subset S_1$, where $\sum_{k=1}^{l} p_{i_k} \leq s^*$ and $w_{i_1}/p_{i_1} \geq w_{i_2}/p_{i_2} \geq \cdots \geq w_{i_l}/p_{i_l}$.

Similarly as the analysis in Lemma 1, we define i and θ as

$$\sum_{j=1}^{i-1} p_j + \sum_{k=1}^{l} p_{i_k} + \theta p_i = s^*, 0 \leq \theta < 1, 1 \leq i \leq i_l - 1.$$

Again we split job J_i into J_i' and J_i'' in the same way as we did above. Let $S_1 = \{J_1, \cdots, J_{i-1}, J_i'\} \cup J^l$, and $S_2 = \{J_i'', J_{i+1}, \cdots, J_n\} \setminus J^l$. Then for this schedule (where J_i is split) we have the total weighted completion time as

$$\begin{aligned} Z_2 = & \sum_{j=1}^{i-1} (\sum_{k=1}^{j} p_k) + \theta w_i (\sum_{k=1}^{i-1} p_k + \theta p_i) + \sum_{k=1}^{l} w_{i_k} (\sum_{j=1}^{i-1} p_j + \theta p_i + \sum_{r=1}^{k} p_{i_r}) \\ & + (1-\theta) w_i (\sum_{j=1}^{i} p_j + \sum_{k=1}^{l} p_{i_k} + f(\sum_{j=1}^{i-1} p_j + \theta p_i + \sum_{k=1}^{l} p_{i_k})) \\ & + \sum_{j>i, j \bar{\in} J^l}^{n} w_j (\sum_{k=1, k \bar{\in} J^l}^{j} p_k + \sum_{k=1}^{l} p_{i_k} + f(\sum_{k=1}^{i-1} p_k + \theta p_i + \sum_{k=1}^{l} p_{i_k})). \end{aligned}$$

Lemma 4. *If the optimal schedule π^* assigns the jobs in J^l before time s^*, then $Z^* \geq Z_2$.*

Proof. Let R_1 denote the set of jobs scheduled before maintenance in the optimal schedule π^* and $R_2 = J \setminus R_1$. Clearly the order of the jobs in R_1 has the following pattern $(B_l, J_{i_l}, B_{l-1}, J_{i_{l-1}}, \cdots, B_1, J_{i_1}, B_0)$, where $B_i, i = 0, 1, \cdots, l$ denotes a set of jobs or an empty set. Let $\{J_{a_i}, i = 1, 2, \cdots, q\}$ be the jobs in $B_i, i = 0, 1 \cdots, l-1$. For each job J_{a_i} we iteratively increase its weight to $w_{a_i} + \triangle w_{a_i}$ until $(w_{a_i} + \triangle w_{a_i})/p_{a_i} = w_{i_l}/p_{i_l}$. Let $J_{a_i}^{'}$ denote the corresponding job of J_{a_i} with a new weight and $B_i^{'}, i = 0, 1, 2, \cdots, l-1$ denote the corresponding set of B_i, for $i = 0, 1, \cdots, l-1$, with new weights.

Let $\pi^{'}$ be the resulting schedule after increasing the job weights. Its objective value is Z'. Those jobs with new weights move to the front in R_1 according to a WSPT order. By some simple calculation (details omitted) we have $Z^* \geq Z' - \sum_i \triangle w_{a_i} c_{a_i}$, where c_{a_i} denotes the completion time of J_{a_i} in π^*.

Let $\pi_1 = (J_1, \cdots, J_{i-1}, J_i^{'}, J_{i_l}, \cdots, J_{i_1}, VM, J_i^{''}, \cdots, J_{a_1}, \cdots, J_{a_q}, \cdots)$ and $\pi_2 = (J_1, \cdots, J_{i-1}, J_i^{'}, J_{i_l}, \cdots, J_{i_1}, VM, J_i^{''}, \cdots, J_{a_1}^{'}, \cdots, J_{a_q}^{'}, \cdots)$ be two schedules, where VM denotes the (variable) maintenance. The corresponding objective values are $Z(\pi_1)$ and $Z(\pi_2)$, respectively. With simple calculation we have $Z(\pi_1) \leq Z(\pi_2) - \sum_i \triangle w_{a_i} c_{a_i}^{'}$, where $c_{a_i}^{'}$ denotes the completion time of $J_{a_i}^{'}$ in π_2. Since $Z(\pi_2) \leq Z'$, we obtain $Z(\pi_1) \leq Z^* + \sum_i \triangle w_{a_i} c_{a_i} - \sum_i \triangle w_{a_i} c_{a_i}^{'}$. Moreover, with $c_{a_i} \leq c_{a_i}^{'}$ we get $Z(\pi_1) \leq Z^*$, which shows that π_1 is an optimal schedule for J' and thus the lemma is proved.

We define critical jobs that are split in Lemmas 1 and 4. More precisely these jobs can be recursively defined below.

- Job J_{i_1} satisfies: $\sum_{j=1}^{i_1-1} p_j \leq s_d < \sum_{j=1}^{i_1} p_j$.
- For $k = 2, \ldots, m$, job J_{i_k} satisfies: $\sum_{j=1}^{i_k-1} p_j \leq s_d - \sum_{j=1}^{k-1} p_{i_j} < \sum_{j=1}^{i_k} p_j$.

Here m is 1 if $i_1 = 1$, otherwise it is determined by $p_1 + \sum_{k=1}^{m} p_{i_k} > s_d$ and $p_1 + \sum_{k=1}^{m-1} p_{i_k} \leq s_d$.

Now we are ready to present the first algorithm. In the description we again use (S_1, S_2) to define a schedule, meaning that

- Jobs in S_1 are scheduled (in a WSPT order) before the maintenance.
- The maintenance starts immediately when the jobs of S_1 are completed, i.e., it starts at time $\sum_{J_i \in S_1} p_i$.
- Jobs in S_2 are scheduled (in a WSPT order too) immediately after the maintenance.

Algorithm H_1.

Step 0. Let J_0 be a dummy job with processing time of zero. It is always in the front in a WSPT order.

Step 1. For $i = 0, 1, \cdots, i_1 - 1$, construct a number of schedules $\pi_0(i)$, where $S_1 = \{J_0, J_1, \cdots, J_i\}$ and $S_2 = \{J_{i+1}, \cdots, J_n\}$.

Step 2. For $l = 1, \ldots, m-1$, and for $i = 0, 1, \ldots, i_l - 1$, construct a number of schedules $\pi_{i_l}(i)$, where $S_1 = \{J_0, J_1, \cdots, J_i, J_{i_l}, J_{i_{l-1}}, J_{i_1}\}$ and $S_2 = \{J_{i+1}, J_{i+2}, \cdots, J_{i_l-1}, J_{i_l+1}, \cdots, J_n)\}$.

Step 3. Construct a schedule $\pi_{i_m}(0)$, where $S_1 = \{J_0, J_{i_m}, J_{i_{m-1}}, \cdots, J_{i_1}\}$ and $S_2 = \{J_1, J_2, \cdots, J_{i_m-1}, J_{i_m+1}, \cdots, J_n\}$.

Step 4. Output the schedule with the minimum total weighted completion time among all schedules above and denote it by π.

Theorem 2. *For a nondecreasing and concave function $f(s)$, Algorithm H_1 is a 2-approximation algorithm with $O(n^2)$ time and the bound is tight.*

Proof. Again let s^* be the starting time of the maintenance in the optimal schedule. To prove this theorem, by Lemma 3 we only need to consider the case that the job J_{i_1} is processed before the maintenance in the optimal schedule.

Divide $[p_{i_1}, \sum_{j=1}^{i_2-1} p_j + p_{i_1}]$ into subintervals $[\sum_{j=1}^{l-1} p_j + p_{i_1}, \sum_{j=1}^{l} p_j + p_{i_1}], l = 1, 2, \cdots, i_2 - 1$. Similarly, if $s^* \in [\sum_{j=1}^{l-1} p_j + p_{i_1}, \sum_{j=1}^{l} p_j + p_{i_1}], l = 1, 2, \cdots, i_2 - 1$, consider schedules

$$\pi_{i_1}(l-1) = (J_1, J_2, \cdots, J_{l-1}, J_{i_1}, VM, J_l, \cdots, J_n), l = 1, \cdots, i_2 - 1,$$

$$\pi_{i_1}(l) = (J_1, J_2, \cdots, J_l, J_{i_1}, VM, J_{l+1}, \cdots, J_n), l = 1, \cdots, i_2 - 1.$$

From Lemma 2, we have

$$
\begin{aligned}
Z^* \geq & \sum_{j=1}^{l-1} w_j(\sum_{k=1}^{j} p_k) + \theta w_l(\sum_{j=1}^{l-1} p_j + \theta p_l) + w_{i_1}(\sum_{k=1}^{l-1} p_k + p_{i_1} + \theta p_l) \\
& + (1-\theta) w_l(\sum_{k=1}^{l} p_k + p_{i_1} + f(\sum_{k=1}^{l-1} p_k + p_{i_1} + \theta p_l)) \\
& + \sum_{j>l, j \neq i_1}^{n} w_j(\sum_{k=1, k\neq i_1}^{j} p_k + f(\sum_{k=1}^{1-1} p_k + p_{i_1} + \theta p_l) + p_{i_1}).
\end{aligned}
$$

Thus

$$Z^* - \theta Z(\pi_{i_1}(l)) \geq (1-\theta) Z(\pi_{i_1}(l-1)) + (\theta^2 - \theta) Z(\pi_{i_1}(l)),$$

i.e.,

$$\min\{Z(\pi_{i_1}(l)), Z(\pi_{i_1}(l-1))\} \leq \frac{4}{3} Z^*.$$

If $s^* \in (\sum_{j=1}^{i_2-1} p_j + p_{i_1}, s_d)$,

$$
\begin{aligned}
& Z(\pi_{i_1}(i_2 - 1)) \\
= & \sum_{j=1}^{i_2-1} w_j(\sum_{k=1}^{j} p_k) + w_{i_1}(\sum_{k=1}^{i_2-1} p_k + p_{i_1}) + w_{i_2}(f(\sum_{k=1}^{i_2-1} p_k + p_{i_1}) + \sum_{k=1}^{i_2} p_k + p_{i_1}) \\
& + \sum_{j>i_1, j\neq i_1} w_j(\sum_{k=1, k\neq i_1}^{j} p_k + f(\sum_{k=1}^{i_2-1} p_k + p_{i_1}) + p_{i_1}).
\end{aligned}
$$

From Lemma 2 we have

$$
\begin{aligned}
Z^* \geq & \sum_{j=1}^{i_2-1} w_j(\sum_{k=1}^{j} p_k) + \theta w_{i_2}(\sum_{k=1}^{i_2-1} p_k + \theta p_{i_2}) + w_{i_1}(\sum_{k=1}^{i_2-1} p_k + \theta p_{i_2} + p_{i_1}) \\
& + (1-\theta) w_{i_2}(f(\sum_{k=1}^{i_2-1} p_k + \theta p_{i_2} + p_{i_1}) + \sum_{k=1}^{i_2} p_k + p_{i_1}) \\
& + \sum_{j>i_2, j\neq i_1}^{n} w_j(\sum_{k=1, k\neq i_1}^{j} p_k + f(\sum_{k=1}^{i_2-1} p_k + \theta p_{i_2} + p_{i_1}) + p_{i_1}).
\end{aligned}
$$

Similarly we can claim that J_{i_2} should be processed before the maintenance in the optimal schedule if $2Z^* < Z(\pi_{i_1}(i_2 - 1))$, i.e., it fails to provide a feasible schedule with an upper bound better than 2.

Continuing the same argument, $J_{i_1}, J_{i_2}, \cdots, J_{i_m}$ should be processed before the maintenance in the optimal schedule if it fails to provide a feasible solution with an upper bound better than 2. Note that either $J_{i_m} = J_1$ or $p_1 + \sum_{k=1}^{m} p_{i_k} > s_d$. Both cases yield a contradiction if we restrict that $Z(\pi_{i_m}(0)) > 2Z^*$ and $J_{i_1}, J_{i_2}, \cdots, J_{i_m}$ are assigned before the maintenance in the optimal schedule.

The analysis above shows that Algorithm H_1 is a 2-approximation algorithm. It is not hard to see that the algorithm actually needs $O(n^2)$ time.

To show the bound is tight, let us consider the following instance. $J = \{J_1, J_2\}$, where $p_1 = w, w_1 = w - 1, p_2 = w - 1, w_2 = w - 2, s_d = w - 1$, and $f(s) = w^2 + 2s$. The schedule by our algorithm is $\pi = (VM, J_1, J_2)$, while an optimal schedule is $\pi^* = (J_2, VM, J_1)$.

$$\frac{Z(\pi)}{Z^*} = \frac{2w^3 - 6w + 2}{w^3 + 4w^2 - 10w + 5} \to 2(w \to \infty)$$

4 A Better Approximation Algorithm

In the section, we present a $(1 + \sqrt{2}/2 + \epsilon)$-approximation algorithm. Let i_1 be determined as in the previous section. For a given schedule π let $Z(\pi)$ be its objective value.

Algorithm H_2

Step 1. Let J_0 be a dummy job with processing time of zero. For $i = 0, 1, \cdots, i_1 - 1$, construct a set of schedules: $\pi_0(i) = (J_0, J_1, \cdots, J_i, VM, J_{i+1}, \cdots, J_n)$.

Step 2. Use the FPTAS in [3] to solve the problem $1|d_i = s_d| \sum_i w_i u_i$ in which $d_i = s_d$ denotes the common deadline for all $J_i \in J$. Let $S_1 = \{i \in J | u_i = 0\}, S_2 = J \setminus S_1$. Construct a schedule $\pi_0(i_1) = (S_1, VM, S_2)$, where the jobs in S_1 and the jobs in S_2 follow a WSPT order.

Step 3. Output the schedule $\pi = \arg \min_{\pi_0(i), i=0,1,\cdots,i_1} \{Z(\pi_0(i))\}$.

Theorem 3. *If $f(s)$ is a nondecreasing and concave function, Algorithm H_2 is a $(1 + \sqrt{2}/2 + \epsilon)$-approximation running in $O(n^2/\epsilon)$ time.*

The proof is similar as Theorem 2 but still involved. Due to the page limit we omit the proof and put it to the full version of this paper.

5 Concluding Remarks

In this paper we consider the problem of scheduling a variable machine maintenance along with jobs on a single machine to minimize the total weighted completion time. We show that the problem is weakly NP-hard. We provide two approximation algorithms, assuming that the duration of the maintenance is a nondecreasing and concave function of its starting time.

It is not difficult to prove that the problem with an arbitrary duration function $f(s)$ of the maintenance cannot be approximated within any constant factor unless $P = NP$. If $f(s)$ is nondecreasing it admits an FPTAS which is only of theoretical interests. We will present the results in the full version of this paper.

References

1. Agnetis, A., Mirchandani, P.B., Pacciarelli, D., Pacifici, A.: Scheduling problems with two competing agents. Operations Research 52, 229–242 (2004)
2. Fu, B., Huo, Y., Zhao, H.: Exponential inapproximability and FPTAS for scheduling with availability constraints. Theoretical Computer Science 410, 2663–2674 (2009)
3. Gens, G., Levner, E.: Fast approximation algorithm for job sequencing with deadlines. Discrete Applied Mathematics 3, 313–318 (1981)
4. Graham, R.L., Lawler, E.L., Lenstra, J.K., Rinnooy Kan, A.H.G.: Optimization and approximation in deterministic sequencing and scheduling: A survey. Annals of Discrete Mathematics 5, 287–326 (1979)
5. Kacem, I.: Approximation algorithm for the weighted flow-time minimization on a single machine with a fixed non-availability interval. Computers and Industrial Engineering 54, 401–410 (2008)
6. Kellerer, H., Kubzin, M.A., Strusevich, V.A.: Two simple constant ratio approximation algorithms for minimizing the total weighted completion time on a single machine with a fixed non-availability interval. European Journal of Operational Research 199, 111–116 (2009)
7. Kellerer, H., Strusevich, V.A.: Fully polynomial approximation schemes for a symmetric quadratic knapsack problem and its scheduling applications. Algorithmica (2009), doi:10.1007/s00453-008-9248-1
8. Kubzin, M.A., Strusevich, V.A.: Planning machine maintenance in two-machine shop scheduling. Operations Research 54, 789–800 (2006)
9. Lee, C.-Y., Chen, Z.: Scheduling jobs and maintenance activities on parallel machines. Naval Research Logistics 47, 145–165 (2000)
10. Ma, Y., Chu, C., Zuo, C.: A survey of scheduling with deterministic machine availability constraints. Computers and Industrial Engineering (2009), doi:10.1016/j.cie.2009.04.014
11. Megow, N., Verschae, J.: Short note on scheduling on a single machine with one non-availability period (2009) (manuscript)
12. Mosheiov, G., Sarig, A.: Scheduling a maintenance activity to minimize total weighted completion-time. Computers and Mathematics with Applications 57, 619–623 (2009)
13. Mosheiov, G., Sidney, J.B.: Scheduling a deteriorating maintenance activity on a single machine. Journal of the Operational Research Society (2009), doi:10.1057/jors.2009.5
14. Schmidt, G.: Scheduling with limited machine availability. European Journal of Operational Research 121, 1–15 (2000)
15. Wang, G., Sun, H., Chu, C.: Preemptive scheduling with availability constraints to minimize total weighted completion times. Annals of Operations Research 133, 183–192 (2005)

Bounded Parallel-Batch Scheduling on Unrelated Parallel Machines

Cuixia Miao[1,2,⋆,⋆⋆], Yuzhong Zhang[1], and Chengfei Wang[1]

[1] School of Operations Research and Management Sciences, Qufu Normal University, Rizhao, Shandong 276826, China
[2] School of Mathematical Sciences, Qufu Normal University, Qufu, Shandong 273165, China
miaocuixia@126.com

Abstract. In this paper, we consider the bounded parallel-batch scheduling problem on unrelated parallel machines. Problems $R_m|B|F$ are NP-hard for any objective function F. For this reason, we discuss the special case with $p_{ij} = p_i$ for $i = 1, 2, \cdots, m$, $j = 1, 2, \cdots, n$. We give optimal algorithms for the general scheduling to minimize total weighted completion time, makespan and the number of tardy jobs. And we design pseudo-polynomial time algorithms for the case with rejection penalty to minimize the makespan and the total weighted completion time plus the total penalty of the rejected jobs, respectively.

Keywords: Parallel-batch scheduling; Unrelated parallel machines; Rejection penalty; Pseudo-polynomial time.

1 Introduction

The parallel-batch scheduling is motivated by burn-in operations in semiconductor manufacturing. Lee et al. [10] provided a background description. Webster and Baker [14] presented an overview of algorithms and complexity results for scheduling batch processing machines. Brucker et al. [2] gave a thorough discussion of the scheduling problem of the batch machine under various constraints and objective functions. Deng and Zhang [6] proved that the problem $1|r_j, B \geq n|\sum w_j C_j$ is NP-hard, they further showed that several important special cases of the problem can be solved in polynomial time. Li et al. [11] presented a polynomial time approximation scheme (PTAS) for the problem of minimizing total weighted completion time on identical parallel unbounded batch machines. Zhang et al. [15] had some results for the parallel-batch scheduling on identical parallel machines to minimize makespan and minimize the maximum lateness. Under the *on-line* setting, Chen et al. [4], for the single batch machine problem

⋆ This work was supported by the National Natural Science Foundation (No.10671108), the Dr.Foundation of Shandong Province (No. 2007BS01014) and the Foundation of Qufu Normal University (No.XJ0714).

⋆⋆ Corresponding author.

B. Chen (Ed.): AAIM 2010, LNCS 6124, pp. 220–228, 2010.

to minimize total weighted completion time, provided an on-line algorithm with $\frac{10}{3}$-competitive for the unbounded model and $4+\epsilon$-competitive on-line algorithm for the bounded model.

In classical scheduling literatures, all jobs must be processed and no rejection is allowed. In the real applications, however, this may not be true. Due to the limited resources, the scheduler can have the option to reject some jobs. The machine scheduling problem with rejection was first considered by Bartal et al. [1]. They studied both the off-line and the on-line versions of scheduling with rejection on identical parallel machines. The objective is to minimize the sum of the makespan of the accepted jobs and the total penalty of the rejected jobs. Subsequently, the scheduling with rejection was extensively studied in the last decade. Engels et al. [7] considered the single machine scheduling with rejection to minimize the sum of weighted completion times of scheduled jobs and total rejection penalty of the rejected jobs. Epstein et al. [8] considered on-line scheduling problem of unit-time jobs with rejection to minimize the total completion time. Cheng and Sun [5] considered the single-machine scheduling problem with deteriorating and rejection. Miao and Zhang [13] studied the on-line scheduling with rejection on identical machines.

Cao and Yang [3] presented a PTAS for the combined model of the above two scheduling models (parallel-batch and rejection) where jobs arrive dynamically. The objective is to minimize the sum of the makespan of the accepted jobs and the total penalty of the rejected ones. Lu et al. [12] also considered the bounded parallel-batch scheduling problem with rejection on one single batch machine.

We address the bounded parallel-batch scheduling problem on unrelated parallel machines. We discuss one special case in this paper. In section 3, we give optimal algorithms for minimizing the makespan, total weighted completion time and the number of tardy jobs. In section 4, we consider the case with rejection, design pseudo-polynomial time dynamic programming algorithms for the makespan plus the total penalty of the rejected jobs and the total weighted completion time plus the total penalty of the rejected jobs. We conclude the paper in the last section.

2 Model and Notation

The parallel-batch processing machine is a machine that can process up to B jobs simultaneously as a batch, and the processing time of the batch is given by the processing time of the longest job in the batch. All jobs contained in the same batch start and complete at the same time. Once processing of a batch is initiated, it can not be interrupted and other jobs cannot be introduced into the batch until processing is completed.

We are given a set $J = \{J_1, \cdots, J_n\}$ of n independent jobs and a set $M = \{M_1, \cdots, M_m\}$ of m unrelated parallel machines. Each job J_j ($j = 1, \cdots, n$) is available for processing from time zero onwards, it has a due date d_j by which it should ideally be completed, a weight w_j and a rejection penalty $e_j > 0$, the processing time of J_j in machine M_i being p_{ij}, we consider the special with $p_{ij} = p_i$ for $i = 1, \cdots, m$, $j = 1, \cdots, n$ in this paper.

The jobs that are processed together form a batch, which we denote by B, and the processing time of the batch denoted by $p(B)$ is $p(B) = max\{p_{ij}|J_{ij} \in B\}$. For convenience, we denote the batch's weight and completion time by $W(B)$ and $C(B)$, respectively, where $W(B) = \sum_{J_j \in B} w_j$. The aim is to schedule the jobs on the set of m unrelated parallel machines so as to minimize $F_1 \in \{\sum w_j C_j, C_{max}, \sum U_j\}$ or minimize $F_2 \in \{\sum_{J_j \in S} w_j C_j + \sum_{J_j \in \overline{S}} e_j, C_{max}(S) + \sum_{J_j \in \overline{S}} e_j\}$, where C_j, $C_{max} = max_{1 \leq j \leq n}\{C_j\}$ and S denotes the completion time of job J_j, the makespan and the set of accepted jobs, respectively. U_j is 0-1 indicator variable that takes the value 1 if J_j is tardy, i.e., $C_j > d_j$, and the value is 0 if J_j is on time, i.e., $C_j \leq d_j$. Using the 3−field notation of Graham et al. [9], we denote our problem as $R_m|B, p_{ij} = p_i|F_1$ and $R_m|B, rej, p_{ij} = p_i|F_2$, where rej denotes the rejection.

3 The Case with General Parallel-Batch Scheduling

In this section, we discuss the problem $R_m|B, p_{ij} = p_i|\sum w_j C_j$, $R_m|B, p_{ij} = p_i|C_{max}$ and $R_m|B, p_{ij} = p_i|\sum U_j$.

3.1 Minimizing Total Weighted Completion Times

In this subsection, we design an optimal algorithm for problem $R_m|B, p_{ij} = p_i|\sum w_j C_j$.

Algorithm $\mathcal{A}1$

Step 1. Re-index jobs in non-increasing order according to their weights so that $w_1 \geq w_2 \geq \cdots \geq w_n$.

Step 2. Form batches by placing jobs J_{jB+1} through $J_{(j+1)B}$ together in the same batch B_{j+1} for $j = 0, 1, \cdots, [\frac{n}{B}]$, where $[\frac{n}{B}]$ denotes the largest integer smaller than $\frac{n}{B}$.

Step 3. Re-index machines in non-decreasing order according to the processing time of job so that $p_1 \leq p_2 \leq \cdots \leq p_m$.

Step 4. Assign batch B_1 on machine M_1, set $j = 2$.

Step 5. Select machine M_{i_0} so that

$$n_j^{i_0} p_{i_0} = min_{1 \leq i \leq m}\{n_j^i p_i\},$$

where n_j^i $(1 \leq j \leq [\frac{n}{B}]+1, 1 \leq i \leq m)$ denotes the number of batches assigned to M_i among batches $\{B_1, \cdots, B_j\}$. Assign batch B_j on machine M_{i_0}, set $j = j+1$, repeat Step 5 until $j = [\frac{n}{B}] + 1$.

The time complexity of Algorithm $\mathcal{A}1$ is $O(nlogn)$.

Theorem 1. *Algorithm $\mathcal{A}1$ generates an optimal solution to problem $R_m|B, p_{ij} = p_i|\sum w_j C_j$.*

Proof. Let π be any optimal schedule for problem $R_m|B, p_{ij} = p_i|\sum w_j C_j$. To prove this Theorem, we only show that π could be transformed in a schedule π'

under condition $w_1 \geq w_2 \geq \cdots \geq w_n$ which has the following three properties, and the objective value does not increase.
(i). The indexes of jobs in every batch are consecutive.
(ii). If there exist two jobs J_i and J_j with $w_i \geq w_j$, then the completion time of J_i must be earlier than or equal to the completion time of J_j, and this conclusion holds true for two batches B_i and B_j with $W(B_i) \geq W(B_j)$.
(iii). All batches are full except possible the one which contains job J_n.

To show (i), suppose that there are two batches B_x, B_y and three jobs J_j, J_{j+1}, J_{j+2} with $w_j \geq w_{j+1} \geq w_{j+2}$ such that $J_j, J_{j+2} \in B_x$ and $J_{j+1} \in B_y$ in schedule π. We distinguish between two cases:

Case 1: B_x and B_y are on the same machine, w.l.o.g, let the machine be M_k $(1 \leq k \leq m)$.

From $p_{ij} = p_i$ for $i = 1, 2, \cdots, m$, $j = 1, 2, \cdots, n$, we know that $P(B_x) = P(B_y) = p_k$. We also distinguish between two subcases in this case as follows.
Subcase 1.1: $C(B_x) < C(B_y)$
We get a new schedule π' by swapping job J_{j+1} with J_{j+2}. Let F and F' denote the objective value, and f and f' denote the total weighted completion times of the other jobs except J_j, J_{j+1}, J_{j+2} in π and π', respectively. And since the completion times of other jobs are unchanged, thus, $f = f'$ in this subcase. Therefore, we have

$$F = f + w_j C_j + w_{j+1} C_{j+1} + w_{j+2} C_{j+2} = f + w_j C(B_x) + w_{j+1} C(B_y) + w_{j+2} C(B_x),$$

$$F' = f' + w_j C'_j + w_{j+1} C'_{j+1} + w_{j+2} C'_{j+2} = f' + w_j C(B_x) + w_{j+1} C(B_x) + w_{j+2} C(B_y),$$

from which we get $F - F' = (w_{j+1} - w_{j+2})(C(B_y) - C(B_x))$.
Since $C(B_x) < C(B_y)$ and $w_{j+1} \geq w_{j+2}$, then we have $F \geq F'$.
Subcase 1.2: $C(B_x) > C(B_y)$
Similarly, we have $F - F' = (w_j - w_{j+1})(C(B_x) - C(B_y))$ by swapping job J_j with J_{j+1} in this subcase, since $C(B_x) > C(B_y)$ and $w_j \geq w_{j+1}$, then we have $F \geq F'$.

Case 2: B_x and B_y are on different machines.

We also distinguish between two subcases in this case.
Subcase 2.1: $C(B_x) \leq C(B_y)$
Similarly, we get a new schedule π' by swapping job J_{j+1} with J_{j+2}. As the above mentioned, we know that $f = f'$ since $p_{ij} = p_i$ for $i = 1, 2, \cdots, m$, $j = 1, 2, \cdots, n$ in this subcase. Therefore, we have

$$F = f + w_j C(B_x) + w_{j+1} C(B_y) + w_{j+2} C(B_x),$$

$$F' = f' + w_j C(B_x) + w_{j+1} C(B_x) + w_{j+2} C(B_y),$$

from which we get $F - F' = (w_{j+1} - w_{j+2})(C(B_y) - C(B_x))$.

Since $C(B_x) \leq C(B_y)$ and $w_{j+1} \geq w_{j+2}$, then we have $F \geq F'$.

Subcase 2.2: $C(B_x) > C(B_y)$

Similarly, we have $F - F' = (w_j - w_{j+1})(C(B_x) - C(B_y))$ by swapping job J_j with J_{j+1} in this subcase, since $C(B_x) > C(B_y)$ and $w_j \geq w_{j+1}$, then we have $F \geq F'$.

Combining the above two cases, we know that the objective value does not increase after the swap. Thus, the schedule π' is still an optimal schedule. A finite number of repetitions of this procedure yield an optimal schedule of the required form.

To show (ii), assuming that there exist two jobs J_i and J_j with $w_i \geq w_j$, but $C_i > C_j$ in schedule π, then we have that there is a similar argument to that in the previous cases, that is we get $F - F' = (w_i - w_j)(C_i - C_j)$ by swapping job J_i with J_j whether J_i and J_j are on the same machine or not. Since $w_i \geq w_j$ and $C_i > C_j$, then we have $F \geq F'$, i.e., the objective value does not increase. Thus, the schedule π' is still an optimal schedule. A finite number of repetitions of this procedure yield an optimal schedule of the required form.

Similarly, if we view the batch B_i as an aggregate job with weight $W(B_i) = \sum_{J_l \in B_i} w_l$, we could prove the second part of of (ii).

To show (iii), first, we find the batch B_x which contains job J_1 in schedule π, according to (i) and (ii), W.l.o.g, let $B_x = \{J_1, \cdots, J_k\}$. If B_x is not full, i.e., $|B_x| = k < B$. According to (ii) and $w_1 \geq w_2 \geq \cdots \geq w_k \geq w_{k+1} \geq \cdots \geq w_B \geq \cdots \geq w_n$, we know that the completion times of $J_{k+1}, J_{k+2}, \cdots, J_B$ must be late than or equal to $C_1 = C(B_x)$, the completion time of job J_1. Thus, if we move the remaining jobs $J_{k+1}, J_{k+2}, \cdots, J_B$ from other batches to B_x to get a schedule π', then B_x is full, and this procedure can not increase the objective value. Now, by applying this argument recursively, we have that all batches are full except possible the one which contained job J_n.

In a word, from the discussion above, we know that the result of the allocation of the optimal schedule π' is consistent with that of Algorithm $\mathcal{A}1$. Thus, Algorithm $\mathcal{A}1$ generates an optimal solution to problem $R_m|B, p_{ij} = p_i|\sum w_j C_j$. □

Assume $w_1 \geq w_2 \geq \cdots \geq w_n$, let $H = \{J_1, J_2, \cdots, J_j\}$ $(1 \leq j \leq n)$ be the subset of J, we can get the following conclusion.

Theorem 2. *The maximum completion time of H is equal to the completion time of the batch containing job J_j which is the last batch assigned by Algorithm $\mathcal{A}1$.*

Proof. To show this, we will assume otherwise and derive the contradiction. Let $C_{max}(H)$ be the makespan generated by Algorithm $\mathcal{A}1$ and $C_j(H)$ be the completion time of J_j among H, then the completion of the batch containing job J_j is also $C_j(H)$. If $C_{max}(H) \neq C_j(H)$, then there exists i $(1 \leq i < j)$ such that $C_{max}(H) = C_i(H)$, thus, $C_j(H) < C_i(H)$. Case 1: if job J_j and J_i are in the same machine, then $C_i(H) \leq C_j(H)$ according to Algorithm $\mathcal{A}1$ and $w_i \geq w_j$. A contradiction. Case 2: if job J_j and J_i are in different machines. The completion time of job J_i will decrease if we move job J_i to the location of J_j, this contradicts Algorithm $\mathcal{A}1$, in which, the completion time of job J_i is minimum. Thus, $C_{max}(H) = C_j(H)$. This completes the proof of Theorem 2. □

3.2 Minimizing Makespan and the Number of Tardy Jobs

We give optimal algorithms for problem $R_m|B, p_{ij} = p_i|C_{max}$ and $R_m|B, p_{ij} = p_i|\sum U_j$ in this subsecton.

Algorithm $\mathcal{A}2$

Step 1. Give an arbitrary order of jobs.

From **Step 2** to **Step 5** are identical with those in Algorithm $\mathcal{A}1$.

Theorem 3. *Algorithm $\mathcal{A}2$ generates an optimal solution to problem $R_m|B, p_{ij} = p_i|C_{max}$.*

Algorithm $\mathcal{A}3$

Step 1. Re-index jobs in non-decreasing due dates order so that $d_1 \le d_2 \le \cdots \le d_n$.

From **Step 2** to **Step 5** are identical with those in Algorithm $\mathcal{A}1$.

Theorem 4. *Algorithm $\mathcal{A}3$ generates an optimal solution to problem $R_m|B, p_{ij} = p_i|\sum U_j$.*

The proofs of Theorem 3 and Theorem 4 are similar to the proof of Theorem 1.

4 The Case with Rejection

In this section, we discuss the bounded parallel-batch schedule problem on unrelated parallel machines with rejection. W.l.o.g, we assume that the job parameters are integral, unless stated otherwise. Each job J_j is either rejected with a rejection penalty e_j having to be paid, or accepted and processed on one of the m unrelated parallel machines. We design pseudo-polynomial time dynamic programming algorithms for $R_m|B, rej, p_{ij} = p_i|C_{max}(S) + \sum_{J_j \in \overline{S}} e_j$ and $R_m|B, rej, p_{ij} = p_i|\sum_{J_j \in S} w_j C_j + \sum_{J_j \in \overline{S}} e_j$, respectively.

4.1 Problem $R_m|B, rej, p_{ij} = p_i|\sum_{J_j \in S} w_j C_j + \sum_{J_j \in \overline{S}} e_j$

Lemma 1. *There exists an optimal schedule for $R_m|B, rej, p_{ij} = p_i|\sum_{J_j \in S} w_j C_j + \sum_{J_j \in S} e_j$ in which the accepted jobs are assigned to the unrelated parallel machines by Algorithm $\mathcal{A}1$.*

Assuming that the jobs have been indexed so that $w_1 \ge w_2 \ge \cdots \ge w_n$. To solve our problem, we set up a dynamic program to find the schedule that minimizes the objective function when the total rejection penalty of the rejected jobs is given. Let $F_j(k, C^j_{max}, E)$ denote the optimal value of the objective function satisfying the following conditions.

- The jobs in consideration are $J_1, \cdots, J_j$.
- The number of accepted jobs is exactly k.

- The maximum completion time of the accepted jobs is C^j_{max}.
- The total penalty of the rejected jobs is E.

To get $F_j(k, C^j_{max}, E)$, we distinguish two cases as follows.

Case 1: Job J_j is rejected.

Since J_j is rejected, then the number of accepted jobs among $J_1, \cdots, J_j$ is the same as the number of accepted jobs among $J_1, \cdots, J_{j-1}$. Therefore, $F_j(k, C^j_{max}, E) = F_{j-1}(k, C^{j-1}_{max}, E - e_j) + e_j$.

Case 2: Job J_j is accepted.

In this case, we distinguish two subcases.

Subcase 2.1: $k = hB + 1$

In this subcase, job J_j has to start a new batch. We use Algorithm $\mathcal{A}1$ to get the completion time of J_j, denoted by C_j, which is equal to the makespan C^j_{max} of the accepted jobs among $J_1, \cdots, J_j$ according to Theorem 2. Therefore, $F_j(k, C^j_{max}, E) = F_{j-1}(k-1, C^{j-1}_{max}, E) + w_j C^j_{max}$.

Subcase 2.2: $k \neq hB + 1$

In this subcase, job J_j can be assigned to the last batch which has existed and the makespan does not change according to Algorithm $\mathcal{A}1$, and the completion time of J_j is C^{j-1}_{max} according to Theorem 2. Therefore, $F_j(k, C^j_{max}, E) = F_{j-1}(k-1, C^{j-1}_{max}, E) + w_j C^{j-1}_{max}$.

Now, combining the above two cases, we have the following dynamic programming algorithm.

Algorithm $\mathcal{DPA}1$

The boundary conditions:

$F_1(1, p_1, 0) = w_1 p_1$ and $F_1(k, C^1_{max}, E) = +\infty$ for $(k, C^1_{max}, E) \neq (1, p_1, 0)$.

$F_1(0, 0, e_1) = e_1$ and $F_1(k, C^1_{max}, E) = +\infty$ for $(k, C^1_{max}, E) \neq (0, 0, e_1)$.

The recursive function:

$$F_j(k, C^j_{max}, E) = \begin{cases} min\{F_{j-1}(k, C^{j-1}_{max}, E - e_j) + e_j, \\ \quad F_{j-1}(k-1, C^{j-1}_{max}, E) + w_j C^j_{max}\} & if\ k = hB + 1 \\ min\{F_{j-1}(k, C^{j-1}_{max}, E - e_j) + e_j, \\ \quad F_{j-1}(k-1, C^{j-1}_{max}, E) + w_j C^{j-1}_{max}\} & if\ k \neq hB + 1. \end{cases}$$

The optimal value is given by

$$F^* = min\{F_n(k, C^n_{max}, E) | 0 \leq k \leq n, 0 \leq C^n_{max} \leq np_m, 0 \leq E \leq \sum_{j=1}^{n} e_j\}.$$

Theorem 5. *Problem $R_m|B, rej, p_{ij} = p_i| \sum_{J_j \in S} w_j C_j + \sum_{J_j \in \overline{S}} e_j$ can be solved in $O(mn^3 p_m \sum e_j)$.*

Proof. The correctness of Algorithm $\mathcal{DPA}1$ is guaranteed by the above discussion. Clearly, we have $0 \leq k \leq n$, $0 \leq C^n_{max} \leq np_m$ and $0 \leq E \leq \sum_{j=1}^{n} e_j$. Thus, the recursive function has at most $O(n^3 p_m \sum e_j)$ states. Each iteration takes an $O(m)$ time to execute. Hence, the running time of Algorithm $\mathcal{DPA}1$ is $O(mn^3 p_m \sum e_j)$. □

4.2 Problem $R_m|B, rej, p_{ij} = p_i|C_{max}(S) + \sum_{J_j \in \overline{S}} e_j$

Lemma 2. *There exists an optimal schedule for $R_m|B, rej, p_{ij} = p_i|C_{max}(S) + \sum_{J_j \in \overline{S}} e_j$ in which the accepted jobs are assigned to the unrelated machines by Algorithm $\mathcal{A}2$.*

Assuming an arbitrary order $\{J_1, \cdots, J_n\}$ of jobs. To solve our problem, we set up a dynamic program to find the schedule that minimizes the objective function when the total rejection penalty of the rejected jobs is given. Let $F_j(k, E)$ denote the optimal value of the objective function satisfying the following conditions.

- The jobs in consideration are $J_1, \cdots, J_j$.
- The number of accepted jobs is exactly k.
- The total rejection penalty of the rejected jobs is E.

To get $F_j(k, E)$, we distinguish two cases as follows.

Case 1: Job J_j is rejected.

Since J_j is rejected, then the number of accepted jobs among $J_1, \cdots, J_j$ is the same as the number of accepted jobs among $J_1, \cdots, J_{j-1}$. Therefore, $F_j(k, E) = F_{j-1}(k, E - e_j) + e_j$.

Case 2: Job J_j is accepted.

In this case, we distinguish two subcases.

Subcase 2.1: $k = hB + 1$

Job J_j has to start a new batch in this subcase. The makespan C^j_{max} is equal to the completion time of J_j according to Algorithm $\mathcal{A}1$ and Theorem 2. Therefore, we have $F_j(k, E) = C^j_{max} + E$.

Subcase 2.2: $k \neq hB + 1$

In this subcase, job J_j can be assigned to the last batch which has existed and the makespan does not change after inserting J_j (Algorithm $\mathcal{A}1$), then we have $F_j(k, E) = F_{j-1}(k - 1, E)$.

Now, combining the above two cases, we have the following dynamic programming algorithm.

Algorithm $\mathcal{DPA}2$

The boundary conditions:

$F_1(1, 0) = p_1$ and $F_1(k, E) = +\infty$ for $(k, E) \neq (1, 0)$.

$F_1(0, e_1) = e_1$ and $F_1(k, E) = +\infty$ for $(k, E) \neq (0, e_1)$.

The recursive function:

$$F_j(k, E) = \begin{cases} min\{F_{j-1}(k, E - e_j) + e_j, C^j_{max} + E\} & if \;\; k = hB + 1 \\ min\{F_{j-1}(k, E - e_j) + e_j, F_{j-1}(k - 1, E)\} & if \;\; k \neq hB + 1. \end{cases}$$

The optimal value is given by

$$F^* = min\{F_n(k, E) | 0 \leq k \leq n, 0 \leq E \leq \sum_{j=1}^{n} e_j\}.$$

Theorem 6. *Problem $R_m|B, rej, p_{ij} = p_i|C_{max}(S) + \sum_{J_j \in \overline{S}} e_j$ can be solved in $O(mn^2 \sum e_j)$.*

5 Conclusion

In this paper, we considered the bounded parallel-batch scheduling problem on unrelated parallel machines under $p_{ij} = p_i$ for $i = 1, \cdots, m$, $j = 1, \cdots, n$. We gave optimal algorithms for the general scheduling to minimize total weighted completion time, makespan and the number of tardy jobs, and designed pseudo-polynomial time algorithms for the case with rejection to minimize the makespan and the total weighted completion time plus the total penalty of the rejected jobs.

References

1. Bartal, Y., Leonardi, S., Marchetti-Spaccamela, A., Sgall, J., Stougie, L.: Multiprocessor scheduling with rejection. SIAM J. Discrete Math. 13, 64–78 (2000)
2. Brucker, P., Gladky, A., Hoogeveen, H., Kovalyov, M.Y., Potts, C.N., Tautenhahn, T., van de Velde, S.L.: Scheduling a batching machine. Journal of Scheduling 1, 31–54 (1998)
3. Cao, Z.G., Yang, X.G.: A PTAS for parallel batch scheduling with rejection and dynamic job arrivals. Theoretical Computer Science 410(27-29), 2732–2745 (2009)
4. Chen, B., Deng, X.T., Zang, W.N.: On-line scheduling a batch processing system to minimize total weighted job completion time. Journal of Combinatorial Optimization 8, 85–95 (2004)
5. Cheng, Y.S., Sun, S.J.: Scheduling linear deteriorating job with rejection on a single machine. European Journal of Operational Research 194, 18–27 (2009)
6. Deng, X.T., Zhang, Y.Z.: Minimizing mean response time in batch processing system. In: Asano, T., Imai, H., Lee, D.T., Nakano, S.-i., Tokuyama, T. (eds.) COCOON 1999. LNCS, vol. 1627, pp. 231–240. Springer, Heidelberg (1999)
7. Engels, D.W., Karger, D.R., Kolliopoulos, S.G., Sengupta, S., Uma, R.N., Wein, J.: Techniques for scheduling with rejection. Journal of Algorithms 49, 175–191 (2003)
8. Epstein, L., Noga, J., Woeginger, G.J.: On-line scheduling of unit time jobs with rejection: Minimizing the total completion time. Operations Research Lettars 30, 415–420 (2002)
9. Graham, R.L., Lawler, L.J.K., Rinnooy Kan, A.H.G.: Optimization and approximation in deterministic sequencing and scheduling: A survey. Ann. Disc. Math. 5, 287–326 (1979)
10. Lee, C.-Y., Uzsoy, R., Martin-Vega, L.A.: Efficient algorithms for scheduling semiconductor burn-in operations. Operations Research 40, 764–775 (1992)
11. Li, S.G., Li, G.J., Wang, X.H.: Minimizing total weighted completion time on parallel unbounded batch machines. Journal of Software 17, 2063–2068 (2006)
12. Lu, L.F., Cheng, T.C.E., Yuan, J.J., Zhang, L.Q.: Bounded single-machine parallel-batch scheduling with release dates and rejection. Computer and Operations Research 36, 2748–2751 (2009)
13. Miao, C.X., Zhang, Y.Z.: On-line scheduling with rejection on identical parallel machines. Journal of System Science and Complexity 19, 431–435 (2006)
14. Webster, S., Baker, K.R.: Scheduling groups of jobs on a single machine. Operations Research 43, 692–703 (1995)
15. Zhang, Y.Z., Wang, Z.Z., Wang, C.Y.: The "transform lemma" and its application in batch scheduling. Journal of Systems Science and Mathematical Sciences 22(3), 328–333 (2002)

Exact Algorithms for Coloring Graphs While Avoiding Monochromatic Cycles

Fabrice Talla Nobibon[1], Cor Hurkens[2], Roel Leus[1], and Frits C.R. Spieksma[1]

[1] University of Leuven, Operations Research Group, Naamsestraat 69, B-3000 Leuven, Belgium
Fabrice.TallaNobibon@econ.kuleuven.be, Roel.Leus@econ.kuleuven.be, Frits.Spieksma@econ.kuleuven.be

[2] Eindhoven University of Technology, Department of Mathematics and Computer Science. P.O. Box 513, 5600 MB Eindhoven, the Netherlands
wscor@win.tue.nl

Summary. We consider the problem of deciding whether a given directed graph can be vertex partitioned into two acyclic subgraphs. Applications of this problem include testing rationality of collective consumption behavior, a subject in micro-economics. We identify classes of directed graphs for which the problem is easy and prove that the existence of a constant factor approximation algorithm is unlikely for an optimization version which maximizes the number of vertices that can be colored using two colors while avoiding monochromatic cycles. We present three exact algorithms, namely an integer-programming algorithm based on cycle identification, a backtracking algorithm, and a branch-and-check algorithm. We compare these three algorithms both on real-life instances and on randomly generated graphs. We find that for the latter set of graphs, every algorithm solves instances of considerable size within few seconds; however, the CPU time of the integer-programming algorithm increases with the number of vertices in the graph while that of the two other procedures does not. For every algorithm, we also study empirically the transition from a high to a low probability of YES answer as function of a parameter of the problem. For real-life instances, the integer-programming algorithm fails to solve the largest instance after one hour while the other two algorithms solve it in about ten minutes.

1 Introduction

Consider the following problem. Given is a finite, directed graph $G = (V, A)$. The goal is to partition the vertices of G into two subsets such that each subset induces an acyclic subgraph. We refer to this problem as the *acyclic 2-coloring problem.* Notice that the acyclic 2-coloring problem is defined for a directed graph. The counterpart for undirected graphs is named *partition into two forests* and is known to be NP-complete [16]. This problem is neither a special case nor a generalization of the problem for directed graphs.

In this paper, we describe applications of the acyclic 2-coloring problem and identify classes of directed graphs for which the problem is easy. For an arbitrary

B. Chen (Ed.): AAIM 2010, LNCS 6124, pp. 229–242, 2010.

directed graph, the problem is NP-complete and we prove that the existence of a constant factor approximation algorithm is unlikely for an optimization version which maximizes the number of vertices that can be colored using two colors while avoiding monochromatic cycles. We develop and implement three exact algorithms, namely an integer-programming (IP) algorithm based on cycle identification (in the rest of this text, we also refer to this algorithm as cycle-identification algorithm), a backtracking algorithm and a branch-and-check algorithm. We compare them based on their CPU time, both for real-life instances coming from micro-economics and for randomly generated graphs. We find that every algorithm solves random graphs of considerable size within few seconds. The CPU time of the cycle-identification algorithm increases with the number n of vertices in the graph while the running time of both the backtracking algorithm and the branch-and-check algorithm does not increase. Further, for every algorithm we study empirically the phase transition of the problem as function of the number of arcs divided by the number of vertices. When applying the three algorithms to real-life instances, on the other hand, we find that the cycle-identification algorithm takes more time than the two other procedures and is not able to decide the largest instance after one hour, while the backtracking algorithm solves it in less than five minutes and the branch-and-check algorithm in about ten minutes.

This paper is organized as follows. In Section 2, we motivate the study of this problem, present a literature review and examine the complexity of the problem. The algorithms are described in Section 3, followed by computational results in Section 4. We conclude in Section 5.

2 Motivation and Notation

In this section, we first explain our motivation for studying this problem and present a brief literature review. Subsequently, we describe some notation and definitions that will be used throughout this paper and finally, we study the complexity of the problem and present some properties.

2.1 Motivation and Literature Review

Our motivation to consider this problem comes from an application in the *study of rationality* of consumption behavior, a field in micro-economics. We now shortly elaborate on this application. Suppose that there is a dataset S consisting of ℓ observations, each observation i being identified by a pair (p^i, x^i) of prices and quantities with $p^i = (p^i_1, \dots, p^i_k)$ and $x^i = (x^i_1, \dots, x^i_k)$, where k denotes the number of goods. The dataset S may, for instance, describe the expenditures of an economic entity, such as a household, over a certain period of time. Economic theory has developed a number of properties that reflect rationality of the dataset (see Varian [15] for an overview). As example of such a property, we mention the Strong Axiom of Revealed Preference (SARP); a dataset S may or may not satisfy SARP. Clearly, a relevant question is how to test whether a given dataset satisfies SARP. It has been shown ([15]) that this question can be answered using graph theory. A directed graph G with ℓ vertices is built by considering each observation i as a vertex. Further, there is an arc from vertex i to vertex j if and only if $p^i x^i \geq p^i x^j$, where $p^i x^i$ is the scalar product. The dataset S satisfies SARP if and only if G is acyclic.

Recently, testing rationality of observed consumption behavior has been extended to households consisting of multiple members (see Cherchye et al. [4]). Deb [8] shows that the problem of testing whether observed data of two-member household consumption behavior satisfies the so-called Generalized Axiom of Revealed Preferences (GARP) is equivalent to an acyclic 2-coloring problem for a directed graph G built from the data. The problem of testing whether observed data of two-member household consumption behavior satisfies the so-called Collective Axiom of Revealed Preferences (CARP) is considered by Cherchye et al. [5] and by Talla Nobibon et al. [12]. The former paper solves an integer programming model, the latter paper describes heuristics that attempt to find a solution to the acyclic 2-coloring problem for specific directed graphs. Clearly, the method described in this paper can be used to color graphs arising either from testing GARP or from testing CARP.

To the best of our knowledge, Deb [8, 7] is the first to explicitly address the acyclic 2-coloring problem. He proves that the problem is NP-hard, and extends the results of Chen [3] for undirected graphs by computing an upper bound on the acyclic chromatic number.

2.2 Notation and Definitions

We denote by $G = (V, A)$ a finite directed graph with $|V| = n$ vertices and $|A| = m$ arcs. For a vertex $p \in V$, the *outdegree* of p is the number of arcs leaving p while the *indegree* of p is the number of incoming arcs to p. The *degree* of p is the sum of its outdegree and its indegree. For ease of exposition, we will use pq to represent the arc $p \to q$. A sequence of vertices $[v_0, v_1, \ldots, v_l]$ is called a *chain* of length l if $v_{i-1}v_i \in A$ or $v_i v_{i-1} \in A$ for $i = 1, \ldots, l$. G is *connected* if between any two vertices there exists a chain in G joining them. A sequence of vertices $[v_0, v_1, \ldots, v_l]$ is called a *path* from v_0 to v_l if $v_{i-1}v_i \in A$ for $i = 1, \ldots, l$. A *vertex-induced subgraph* (subsequently called *induced subgraph* in this text) is a subset of vertices of G together with all arcs whose endpoints are both in this subset. A *strongly connected component* (SCC) of G is a maximal induced subgraph $S = (V(S), A(S))$ where for every pair of vertices $p, q \subset V(S)$, there is a path from p to q and a path from q to p. A sequence of vertices $[v_0, v_1, \ldots, v_l, v_0]$ is called a *cycle* of length $l + 1$ in $G = (V, A)$ if $v_{i-1}v_i \in A$ for $i = 1, \ldots, l$ and $v_l v_0 \in A$. A graph is *acyclic* if it contains no cycle; otherwise it is *cyclic*. A k-*coloring* of the vertices of G is a partition $V_1, V_2, \ldots, V_k$ of V; the sets V_j $(j = 1, \ldots, k)$ are called *color classes*. Given a k-coloring of G, a cycle $[v_0, v_1, \ldots, v_l, v_0]$ in G is *monochromatic* if there exists $i \in \{1, \ldots, k\}$ such that v_0, v_1, ..., $v_l \in V_i$. In this paper, we use the notions vertex coloring and vertex partition of a graph interchangeably.

Given an integer k, an *acyclic k-coloring* of G is a k-coloring in which the subgraph induced by each color class is acyclic. The *acyclic chromatic number* $a(G)$ of G is the smallest k for which G has an acyclic k-coloring. The *directed line graph* LG of G has $V(LG) \equiv A(G)$ and a vertex (u, v) is adjacent to a vertex (w, z) if $v = w$. An arc $pq \in A$ is called a *single arc* if the arc $qp \notin A$. We define the 2-*undirected graph* $G_2 = (V, E)$ associated with G as the undirected graph obtained from G by deleting all single arcs and transforming a pair of arcs forming a cycle of length 2 into an edge (undirected arc); more precisely, $\{v_1, v_2\} \in E$ if and only if $v_1v_2 \in A$ and $v_2v_1 \in A$. We define the *single directed graph* $G_s = (V, A_s)$ of G as the subgraph of G containing only single arcs; more precisely, for a given pair of vertices v_1 and v_2 in V, $v_1v_2 \in A_s$ if and only if $v_1v_2 \in A$ and $v_2v_1 \notin A$. In the rest of this paper, $G = (V, A)$ is a given directed graph, G_2 is its associated 2-undirected graph and G_s its single directed graph.

2.3 Complexity and Properties of the Problem

The acyclic 2-coloring problem is explicitly defined as the following decision problem.
INSTANCE: A finite directed graph $G = (V, A)$.
QUESTION: Does G have an acyclic 2-coloring?

There are classes of directed graphs for which the acyclic 2-coloring problem is always a YES instance. This is the case for the class of directed acyclic graphs, the class of line graphs (see Lemma 1 in [12]) and the class of partial directed line graphs [2]. For an arbitrary directed graph G, Deb [7] has shown that the acyclic 2-coloring problem is NP-complete.

Further, if we consider an optimization formulation, Max-A2C, of the acyclic 2-coloring problem which maximizes the number of vertices of G that can be colored using two colors such that each subgraph induced by a color class is acyclic, we prove the following non-approximability result in [13].

Theorem 1. *There exists an $\epsilon > 0$ such that Max-A2C cannot be approximated in polynomial time with ratio n^ϵ unless $P = NP$.*

The following properties of the problem are used in the next section to build exact algorithms.

Proposition 1. *If the set V of vertices of a given graph G can be partitioned into two sets, RED and BLUE, such that G_2 is bipartite with all the vertices in RED on one side and those in BLUE on the other side; and the single directed graphs induced by RED, $G_s(RED)$, and by BLUE, $G_s(BLUE)$, respectively, are acyclic then G is a YES instance of the acyclic 2-coloring problem; otherwise G is a NO instance.*

Proposition 2. *If G_2 is not a bipartite graph then G is a NO instance of the acyclic 2-coloring problem, while if G_2 is a bipartite graph and G_s is an acyclic graph, then G is a YES instance.*

3 Algorithms

In this section, we describe three exact algorithms for solving the acyclic 2-coloring problem, namely a *cycle-identification* algorithm, a *backtracking* algorithm and a *branch-and-check* (B&C) algorithm. We also present two dominance rules used to reduce the size of the initial graph.

3.1 Cycle-Identification Algorithm

This algorithm is based on the following IP formulation of the problem where we have a binary decision variable x_i ($i = 1, \ldots, n$) which equals 1 if vertex i is colored red and 0 if it is colored blue. We are looking for a coloring x_i with a maximum number of red vertices for which there is no monochromatic cycle. Notice that any other objective function can be chosen. For each oriented cycle $\mathcal{C}$ in G, there is a pair of constraints $1 \leq \sum_{i \in \mathcal{C}} x_i \leq |\mathcal{C}| - 1$, where $|\mathcal{C}|$ is the number of vertices in $\mathcal{C}$. A formal description of the cycle identification algorithm is given by CycleId(G).

CycleId(G)

```
1: solve a restricted IP problem containing only a subset of constraints
2: if there exists a feasible solution
3:    for each color class, search for a monochromatic cycle
4:    if monochromatic cycle found
5:       add the corresponding pair of constraints to the restricted IP problem
6:       solve the IP problem again and goto 2
7:    else return YES
8: else return NO
```

3.2 Backtracking Algorithm

An "ordinary" backtracking algorithm for solving the acyclic 2-coloring problem is an adaptation of the well known backtracking algorithm for graph coloring on undirected graphs. It would work as follows: it successively colors the vertices of G either red or blue and each time a new vertex is colored, the corresponding color class is checked to see whether it is acyclic; otherwise the color of the last vertex is switched and its new color class is then checked. If it is not acyclic, the algorithm backtracks.

In this paper, we propose a backtracking algorithm based on Proposition 1. The key difference between our algorithm and an ordinary backtracking algorithm is that the backtracking algorithm described here can anticipate a NO conclusion earlier without having to color many vertices. A formal description of the backtracking algorithm is given by BT(RED, $BLUE$, G) with $RED = \emptyset$ and $BLUE = \emptyset$ at the beginning. In the description, the function bipartite(RED, $BLUE$, G_2) returns YES if G_2 is bipartite given that the vertices in RED are on one side and those in BLUE are on the other side; otherwise it returns NO. $G_s(A)$ denotes the single directed graph induced by a set A.

```
BT(RED, BLUE, G)
1: if V = RED ∪ BLUE, then return YES
2: choose a vertex p in V \ {RED ∪ BLUE}
3: RED = RED ∪ {p}
4: if bipartite(RED, BLUE, G2) == YES and Gs(RED) is acyclic then
5:    if BT(RED, BLUE, G) == YES then return YES
6: RED = RED \ {p}, BLUE = BLUE ∪ {p}
7: if bipartite(RED, BLUE, G2) == YES and Gs(BLUE) is acyclic then
8:    if BT(RED, BLUE, G) == YES then return YES
9: return NO
```

3.3 Branch-and-Check Algorithm

This B&C algorithm is based on Proposition 2. Like the backtracking algorithm, it is an explicit enumeration algorithm where at each node we check some conditions and decides whether to proceed or to stop. Unlike the backtracking algorithm, however, the B&C algorithm is an implicit coloring algorithm which branches on an arc, and the directed graph obtained at every child node is different from the graph of the parent node.

We now explain how to construct two new graphs from G. Let $p, q \in V$ be two adjacent vertices in G_s such that there is a cycle in G_s containing the arc pq. Consider the directed graphs $H^{pq} = (V'', A'')$ and $F^{pq} = (V', A')$ defined as follows. For H^{pq}, $V'' = V$ and $A'' = A \cup \{qp\}$. For F^{pq}, V' contains V and two vertices (pq_1) and (pq_2); that is $V' = V \cup \{(pq_1), (pq_2)\}$ and A' is built as follows. First, every arc in $A \setminus \{pq\}$ is an arc in A'. Second, for every single incoming arc ap into p, add an arc $a(pq_2)$ in A'. Third, for every single outgoing arc qa out of q, add an arc $(pq_2)a$ in A' and finally, add the following arcs: $p(pq_1)$, $(pq_1)p$, $q(pq_1)$, $(pq_1)q$, $(pq_1)(pq_2)$, $(pq_2)(pq_1) \in A'$. H^{pq} corresponds to a setting where p and q receive different colors, whereas F^{pq} represents the setting where p and q have the same color in any feasible coloring.

Proposition 3. *Let p and q be two adjacent vertices contained in a cycle in G_s. F^{pq} or H^{pq} is a YES instance of the acyclic 2-coloring problem if and only if G is a YES instance.*

The proof of this result can be found in [13]. A formal description of the B&C algorithm is given by BnC(G).

```
BnC(G)
1: determine G2, Gs
2: if G2 not bipartite, then return NO
3: if Gs acyclic, then return YES
4: choose pq on a cycle in Gs
5: determine H^pq, F^pq
6: if BnC(H^pq) == YES then return YES
7: else return BnC(F^pq)
```

In [13], we prove the following result about the correctness of the B&C algorithm.

Theorem 2. Correctness of the Branch-and-check algorithm
Suppose that the B&C algorithm is run on G. Then, its execution terminates after a finite number of iterations and the decision corresponds to the decision for the original graph G.

3.4 Refinements

We present two dominance rules used to reduce the size of the considered graph. The first rule removes the vertices of G with outdegree or indegree less than or equal to one. The second rule aims to identify and remove all single arcs not involved in any cycles in G_s. It proceeds as follows. The vertices of G_s are partitioned into SCCs and single arcs between two distinct SCCs are deleted. Notice that if either the first rule or the second rule removes at least one arc or at least one vertex, then the repeated application of the other rule may further remove new arcs or vertices.

3.5 Implementation Issues

Bipartiteness test, acyclicness test and strongly connected components. An adapted breadth-first-search algorithm [6] is implemented to check whether a given graph is bipartite while a topological ordering algorithm [1] is used for testing acyclicness. Tarjan's algorithm [14] is used to identify the SCCs of a graph.

Cycle identification algorithm. The restricted IP problem contains only pairs of constraints coming from cycles of length 2. The IP problems are solved using CPLEX. Throughout the algorithm, we use the Floyd-Warshall algorithm [6,1] to find (if it exists) a monochromatic cycle with the smallest number of vertices in the single directed graphs induced by every color class.

Backtracking algorithm

Branching strategy: This involves the selection of a vertex $p \in V$ which is neither in RED nor in BLUE. We investigate two choices: the first one is simply the first uncolored vertex found while the second choice is an uncolored vertex with the highest degree; with ties broken arbitrarily.

Propagation rule: Suppose a new vertex p is added to RED (BLUE). Then for any vertex q which is such that pq and qp exist, if q is not yet in BLUE (RED) then we add q in BLUE (RED). The procedure is repeated for any new vertex added to either RED or BLUE.

Node selection: Our goal is to color all the vertices as soon as possible (provided it is possible). Therefore, we use a *depth-first-search* strategy.

Branch-and-check algorithm

Branching strategy: We select a single arc pq which is such that there is a cycle in G_s containing that arc. Before choosing the arc pq, the graph G_s is first reduced by deleting all single arcs linking vertices of the same connected component in G_2 with different colors obtained from the bipartiteness test, and a single arc between vertices of the same connected component in G_2 with the same color is not considered for branching. We investigate two different choices of pq. The first choice is the first arc pq found that meets the above restriction. The second choice is an arc pq with p having the highest degree possible, breaking ties arbitrarily. In both cases, if in addition there is no path in G_s from p to q other than the arc pq, we define a simplified version of $F^{pq} = (V', A')$ by merging p and q. V' contains a vertex (pq) and all vertices in V except p and q such that $|V'| = |V| - 1$ while A' is built as follows. First, every arc $ab \in A$ with $a, b \notin \{p, q\}$ is an arc in A'. Second, for every single incoming arc ax to x with $x \in \{p, q\}$, (respectively every single outgoing arc xa from x), add an arc $a(pq)$ (respectively $(pq)a$) in A' while avoiding the repetition of arcs.

Branch-pruning criterion: This branch-pruning criterion considers each connected component of G_2 and the coloring of its vertices given by the bipartiteness test. If there exists a color class in a connected component which is such that the induced single directed graph is cyclic, then any graph built at a child node of that node is a NO instance of the problem and that node is pruned.

Node selection: We again use a *depth-first-search* strategy since we wish to reach a node with a YES answer as soon as possible (provided it exists).

4 Computational Experiments

All algorithms have been coded in C using Visual Studio C++ 2005; all the experiments were run on a Dell Optiplex 760 personal computer with Pentium R processor with 3.16 GHz clock speed and 3.21 GB RAM, equipped with Windows XP. CPLEX 10.2 was used for solving the IP problems.

4.1 Data

The three algorithms were tested on real-life graphs [5, 12] and on randomly generated graphs with n vertices, where $n = 50$, 100, 200, 500 and 1 000. Table 1 reports some properties of the real-life instances.

The random graphs are generated using a two-phase procedure. Each graph is connected and cyclic. To diversify the instances, we vary the density D of the graph, which is the number of arcs divided by $n(n-1)$. During the first phase, for every n, 400 graphs are randomly generated with 40 different densities, starting from a lower bound of 2.5% for $n = 50$, 1.5% for $n = 100$, 1% for $n = 200$ and 0.5% for $n = 500$ and $n = 1\,000$; and increased with a step of 0.5%. For

Table 1. Properties of the real-life graphs

Instance	1	2	3	4	5	6	7	8	9	10	11	12
n	22	48	68	95	118	139	226	279	294	410	755	4384
m	53	169	297	513	699	985	1979	2012	2427	3660	10113	124321
density	11.47	7.49	6.52	5.74	5.06	5.14	3.89	2.59	2.82	2.18	1.78	0.65

every n, the lower bound is obtained by taking the first multiple of 0.5 greater than or equal to the smallest density for which a connected and cyclic graph can be built. For each value of D, 10 directed graphs with $\lceil D \times (n^2 - n) \rceil$ arcs are generated, leading to $400 \times 5 = 2\,000$ graphs.

After preliminary computation on the graphs of the first phase, we identify for every n a *critical interval* containing densities with at least one YES instance and at least one NO instance; we then generate additional graphs with the densities given in Table 2. For every density, 100 graphs are generated, leading to $1\,151 \times 100 = 115100$ graphs for the second phase. Both the real-life instances and the random graphs can be found at http://www.econ.kuleuven.be/public/NDBAC96/acyclic_coloring.htm

Table 2. Densities of the graphs generated in the second phase

n	density (D)			
	from	to	step	total
50	8%	15.75%	0.25%	32
100	3.05%	8.95%	0.05%	119
200	2.01%	3.99%	0.01%	199
500	0.8%	1.498%	0.002%	350
1 000	0.3%	1.2%	0.002%	451

4.2 Computational Results

In this section, we examine different implementations of every algorithm. The three algorithms are subsequently compared based on their best implementation. Finally, we study the phase transition [10, 11] of the acyclic 2-coloring problem as function of the number of arcs divided by n.

Comparison of different implementations of every algorithm. Different implementations of every algorithm are compared based on their average CPU time for the set of 50-vertex graphs generated during the first phase; a time limit of ten minutes is used. For more details, we refer to [13].

Cycle-identification algorithm: We compare four implementations. The first one is CycleId(G), where if a monochromatic cycle is found for one color class, we do not search for a monochromatic cycle in the other color class. The second implementation is similar to the first one, except that irrespective of finding a monochromatic cycle in the first color class, we search for a monochromatic cycle in the second color class. The third (fourth) implementation considers the first (second) implementation with in addition the use of dominance rules. Based on the comparison of CPU time, we find that the last implementation is better than the others, and we use it for the rest of our experiments.

Backtracking algorithm: Four implementations are examined. The first one is the pseudocode BT(RED, $BLUE$, G) with in addition the use of the propagation rule; we branch on the first uncolored vertex found. The second implementation is similar to the first one, but we choose an uncolored vertex with the highest degree. The third (fourth) implementation is the first (second) one with the use of dominance rules. We find that the last implementation outputs better results.

B&C algorithm: We compare six implementations. The first one is the B&C algorithm as described by the pseudocode BnC(G), with in addition the use of the branch-pruning criterion and the arc pq selected is the first arc found. The second implementation is similar to the first one, except that we choose an arc pq with vertex p having the highest degree possible. The third (fourth) implementation considers the first (second) implementation with dominance rules applied at the root node. The fifth (sixth) implementation considers the first (second) implementation with dominance rules at every node of the branching tree. We find that the fourth implementation needs less time than the others.

Comparison of the three algorithms. We have compared the three algorithms both on random graphs and on real-life instances [13].

Random graphs: In Figure 1 we plot, for every n, the average CPU time of every algorithm as function of the number of arcs divided by n. Figure 1(a) shows the average CPU time for the 50-vertex graphs. The B&C algorithm (BnC) usually reports a higher CPU time than the other algorithms. However, the highest average CPU time is less than 1.2 seconds. The cycle-identification algorithm (CycleId) usually uses, on average, the smallest CPU time. For 100-vertex graphs (Figure 1(b)), we see that the average CPU time of CycleId is usually between that of BnC and that of BT, with BT using, in most cases, the smallest average time. For the large graphs (with more than 100 vertices, see Figures (c), 1(d) and 1(e)), the average CPU time reported for CycleId increases with n, while those of BnC and BT are stable, close to each other and usually below one second.

Real-life instances: Table 3 reports the CPU time of every algorithm when applied to real-life instances with a time limit of one hour. We see that CycleId is not able to decide the largest instance within the time limit; and although it solves the remaining instances, the reported CPU time is usually higher than that of the other algorithms. As for the other two algorithms, BT reports the

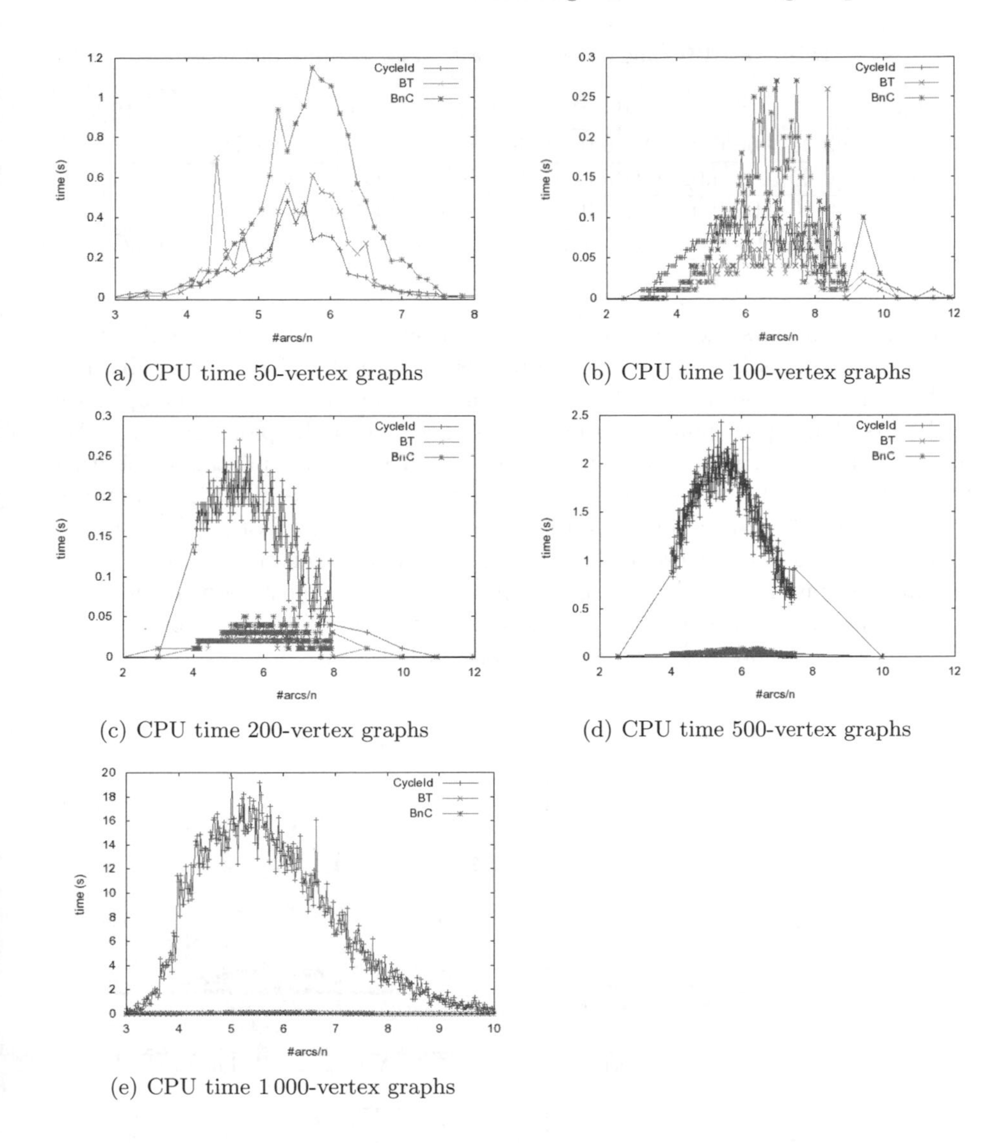

(a) CPU time 50-vertex graphs

(b) CPU time 100-vertex graphs

(c) CPU time 200-vertex graphs

(d) CPU time 500-vertex graphs

(e) CPU time 1 000-vertex graphs

Fig. 1. Average CPU time of every algorithm for random graph instances

best CPU time for 6 instances out of 12 while BnC reports the best CPU time for 9 instances out of 12. For the largest instance, however, BT spends less than five minutes, compared to about ten minutes for BnC.

Phase transition. We investigate the transition from a high probability to a low probability of YES answer as function of the following parameter of the problem: *the number of arcs divided by n*. Further, for every algorithm, we show how the average CPU time varies as function of that parameter.

Table 3. CPU time of every algorithm for real-life instances

Instance	1	2	3	4	5	6	7	8	9	10	11	12
CycleId	0.00	0.00	0.01	0.03	0.06	0.23	1.03	0.27	1.41	1.81	29.75	3930.47*
BT	0.00	0.00	0.02	0.03	0.06	0.09	0.36	0.28	0.31	0.28	3.45	283.72
BnC	0.00	0.00	0.01	0.02	0.05	0.09	0.59	0.05	0.28	0.11	3.84	612.41

* means that the time limit was exceeded without any decision found.

Figure 2 presents the probability of YES answer as well as the CPU time of every algorithm as function of the parameter. Figure 2(a) shows the probability of YES answer. The plots in Figure 2(a) are Bézier approximations [9] of the real plots. This approximation is mainly used to render the plots smoother. For every n, the plot has three regions. In the first region, with parameter between 0 and 3, almost all the instances have YES answer. The second region between 3 and 8, called *critical interval*, contains densities for which both YES instances and NO instances are present. The last region, with parameter greater than 8, contains graphs for which the probability of YES is almost zero. Overall, we remark that the *threshold* value of the parameter for which the probability of YES answer is equal to $\frac{1}{2}$ is almost the same for every n.

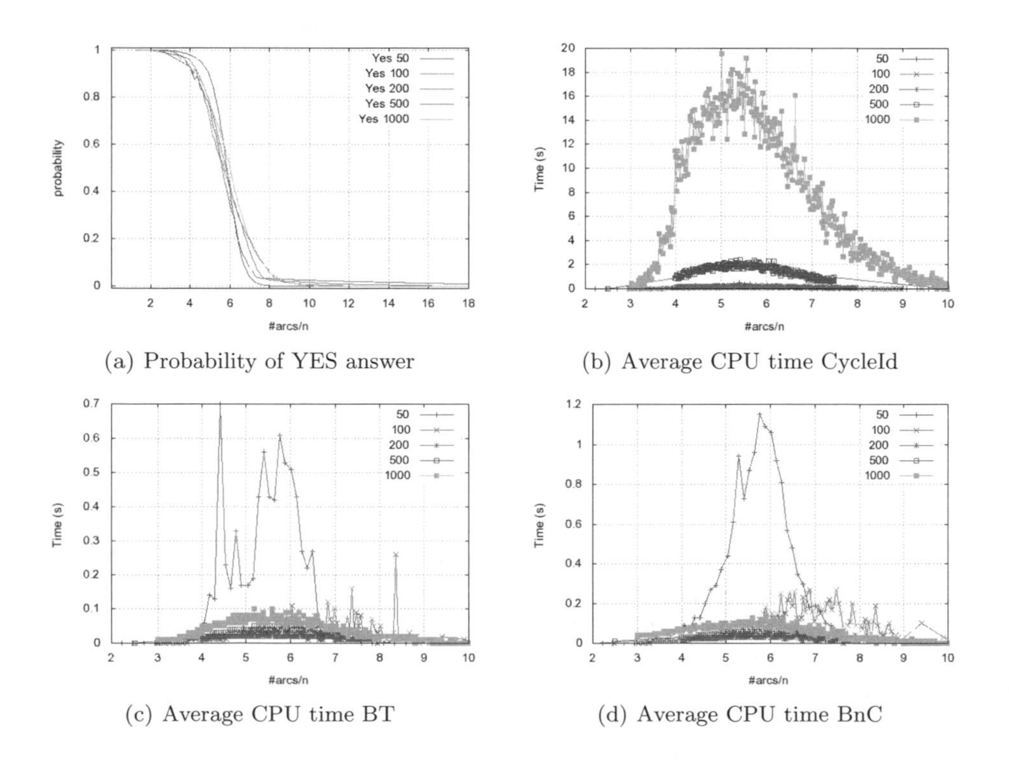

(a) Probability of YES answer

(b) Average CPU time CycleId

(c) Average CPU time BT

(d) Average CPU time BnC

Fig. 2. Probability of YES answer and average CPU time of every algorithm

The plots in Figures 2(b), 2(c) and 2(d) are obtained using the data that are used to generate the plots in Figure 1, but here the plots are grouped by algorithm. Figure 2(b) plots the average CPU time of CycleId for every n. The plots respect the three regions described above. For the first and the third region, the CPU time is very close to zero while in the critical interval, we have a non-negligible CPU time, showing an easy-hard-easy transition. Further, CycleId has a CPU time which increases with n, which probably occurs simply because when n increases the IP problem becomes more difficult to solve. Figure 2(c) plots the average CPU time of BT for every n. The easy-hard-easy transition is also observed here. However, unlike CycleId, BT spends more time in deciding 50-vertex and 100-vertex instances in the critical interval than in deciding instances with more vertices. This decrease in CPU time as n increases stops beyond $n = 200$. In Figure 2(d), the plots of the average CPU time of BnC for every n exhibit characteristics similar to those observed for BT. A possible explanation for this decrease in CPU time is the following: when n increases, the size (number of edges) of the undirected graph G_2 will increase, making the bipartiteness test used by both BT and BnC more efficient in detecting NO instances. At the same time, both the propagation rule (used by BT) and the branch-pruning criterion (used by BnC) become stronger, reducing the number of possible nodes to investigate in order to arrive at a YES answer.

In general for every n, the highest CPU time is obtained for values of the parameter around the threshold value. Further, for a value of the parameter in the critical region, very few instances have CPU time greater than a seconds, clearly indicating that difficult instances are hard to find. This also leads to high variability of the CPU time.

5 Summary and Conclusions

This text studies the problem of coloring the vertices of a directed graph using two colors such that no monochromatic cycle occurs. We were motivated to consider this problem by an application in the study of rationality of consumption behavior in households with multiple members. Our work confirms that the problem is NP-complete and that the existence of a constant factor approximation algorithm is unlikely for an optimization version which maximizes the number of vertices that can be colored using two colors while avoiding monochromatic cycles. We present a integer-programming algorithm based on cycle identification, a backtracking algorithm and a branch-and-check algorithm to solve the problem exactly. We compare them, based on their CPU time, both on real-life instances and on random graphs. For the latter set, graphs with up to 1000 vertices are solved in few seconds by every algorithm. We also study the phase transition of the problem. For real-life instances, the backtracking algorithm and the branch-and-check algorithm solve all instances, the largest taking about ten minutes.

References

[1] Ahuja, R.K., Magnanti, T.L., Orlin, J.B.: Network Flows: Theory, Algorithms, and Applications. Prentice-Hall, Englewood Cliffs (1993)
[2] Apollonio, N., Franciosa, P.G.: A characterization of partial directed line graphs. Discrete Mathematics 307, 2598–2614 (2007)
[3] Chen, Z.: Efficient algorithm for acyclic colorings of graphs. Theoretical Computer Science 230, 75–95 (2000)
[4] Cherchye, L., De Rock, B., Vermeulen, F.: The collective model of household consumption: a nonparametric characterization. Econometrica 75, 553–574 (2007)
[5] Cherchye, L., De Rock, B., Sabbe, J., Vermeulen, F.: Nonparametric tests of collectively rational consumption behavior: an integer programming procedure. Journal of Econometrics 147, 258–265 (2008)
[6] Cormen, T.H., Leiserson, C.E., Rivest, R.L., Stein, C.: Introduction to Algorithms. The MIT Press, Cambridge (2005)
[7] Deb, R.: Acyclic partitioning problem is NP-complete for $k = 2$. Private communication. Yale University, United States (2008)
[8] Deb, R.: An efficient nonparametric test of the collective household model. Working paper. Yale University, United States (2008)
[9] Farin, G.: Class A Bézier curves. Computer Aided Geometric Design 23, 573–581 (2006)
[10] Hogg, T.: Refining the phase transition in combinatorial search. Artificial Intelligence 81, 127–154 (1985)
[11] Monasson, R., Zecchina, R., Kirkpatrick, S., Selman, B., Troyansky, L.: Determining computational complexity from characteristic 'phase transitions'. Nature 400, 133–137 (1999)
[12] Talla Nobibon, F., Cherchye, L., De Rock, B., Sabbe, J., Spieksma, F.C.R.: Heuristics for deciding collectively rational consumption behavior. Research report ces08.24, University of Leuven (2008)
[13] Talla Nobibon, F., Hurken, C., Leus, R., Spieksma, F.C.R.: Coloring graphs to avoid monochromatic cycles. University of Leuven (2009) (manuscript)
[14] Tarjan, R.E.: Depth-first search and linear graph algorithms. SIAM Journal on Computing 2, 146–160 (1972)
[15] Varian, H.: Revealed preference. In: Samuelsonian Economics and the 21st Century (2006)
[16] Wu, Y., Yuan, J., Zhao, Y.: Partition a graph into two induced forests. Journal of Mathematical Study 1, 1–6 (1996)

Randomized Approaches for Nearest Neighbor Search in Metric Space When Computing the Pairwise Distance Is Extremely Expensive

Lusheng Wang[1], Yong Yang[1], and Guohui Lin[2]

[1] Department of Computer Science, City University of Hong Kong, Hong Kong
[2] Department of Computing Science, University of Alberta, Edmonton, Alberta T6G 2E8, Canada
cswangl@cityu.edu.hk, yongyang2@student.cityu.edu.hk, guohui@ualberta.ca

Abstract. Finding the closest object for a query in a database is a classical problem in computer science. For some modern biological applications, computing the similarity between two objects might be very time consuming. For example, it takes a long time to compute the edit distance between two whole chromosomes and the alignment cost of two 3D protein structures. In this paper, we study the nearest neighbor search problem in metric space, where the pair-wise distance between two objects in the database is known and we want to minimize the number of distances computed on-line between the query and objects in the database in order to find the closest object. We have designed two randomized approaches for indexing metric space databases, where objects are purely described by their distances with each other. Analysis and experiments show that our approaches only need to compute $O(\log n)$ objects in order to find the closest object, where n is the total number of objects in the database.

1 Introduction

Finding the closest object for a query in a database is a classical problem in computer science. For some modern biological applications, computing the similarity between two objects might be very time consuming. For example, it takes a long time to compute the edit distance between two whole chromosomes [1,2] and the alignment cost of two 3D protein structures [3]. With the rapid development of biological sequence and structure databases, efficient methods for finding the closest object to a query in those biological databases are desperately required.

Here we study the nearest neighbor search problem in metric space, where the pair-wise distance between two objects in the database is known (pre-computed) and the distance function forms a metric space. That is, for any objects o_i, o_j and o_k, the distance function satisfies the following conditions:

1. $d(o_i, o_j) \geq 0$, and $d(o_i, o_j) = 0$ if and only if $o_i = o_j$;
2. $d(o_i, o_j) = d(o_j, o_i)$,
3. $d(o_i, o_j) \leq d(o_i, o_k) + d(o_k, o_j)$.

B. Chen (Ed.): AAIM 2010, LNCS 6124, pp. 243–252, 2010.

We want to minimize the number of distances computed on-line between the query and the objects in the database in order to find the closest object. We have designed two randomized approaches for indexing metric space databases, where objects are purely described by their distances from each other.

Lots of work have been done for the nearest neighbor search problem. Some methods organize the database in the preprocessing phase. The M-tree method is first introduced in [4]. It splits the data-set into spheres represented by nodes in the M-tree. Each node can hold a limited number of objects, and is led by a routing object (pivot). Every pivot maintains a covering radius that is the largest distance between itself and all the data objects in its covering tree, which is the subtree rooted at the node led by this pivot. During the M-tree construction stage, when a new data object arrives, its distances to pivots stored in the root node are computed and it is inserted into the most suitable node (i.e., led by the closest pivot from the root node) or the node that minimizes the increase of the covering radius, and subsequently descended down to a leaf node. If the size of a node reaches the pre-defined limit, this node is split into two new nodes with their pivots elected accordingly and inserted into their parent node. In the query stage, M-tree uses the covering radius to perform triangle inequality pruning. SA-Tree [5] uses spatial approximation inspired by the Voronoi diagram to reduce the number of visited subtrees during the query stage.

Non-hierarchical search methods [6,7,8,9,10,11] select in the preprocessing phase a set of pivots and compute the distance between each pivot and every database object; during query processing, all the pivots are used collectively to prune objects from the k-nn candidate list using the triangle inequality. That is, any object whose estimated lower bound is larger than the k-th smallest estimated upper bound is deleted. Approximating and Eliminating Search Algorithm (AESA) was proposed by Vidal [12]. Assume that all the pairwise distances in the database are known. The algorithm randomly chooses an object in the database, and computes the distance between the object and the query. If it is better than the current result, update the current result and the minimum distance. When a new sample is obtained, the symmetry property is used to eliminate all the objects that are impossible to be the result, and a lower bound for each remaining object can be obtained. The object with the smallest lower bound in the remaining database is selected as the next sample. Other methods such as Burkhard-Keller Tree, Fixed Queries Trees and Bisector Tree can be found in [13].

Our analysis and experiments show that our approaches only need to compute $O(\log n)$ objects in order to find the closest object, where n is the number of objects in the database.

2 Preliminary

In a metric database, the distribution of the objects plays a crucial role in designing indices. The philosophy of indexing is to put similar objects in one group or creating a representative for a group of similar objects. For some bad distributions, it is impossible to give an indexing so that the query can take less than $O(n)$ computation of pairwise distance.

A bad example: Let $U = \{o_1, o_2, \ldots, o_n\}$ be a database containing n objects. The distance between any two objects o_i and o_j is defined as follows:

$$d(o_i, o_j) = 1 \text{ if } i \neq j; \tag{1}$$

$$d(o_i, o_j) = 0 \text{ if } i = j. \tag{2}$$

In this database, object o_i has an equal distance to any other objects in the database. Let q be the query and o be the object that is 0-distance from q. Then, for any other object $o' \in U - \{o\}$, the distance between q and o' is 1. To search o from the database, it does not help much even if we know the distances between q and a set of objects $\{o_{i_1}, o_{i_2}, \ldots, o_{i_k}\}$, where $o \neq o_{i_j}$ for $j = 1, 2, \ldots, k$. In this case, we have to compute the distance between q and every object in the database one by one until the 0-distance object is found.

The above example hints us to design indices based on the distribution of objects. Here we assume that we have pre-computed the distance between any pair of objects in the database. For each fixed object o, we sort the objects according to their distances to the fixed object o. Then each object can serve as a representative for a set of objects that are within distance r to o for any given r.

3 The Basic Approach

Let U be the set of all objects in the database. For any objects o_i and o_j, we use $d(o_i, o_j)$ to denote the distance between objects o_i and o_j. In this section, we assume that we have pre-computed $d(o_i, o_j)$ for all pairs of o_i and o_j. For each object o_i, we sort all the objects in U according to their distance to o_i in a non-decreasing order. For each object o_i and a radius r, we define $G(o_i, r) = \{o | d(o, o_i) \leq r\}$ be the set of objects in U that are within distance r from o_i.

Let q be the query and o_q be the object in U that is closest to q. Our basic idea is to randomly and independently select a set of t objects $S = \{s_1, s_2, \ldots, s_t\}$ from U. We then compute the distance $d(q, s_i)$ for $i = 1, 2, \ldots, t$. Let s^* be the object in S with $d(q, s^*) \leq d(q, s_i)$ for any $s_i \in S$. Let $r_S = d(q, s^*)$. The object o_q that is the closest to q must be in $G(q, r_S)$. However, since we have not compared q with all the objects in U, $G(q, r_S)$ is not known. The key idea here is to reduce the size of the set of candidate objects that o_q is in. First, we know that o_q is in $G(s^*, 2r_S)$. Second, for any $s_i \in S - \{s^*\}$, $o_q \in G(s_i, d(q, s_i) + r_S)$. Finally, for any $s_i \in S - \{s^*\}$, o_q is not in $G(s_i, d(q, s_i) - r_S)$. Therefore, we define

$$U(S) = G(s^*, 2r_S) \bigcap_{s_i \in S} G(s_i, d(q, s_i) + r_S) - \bigcup_{s_i \in S} G(s_i, d(q, s_i) - r_S). \tag{3}$$

To demonstrate the advantages of this formula, we use the m-dimensional Euclidean space to illustrate. In the m-dimensional Euclidean space, the volume

The Outside-In Algorithm	
Input	the set of objects U in the database and a query q.
Output	An object in U that is closest to the query q.
1.	$S = \emptyset$;
2.	randomly and independently select t samples from U and add them into S and construct $U(S)$;
3.	set $U = U(S)$ and repeat Step 2 until the size of $U(S)$ is less than or equal to t;
4.	return the object that is closest to q among S and the remaining objects.

Fig. 1. The Outside-In Algorithm

of a ball with radius $2r$ is 2^m times of the volume of a ball with radius r. Assume that the objects are evenly distributed in the space and the number of objects in $G(o, r)$ is proportional to the volume. Then the number of objects in $G(s^*, 2r_S)$ is 2^m times of that in $G(q, r_S)$. In other words, the number of objects in $G(s^*, 2r_S)$ would be significantly larger than that in $G(q, r_S)$. Obviously, when the size of S increases, the sizes of both $G(s^*, 2r_S) \bigcap_{s_i \in S} G(s_i, d(q, s_i) + r_S)$ and $G(s^*, 2r_S) - \bigcup_{s_i \in S} G(s_i, d(q, s_i) - r_S)$ decrease.

Consider the m-dimensional Euclidean space with points $(x_1, x_2, \ldots, x_m)$, where $x_i \in (-L, L)$ for $i = 1, 2, \ldots, m$. For each point $(x_1, x_2, \ldots, x_m)$ in U, x_i is a random number in $(-L, L)$. $G(s, r)$ represents a ball with radius r centering at s. Let S be a set of randomly selected points in the space, q the query and s^* the point with $d(s^*, q) = r_S$.

Lemma 1. *With high probability,*

$$\lim_{|S| \to \infty} G(s^*, 2r_S) \bigcap_{s_i \in S} G(s_i, d(q, s_i) + r_S) = G(q, r_S).$$

Consider the 2-dimensional Euclidean space. The circle $G(q, r_S)$ for $q = (0, 0)$ is bounded within a square formed by the four lines $x = r_S$, $x = -r_S$, $y = r_S$ and $y = -r_S$. Thus, if S contains the four line and circle intersection points, then the size of $G(s^*, 2r_S) \bigcap_{s_i \in S} G(s_i, d(q, s_i) + r_S)$ is upper bounded by the size of the square which is $c \times |G(q, r_S)|$ for some constant c. In m-dimensional Euclidean space, if S contains the $2m$ points that form the $2m$ hyper-planes of an m-dimensional cube, then $G(s^*, 2r_S) \bigcap_{s_i \in S} G(s_i, d(q, s_i) + r_S)$ is upper bounded by $[\Gamma(1 + \frac{m}{2}) / (\frac{\pi}{4})^{\frac{m}{2}}] \times G(q, r_S)$, where $\Gamma(\cdot)$ is the well-known Gamma function. This upper bound is too big for our purpose. Therefore, in practice, we have to use a very large number of sample points for high dimension.

In our algorithm, we always try to use formula (3) to reduce the remaining set of candidate objects. Here r_S is the distance between q and the object in S that is the closest to q. Obviously, $U(S)$ always contains the object o_q that is the closest object to q. The complete algorithm is given in Figure 1.

Note that, in order to make full use of selected samples, in our algorithm, the sample set S contains *all* the sample objects selected in all steps. When the size of S is getting bigger, more and more objects will be eliminated from $U(S)$. Obviously, the size of $U(S)$ depends on the distribution of the objects in U.

3.1 The Size of $G(q, r_S)$

In this section, we will set up the relationship between $|G(q, r_S)|$ and $|S|$. Let $F(o, k)$ be the set of objects in U containing the first $\frac{|U|}{k}$ objects in the sorted list for o, i.e. $|F(o, k)| = \frac{|U|}{k}$. In order to get $|G(q, r_S)| \leq |F(q, k)| = \frac{|U|}{k}$ for any integer $k > 0$, we estimate how many sample objects in S are required.

Lemma 2. *Let B be an integer and $|S| = Bk$. With probability at least $1 - e^{-\frac{B}{2}+1}$, $G(q, r_S) \leq \frac{|U|}{k}$.*

Lemma 2 shows that it is very easy to control the size of $G(q, r_S)$. For example, in order to get $|G(q, r_S)| \leq 0.01|U|$, with probability at least $1 - \frac{1}{|U|}$, we only have to set $k = 100$ and $B = 2 \ln |U| + 2$. The remaining task is to control the size of $U(S)$ defined in (3).

3.2 Experiments on the Number of Required Samples

In order to estimate the relationships between the number of samples and the number of objects in the database, we first do some simulations in the m-dimensional space. We choose to use m-dimensional space due to the following reasons: (1) it is easy to generate objects in this space, (2) it is very fast to compute the distance between objects so that we can study high dimensions, and (3) the performance of the algorithm might be similar if the distribution function $|G(o, r)|$ for objects in U are similar, though the algorithm is designed for the case, where computing distance between objects is extremely expensive.

The set of objects in the database U is randomly generated in a cube of the m-dimensional Euclidean space. The relationships between the number of objects in U and the total number of samples required in order to get o_q for $m = 3, 4, \ldots, 10$ are shown in Figure 2. We also did experiments for $m = 11, 12, \ldots, 19$, the relationships are very similar with $m = 3, 4, \ldots, 10$.

We can see that with the increase of the number of objects in U, the total number of samples required increases fast at the beginning and then grows very slowly after the number of samples reach a certain constant number for each fixed m. The constant numbers of samples, C_m, for different m are listed in Table 1.

From Lemma 2 and the experimental results, we know that for an m-dimensional Euclidean space, when the number of samples in S reaches a certain constant number C_m, the size of $U(S)$ for each round is approximately

$$c \times |G(q, r_S)| \tag{4}$$

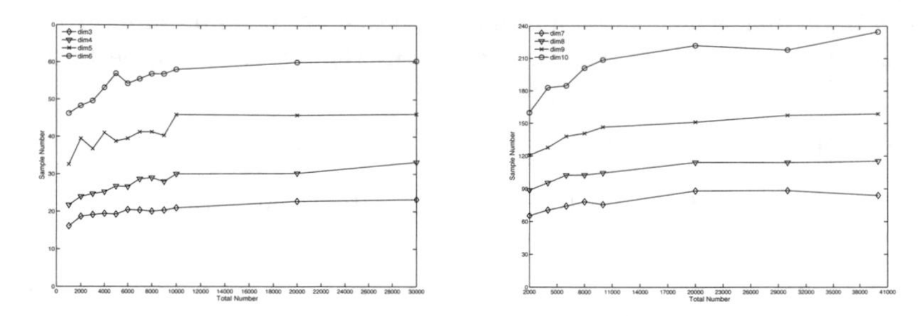

(a) The number of samples required for $m = 3, 4, 5$, and 6.

(b) The number of samples required for $m = 7, 8, 9$, and 10.

Fig. 2. The relationship between number of samples and the number of objects in U

Table 1. The constant C_m

m	3	4	5	6	7	8	9	10	
C_m	18	29	49	59	89	115	158	240	
m	11	12	13	14	15	16	17	18	19
C_m	300	450	640	860	1250	1800	2600	3500	4800

for some constant c. From Lemma 2, we know that at the i-th time that Steps 2-3 are repeated, if we choose a set $S_i(t)$ of $t = Bk$ samples and add them into S, the size of $G(q, r_S)$ satisfies

$$|G(q, r_S)| \leq \frac{G(q, r_{S'})}{k}, \tag{5}$$

where S' is the set of all samples selected up to last round and $S = S' \cup S_i(t)$ is the set of all samples selected up to the current round. From (4) and (5), we know that the size of $U(S)$ is upper bounded by

$$|U(S)| \leq c \times |G(q, r_S)| \leq \frac{c}{k}|G(q, r_{S'})| \approx \frac{|U(S')|}{k}. \tag{6}$$

Thus, for each round, the number of candidate objects that are the closest to q is reduced by $1/k$ times. Therefore, the total number of rounds required is at most $\log_k |U|$. The total number of samples required is at most $C_m + \log_k |U| \times Bk$.

Proposition 1. *For the m-dimensional Euclidean space, the number of samples required to find the closest object to q is $C_m + \log_k |U| \times Bk$.*

Proposition 1 explains why the number of samples required almost remains the same when the number of objects in U is big enough.

The value of C_m: Table 1 shows that C_m increases very fast when m increases. When m changes from i to $i + 1$, the C_{i+1} increases by approximately 30%.

4 The Inside-Out Algorithm

In Figure 1, we use $U(S)$ defined in (3) to keep the candidate objects. $U(S) \cup S$ always contains the desired object o_q. The size of $U(S)$ is reduced from $|U|$ to t in the process. This is a kind of outside-in algorithm. This algorithm needs lots of samples for a high dimensional space. In this section, we propose the Inside-Out approach.

There are two important ideas in this algorithm. First, we randomly select a sample s from U. Second, we keep two sets U_{in} and U_{out} of objects and two radii r_1 and r_2. U_{out} is the set of remaining objects in the database that contains the candidate objects. $r_2 = \min_{s \in S} d(q, s)$ be the smallest distance between the query q and a sample object examined so far. r_1 is the lower bound of the possible distance between q and any object in U_{out}. (Initially, $r_1 = 0$.) We set $r = r_1 + (r_2 - r_1)/3$ as a guess of the smallest distance between q and an object in U_{out}. We use r as the radius in (3) to get $U_{in} = U_{out} \cap_{s \in S} G(s, d(q, s) + r) - \cup_{s \in S} G(s, d(q, s) - r)$, where S is the set of all samples we have selected. Since the value of r is smaller, the size of U_{in} is smaller than that of U_{out}. After obtaining U_{in}, we randomly select samples from U_{in} one by one. For the currently selected sample s, there are two cases. Case 1: we find a sample s in U_{in} such that $d(q, s) < r_2$. In this case, we update r_2 and U_{out} by setting $r_2 = d(q, s)$ and $U_{out} = U_{out} \cap_{s \in S} G(s, d(q, s) + r_2) - \cup_{s \in S} G(s, d(q, s) - r_2)$. Case 2: $d(q, s) \geq r_2$. In this case, we reduce the size of U_{in} by setting $U_{in} = U_{in} \cap G(s, d(q, s) + r) - G(s, d(q, s) - r)$. At the end, either r_2 is updated or U_{in} becomes empty. The latter case, r_1 is updated as r. The process is repeated until U_{out} becomes empty. We then output the sample s in S such that $d(q, s) = \min_{s' \in S} d(q, s')$ as the closest object to q.

There are two circles with radii r_1 and r_2, respectively. The main point is to guess a smaller value of the possible distance between q and the objects and select samples in the smaller circle in the hope that the sample is closer to q. The complete algorithm is given in Figure 3.

4.1 Experiments on the Number of Required Samples

In order to estimate the relationships between the number of samples and the number of objects in the database, we again do some simulations in the m-dimensional space. We run 300 times the Outside-In algorithm since the performance of this algorithm is relatively stable. We run 1000 times the Inside-Out algorithm. To show the performance of our algorithms, we compare our algorithm with the AESA algorithm. We run 1000 times the AESA algorithm.

The set of objects in the database U is randomly generated in a cube of the m-dimensional Euclidean space. We test the cases for $|U| = 10000, 20000, 30000, 40000$, and 50000. The total numbers of samples required in order to obtain o_q for $m = 2, 3, \ldots, 20$ are shown in Table 2. Because of the page limit, we only show the case for $|U| = 50000$. The result for $|U| = 10000, 20000, 30000$, and 40000 are similar to the case for $|U| = 50000$. *Ave. No.* is the average number of samples

The Inside-Out Algorithm

Input the set of objects U in the database and a query q.
Output An object in U that is closest to the query q.

1. $S = \emptyset$; $r_1 = 0$; $r_2 = \infty$; $U_{out} = U$; $U_{in} = U$;
2. randomly select a sample s from U_{out}, $r_2 = d(q, s)$, $s^* = s$, and $S = S \cup \{s\}$;
3. $r = r_1 + (r_2 - r_1)/3$.
4. $U_{out} = U_{out} \cap_{s \in S} G(s, d(q, s) + r_2) - \cup_{s \in S} G(s, d(q, s) - r_2)$.
5. $U_{in} = U_{out} \cap_{s \in S} G(s, d(q, s) + r) - \cup_{s \in S} G(s, d(q, s) - r)$.
6. **while** $U_{in} \neq \emptyset$ **do**
7. randomly select a sample s from U_{in}, compute $d(q, s)$, and $S = S \cup \{s\}$;
8. **if** $d(q, s) \geq r_2$ **then** $U_{in} = U_{in} \cap G(s, d(q, s) + r) - G(s, d(q, s) - r)$;
   else $r_2 = d(q, s)$, $r_1 = 0$, $s^* = s$, goto Step 3.
9. $r_1 = r$.
10. **if** $U_{out} \neq \emptyset$ **then** goto Step 3 **else** return s^*.

Fig. 3. The Inside-Out Algorithm

required to find o_q for the Inside-Out algorithm/AESA algorithm. *Outside-in* is the average number of samples required to find o_q for the outside-in algorithm.

The Inside-Out algorithm and AESA algorithm can find o_q after testing some number of samples (see Ave. First) in Table 2. However, when the first time the algorithm finds o_q, the algorithm does not know that this object o_q is the closest object to q and the algorithm has to test more samples before it can stop. *Ave. First* is the average number of samples required to first get o_q for the Inside-Out algorithm/AESA algorithm. We can see that *Ave. First* is much smaller than *Ave. No.* in some cases. *Worst First* is the biggest number of samples that the Inside-Out/AESA algorithm first finds o_q among the 1000 executions. *Worst No.* is the biggest number of samples required for the Inside-Out/AESA algorithm to stop among the 1000 executions. From Table 2, we can see that for high dimensions, the Inside-Out algorithm is better than AESA algorithm and Outside-In algorithm.

5 Reducing the Space

Both proposed methods assume that we know the pairwise distances for all pairs of objects. Let n be the number of objects in U, then we needs $O(n^2)$ space to store the pairwise distance matrix. When the size of the database is big, the size of the pairwise distance matrix cannot be completely stored in the memory. To reduce the required space, we can decompose the set of objects in U into p groups U_1, U_2, ..., U_p. For each group U_i, we have a pairwise distance matrix D_i of size $O(|U_i|^2)$ to store the pairwise distance between objects in U_i. We also randomly select a set S_g of $O(C_m)$ sample objects from U and construct a (global) distance matrix D of size $|S_g| \times |U|$ to store the distances between every object in S_g and every object in U. In this way, the size of the required space can be dramatically reduced.

Table 2. Number of samples required for 50000 objects. For the Ave. No., Worst No., Ave. First and Worst First, the first number is for the Inside-Out algorithm, and the second number is for the AESA algorithm.

m	Ave. No.	Worst No.	Ave. First	Worst First	Outside-in
2	6/4	9/9	6/4	9/9	17
3	7/5	12/12	7/5	11/10	24
4	9/8	18/20	8/6	15/17	35
5	12/10	27/29	10/8	23/19	50
6	15/14	49/43	11/9	36/30	66
7	20/19	63/58	13/11	42/41	88
8	27/26	102/129	16/12	57/39	125
9	39/38	204/192	18/14	70/52	157
10	54/55	383/404	21/16	112/80	233
11	79/81	513/678	25/18	115/108	335
12	123/127	1278/1393	28/21	138/143	449
13	194/202	1800/1869	35/25	240/211	614
14	301/313	1990/2023	39/26	207/162	871
15	443/459	3363/3543	48/31	798/542	1221
16	748/772	6346/6648	57/34	384/462	1800
17	1035/1062	7314/7561	62/36	551/386	2158
18	1513/1541	9126/9279	77/43	979/1715	2837
19	2267/2308	14783/15181	92/45	1247/724	3764
20	3017/3063	15604/15762	113/52	2660/2002	4791

Table 3. Experiment results for 200000 objects divided into 10 groups. Original Ave. No. is the average number of samples required to get the closest object when the 200,000 objects are stored in one group. Ave. No. (divided) is the average number of samples required to get the closest object when the 200,000 objects are divided into 10 groups. Global samples is the number of samples in S_g for each fixed m (dimension).

m	Original Ave. No.	Ave. No.(divided)	global samples	overhead
2	19	24	8	22%
3	28	32	11	15%
4	38	43	16	15%
5	54	62	22	15%
6	72	95	31	32%
7	95	122	43	29%
8	130	169	60	30%
9	178	235	84	32%
10	247	336	118	36%
11	349	488	165	40%
12	494	683	231	38%
13	724	918	324	27%
14	993	1319	454	33%
15	1289	1709	635	33%
16	1769	2446	889	38%
17	2518	3512	1245	39%
18	3453	4590	1742	33%
19	4932	6580	2439	33%
20	6537	8674	3415	33%

Experiments: To test the algorithm, we use m dimensional Euclidean space. We randomly generate 200,000 objects (points in m dimensional Euclidean space) as the database U. We then randomly divide the 200,000 objects into 10 groups of size 20,000. The results of Outside-In Algorithm are listed in Table 3. We see that the number of samples required is increased by 15%-40% for different m.

Acknowledgements. The work is fully supported by a grant from the Research Grants Council of the Hong Kong Special Administrative Region, China [Project CityU 121207].

References

1. Altschul, S.F., Madden, T.L., Schffer, A.A., Zhang, J., Zhang, Z., Miller, W., Lipman, D.J.: Gapped BLAST and PSI-BLAST: a new generation of protein database search programs. Nucleic Acids Res. 25, 3389–3402 (1997)
2. Pearson, W.R., Lipman, D.J.: Improved tools for biological sequence comparison. Proceedings of the National Academy of Sciences of the United States of America 85(8), 2444–2448 (1988)
3. Guo, F., Wang, L., Yang, Y.: Efficient Algorithms for 3D Protein Substructure Identification. In: The 4th International Conference on Bioinformatics and Biomedical Engineering, iCBBE 2010 (accepted 2010)
4. Ciaccia, P., Patella, M., Zezula, P.: M-tree: An Efficient Access Method for Similarity Search in Metric Spaces. In: Proceedings of the 23rd International Conference on Very Large Data Bases (VLDB 1997), pp. 426–435 (August 1997)
5. Navarro, G.: Searching in Metric Spaces by Spatial Approximation. The VLDB Journal 11, 28–46 (2002)
6. Marvin, S.: The choice of reference points in best-match file searching. Communnications of the ACM 20, 339–343 (1977)
7. Micó, M.L., Oncina, J., Vidal, E.: A new version of the nearest-neighbour approximating and eliminating search algorithm (AESA) with linear preprocessing time and memory requirements. Pattern Recognition Letters 15, 9–17 (1994)
8. Filho, R.F.S., Traina, A.J.M., Traina Jr., C., Faloutsos, C.: Similarity search without tears: the OMNI-family of all-purpose access methods. In: Proceedings of the 17th International Conference on Data Engineering (ICDE 2001), pp. 623–630 (2001)
9. Bustos, B., Navarro, G., Chávez, E.: Pivot selection techniques for proximity searching in metric spaces. Pattern Recognition Letters 24, 2357–2366 (2003)
10. Rico-Juan, J.R., Micó, L.: Comparison of AESA and LAESA search algorithms using string and tree-edit-distances. Pattern Recognition Letters 24, 1417–1426 (2003)
11. Digout, C., Nascimento, M.A., Coman, A.: Similarity search and dimensionality reduction: Not all dimensions are equally useful. In: Lee, Y., Li, J., Whang, K.-Y., Lee, D. (eds.) DASFAA 2004. LNCS, vol. 2973, pp. 831–842. Springer, Heidelberg (2004)
12. Vidal, E.: An algorithm for finding nearest neighbours in (approximately) constant average time. Pattern Recognition Letters 4, 145–157 (1986)
13. Zezula, P., Amato, G., Dohnal, V., Batko, M.: Similarity Search: The Metric Space Approach. Advances in Database Systems, vol. 32. Springer, Heidelberg (2006)
14. Li, M., Ma, B., Wang, L.: On the closest string and substring problems. Journal of the ACM 49, 157–171 (2002)

A Primal-Dual Approximation Algorithm for the k-Level Stochastic Facility Location Problem

Zhen Wang[1], Donglei Du[2], and Dachuan Xu[1]

[1] Department of Applied Mathematics, Beijing University of Technology, 100 Pingleyuan, Chaoyang District, Beijing 100124, P.R. China
{zw,xudc}@bjut.edu.cn

[2] Faculty of Business Administration, University of New Brunswick, Fredericton, NB Canada E3B 5A3
ddu@unb.ca

Abstract. We present a *combinatorial* primal-dual 7-approximation algorithm for the k-level stochastic facility location problem, the stochastic counterpart of the standard k-level facility location problem. This approximation ratio is slightly worse than that of the primal-dual 6-approximation for the standard k-level facility location problem [3] because of the extra stochastic assumption. This new result complements the recent *non-combinatorial* 3-approximation algorithm for the same problem by Wang et al [21].

Keywords: Facility location; Approximation algorithm; Primal-dual.

1 Introduction

In the *k-level facility location problem* (k-FLP), each client must be served by an open path along k level sets of facilities. Approximation algorithms have been proposed for this problem in the literature, including the non-combinatorial 3-approximation algorithm [1,7] and the combinatorial 3.27-approximation algorithm [2]. For the special case of 2-FLP, a better non-combinatorial 1.77-approximation algorithm exists [23]. The k-FLP is a natural extension of the extensively studied *facility location problem* (FLP) [4,6,8,9,10,12,16]. The latter problem has been extended in many different ways [5,17,18,19,24,25]. In particular, the *stochastic facility location problem* and its variants [11,13,15,20,22] are relevant to the problem we are interested in.

The focus of this work is on the *stochastic k-level facility location problem* (k-SFLP), the stochastic counterpart of the above k-FLP where the demands are assumed to be stochastic. A non-combinatorial LP rounding 3-approximation algorithm exists for this problem [21]. Our contribution in this work is to propose a *combinatorial* primal-dual 7-approximation algorithm by integrating the techniques in [3,10,11].

The significance of combinatorial algorithms compared to the LP-based algorithms is well-established in the literature (e.g. [14]) mainly because the former can be implemented much faster than the latter, and hence more practical in

B. Chen (Ed.): AAIM 2010, LNCS 6124, pp. 253–260, 2010.

solving large-size real-life problems. Moreover, combinatorial algorithms require the exploiting of the underlying combinatorial structures, shedding more lights on the special-structure of the investigated problems.

Formally, the k-SFLP can be defined as follows.

Stochastic k-level facility location problem (k-SFLP): Let D be the set of all clients. Let $F = \bigcup_{\ell=1}^{k} F^\ell$ be the set of all facilities, where each F^ℓ is the set of sites where facilities on level ℓ $(1 \le \ell \le k)$ are located and the sets $F^1, \ldots, F^k$ are pairwise disjoint. We refer $p = (i_1, \ldots, i_k)$ $(i_\ell \in F^\ell,\ \ell = 1, \ldots, k)$ to a path of facilities. The set of all possible paths is denoted by P. The shipping cost between site $i \in F$ and client $j \in D$ is equal to c_{ji} which is metric (that is, satisfying the triangle inequality). The cost incurred by assigning client j to path $p = (i_1, \ldots, i_k)$ is equal to $c_{jp} = \sum_{\ell=1}^{k} c_{ji_\ell}$. Assume that there are S scenarios. Facility i has a first-stage opening cost of f_i^0, and recourse cost of f_i^s in scenario $s \in \{1, \ldots, S\}$. Each given scenario s materializes with probability q_s. Let the set of active clients in the s-th scenario be D_s. For any $t \in \{0, 1, \ldots, S\}$ and any subset $P' \subseteq P$, let P'_t denote the path set P' in the t-th scenario and $f(P'_t) = \sum_{i \in \bigcup_{p \in P'} \{i | i \in p\}} f_i^t$.
The problem is to open a set of paths P_0 in the first stage and a set of paths P_s in the second stage for the s-th scenario, and then assign client $j \in D$ to precisely one facility at each of the k levels along open paths, so as to minimize the expected facility cost $f(P_0) + \sum_{s=1}^{S} q_s f(P_s)$ and the expected connection cost $\sum_{s=1}^{S} q_s \sum_{j \in D_s} \min\left\{ \min_{p \in P_0} \{c_{jp}\}, \min_{p \in P_s} \{c_{jp}\} \right\}$. □

We present the main algorithm and its analysis in sections 2 and 3 respectively, then followed by some concluding remarks in Section 4.

2 The Primal-Dual Algorithm

We introduce some notations first.

$$\mathcal{D} = \{(j, s) | j \in D_s, s = 1, \cdots, S\},\ \mathcal{F} = \{(i, t) | i \in F, t = 0, 1, ..., S\},$$
$$\mathcal{P} = \{(p, t) | p \in P, t = 0, 1, ..., S\}, q_0 = 1,$$
$$c_{jp}^{st} = \begin{cases} c_{jp} & \text{if } t = 0 \text{ or } s, \\ +\infty & \text{otherwise.} \end{cases} \tag{2.1}$$

Let x_{jp}^{st} be equal to 1 if client (j, s) is assigned to path (p, t), and 0 otherwise. Let y_i^t be equal to 1 if facility (i, t) is opened, and 0 otherwise.

Our algorithm will be based on the following integer linear program formulation of k-SFLP [21].

$$
(IP) \quad
\begin{array}{ll}
\min & \sum_{(i,t)\in\mathcal{F}} q_t f_i^t y_i^t + \sum_{(p,t)\in\mathcal{P}} \sum_{(j,s)\in\mathcal{D}} q_s c_{jp}^{st} x_{jp}^{st} \\
\text{s.t.} & \sum_{(p,t)\in\mathcal{P}} x_{jp}^{st} \geq 1, \qquad \forall (j,s)\in\mathcal{D} \\
& \sum_{p:i\in p} x_{jp}^{st} \leq y_i^t, \qquad \forall (j,s)\in\mathcal{D}, \forall (i,t)\in\mathcal{F} \\
& x_{jp}^{st}, y_i^t \in \{0,1\}, \qquad \forall (j,s)\in\mathcal{D}, \forall (p,t)\in\mathcal{P}, \forall (i,t)\in\mathcal{F}.
\end{array}
\tag{2.2}
$$

In the above program, the first constraints ensure that client (j,s) is assigned to at least one path. The second constraints guarantee that no assignment of client (j,s) to a path using facility (i,t) is possible unless facility (i,t) is open, for any given client (j,s) and given facility (i,t).

By relaxing the integrality constraints, we get the linear program relaxation of k-SFLP (note that the constraints $x_{jp}^{st} \leq 1$ and $y_i^t \leq 1$ are implied by other constraints, and hence discarded in the relaxation program). The dual linear program is

$$
\begin{array}{ll}
z_{LP} := \max & \sum_{(j,s)\in\mathcal{D}} \alpha_j^s \\
\text{s.t.} & \alpha_j^s - \sum_{i\in p} \beta_{ji}^{st} \leq q_s c_{jp}^{st}, \qquad \forall (j,s)\in\mathcal{D}, \forall (p,t)\in\mathcal{P} \\
& \sum_{(j,s)\in\mathcal{D}} \beta_{ji}^{st} \leq q_t f_i^t, \qquad \forall (i,t)\in\mathcal{F} \\
& \alpha_j^s, \beta_{ji}^{st} \geq 0, \qquad \forall (j,s)\in\mathcal{D}, \forall (i,t)\in\mathcal{F}.
\end{array}
\tag{2.3}
$$

Intuitively, the first constraint in (2.3) suggests that variables α_j^s can be viewed as a budget that client $(j,s)\in\mathcal{D}$ is willing to pay for getting connected, partially for the connection cost and partially for the facility open cost.

Now we are ready to present our algorithm which is a dual ascent method by generalizing the approaches in [3,10,11].

Algorithm 1. *(Primal-dual algorithm)*

Phase 0. (Initialization). For every $(j,s)\in\mathcal{D}$, initialize α_j^s to 0. All facilities are closed and all clients are unfrozen.

Phase 1. (Construction of a dual feasible solution). We introduce the notion of *time* τ starting from 0. We define the following three concepts before constructing a dual feasible solution.

- A facility $(i_l,t)\in\mathcal{F}$ $(i_l\in F^l)$ is temporarily *open* when $\sum_{(j,s)\in\mathcal{D}} \beta_{ji_l}^{st} = q_t f_{i_l}^t$. Denote by $T_{i_l,t}$ the time when facility $(i_l,t)\in\mathcal{F}$ $(i_l\in F^l)$ becomes temporarily open.
- A client $(j,s)\in\mathcal{D}$ *reaches* $(i_l,t)\in\mathcal{F}$ $(i_l\in F^l)$ if for some path $p = (i_1,i_2,\ldots,i_l)$ from i_1 to i_l, all facilities $(i_1,t),(i_2,t)\ldots,(i_{l-1},t)$ are open and $\alpha_j^s = q_s c_{jp}^{st} + \sum_{l'=1}^{l} \beta_{ji_{l'}}^{st}$.
- If, in addition, also (i_l,t) is open, we say that (j,s) *leaves* (i_l,t) or, in case $l=k$, that (j,s) gets connected.

We increase the dual variables α_j^s for all unfrozen clients $(j, s) \in \mathcal{D}$ uniformly at rate q_s. When the client $(j, s) \in \mathcal{D}$ reaches some closed facility $(i_l, t) \in \mathcal{F}$ $(i_l \in F^l)$, the dual variable $\beta_{ji_l}^{st}$ will be increased at the same rate as α_j^s. When (i_l, t) is open, then freeze all the dual variables $\beta_{ji_l}^{st}$, $(j, s) \in \mathcal{D}$. Keep increasing time τ until there is no unfrozen client. The predecessor of (i_l, t) will be the facility in the level $l - 1$ via which (i_l, t) was for the first time reached by a client, i.e.,

$$\mathrm{pred}(i_l, t) := \mathrm{argmin}_{i \in F^{l-1}}\{T_{i,t} + c_{ii_l}^{tt}\}.$$

The predecessor of a temporarily open (i_1, t) $(i_1 \in F^1)$ will be its closest client and we define the time $T_{\mathrm{pred}(i_1,t)} = 0$. As time increases, the following three cases may occur:

- Facility $(i_k, 0)$ is temporarily open. In this case, freeze those unfrozen clients $(j, s) \in \mathcal{D}$ with $\beta_{ji_k}^{s0} > 0$ and connect them to facility $(i_k, 0)$, which is called the *connecting witness* for (j, s). In addition, denote $p(i_k, 0) = (i_1, \ldots, i_k; 0)$ as the associated central path such that

$$i_l = \mathrm{pred}(i_{l+1}, 0), \forall 1 \leq l \leq k - 1,$$

  and $(j, s)_{i_k}$ as the predecessor of $(i_1, 0)$. We call the neighborhood of $(i_k, 0)$ the set of clients contributing to $p(i_k, 0)$, i.e.,

$$N(i_k, 0) = \{(j, s) \in \mathcal{D} \mid \beta_{ji_l}^{s0} > 0 \text{ for some } i_l \in p(i_k, 0)\}.$$

- Facility (i_k, s) $(s = 1, \ldots, S)$ is temporarily open. In this case, freeze those unfrozen clients $(j, s) \in \mathcal{D}$ with $\beta_{ji_k}^{ss} > 0$ and connect them to facility (i_k, s), which is called the *connecting witness* for (j, s). In addition, denote $p(i_k, s) = (i_1, \ldots, i_k; s)$ as the associated central path such that

$$i_l = \mathrm{pred}(i_{l+1}, s), \forall 1 \leq l \leq k - 1,$$

  and $(j, s)_{i_k}$ as the predecessor of (i_1, s). We call the neighborhood of (i_k, s) the set of clients contributing to $p(i_k, s)$, i.e.,

$$N(i_k, s) = \{(j, s) \in \mathcal{D} \mid \beta_{ji_l}^{ss} > 0 \text{ for some } i_l \in p(i_k, s)\}.$$

- If an unfrozen client (j, s) reaches a temporarily open facility (i_k, t), then freeze (j, s) and connect (j, s) to (i_k, t), which is also called the *connecting witness* for (j, s).

When all clients are frozen, the first phase terminates. If several events occur simultaneously, the algorithm executes them in an arbitrary order.

Phase 2. (Construction of integer primal feasible solution). For each $t \in \{0, 1, \ldots, S\}$, let $\tilde{F}^{kt} \subseteq \{(i, t) \mid i \in F^k\}$ be the set of temporarily open facilities ordered according to nondecreasing T-value at level k. Facilities (i_k, t) and (i'_k, t) $((i_k, t), (i'_k, t) \in \tilde{F}^{kt})$ are *dependent* if there exists some client $(j, s) \in \mathcal{D}$ such that $(j, s) \in N(i_k, t) \cap N(i'_k, t)$.

- Open facility (i_k, t) if and only if there is no (i'_k, t) such that (i'_k, t) is already open, $T_{i'_k,t} \le T_{i_k,t}$ and (i_k, t), (i'_k, t) are dependent. Let the subset $\bar{F}^{kt} \subseteq \tilde{F}^{kt}$ be the finally open set, and also open the associated central path $p(i_k, t)$, $(i_k, t) \in \bar{F}^{kt}$.
- Assign client $(j, s) \in \mathcal{D}$ to an open facility $(i_k, t) \in \bar{F}^k$ by the following rule. If there is (i_k, t) such that $(j, s) \in N(i_k, t)$, then (j, s) is connected to (i_k, t) along the associated central path $p(i_k, t)$. Otherwise, (j, s) is connected to the closest open facility (i_k, s) (or $(i_k, 0)$) along path $p(i_k, s)$ (or $p(i_k, s)$). □

3 Analysis

Firstly, we consider the expected facility cost.

Lemma 2

$$\sum_{t=0}^{S} \sum_{(i_k,t)\in \bar{F}^{kt}} \sum_{i_l \in p(i_k,t)} q_t f_{i_l}^t \le 2 \sum_{(j,s)\in\mathcal{D}} \alpha_j^s.$$

Proof. The cost of opening facilities along a central path $p(i_k, t)$ can be bounded by $\sum_{i_l \in p(i_k,t)} f_{i_l}^t \le \sum_{(j,s)\in N(i_k,t)} \alpha_j^s$. On the other hand, it follows from Phase 2 of Algorithm 1 that each $(j, s) \in \mathcal{D}$ contributes to at most two open facilities in level k.

Secondly, we consider the connection cost of client $(j, s) \in \mathcal{D}$.

Lemma 3. *If $(j, s) \in \mathcal{D}$ is assigned to $p(i_k, 0)$ in Phase 2 of Algorithm 1, then we have $c_{jp(i_k)}^{s0} \le 5\alpha_j^s/q_s$.*

Proof. For any client $(j, s) \in \mathcal{D}$, assume that (j, s) is connected to $(i_k, 0)$ by the path $p(i_k, 0) = (i_1, \ldots, i_k)$ in Phase 2 of Algorithm 1. Consider the following two possibilities.

1. Facility $(i_k, 0)$ is the connecting witness for client (j, s). In this case, there exists a path $p_{i_k,0}$ such that $c_{jp_{i_k}}^{s0} \le \alpha_j^s/q_s$, and $T_{i_k,0} \le \alpha_j^s/q_s$. From the triangle inequality, we have

$$c_{jp(i_k,0)}^{s0} = c_{ji_1}^{s0} + \sum_{l=1}^{k-1} c_{i_l i_{l+1}}^{00} \le c_{jp_{i_k,0}}^{s0} + 2\sum_{l=1}^{k-1} c_{i_l i_{l+1}}^{00} \le \alpha_j^s/q_s + 2T_{i_k,0} \le 3\alpha_j^s/q_s.$$

2. Facility $(i_k, 0)$ is not the connecting witness for client (j, s). From Algorithm 1, assume that (i'_k, t') $(t' = 0$ or $s)$ is the connecting witness of client (j, s). Similar to the above case, we have $c_{jp(i'_k,t')}^{st'} \le 3\alpha_j^s/q_s$. According to Phase 1 of Algorithm 1, there exists a path $p_{i'_k,t'}$ such that

$$c_{jp_{i'_k,t'}}^{st'} \le \alpha_j^s/q_s \quad \text{and} \quad T_{i'_k,t'} \le \alpha_j^s/q_s.$$

If $(i'_k, t') \in \bar{F}^k$, one can show that

$$c^{s0}_{jp(i_k,0)} \leq c^{st'}_{jp(i'_k,t')} \leq 3\alpha^s_j/q_s.$$

Otherwise, we have $(i'_k, t') \notin \bar{F}^{kt'}$ and there exists $(i''_k, t') \in \bar{F}^{kt'}$ such that

$$T_{i''_k,t'} \leq T_{i'_k,t'}, \qquad N(i'_k, t') \cap N(i''_k, t') \neq \emptyset.$$

Let $(j_0, s_0) \in N(i'_k, t') \cap N(i''_k, t')$. From Algorithm 1, there exist two paths $\tilde{p}_{i'_k,t'}$ and $\tilde{p}_{i''_k,t'}$ satisfying

$$c^{s_0t'}_{j_0\tilde{p}_{i'_k,t'}} \leq T_{i'_k,t'}, \; c^{s_0t'}_{j_0\tilde{p}_{i''_k,t'}} \leq T_{i''_k,t'}.$$

So we have that

$$c^{s0}_{jp(i_k,0)} \leq c^{st'}_{jp(i''_k,t')} \leq c^{st'}_{jp(i'_k,t')} + c^{s_0t'}_{j_0\tilde{p}_{i'_k,t'}} + c^{s_0t'}_{j_0\tilde{p}_{i''_k,t'}} \leq 3\alpha^s_j/q_s + 2T_{i'_k,t'} \leq 5\alpha^s_j/q_s.$$

Summarizing the above two cases, we obtain the desired result.

Similar to Lemma 3, we have

Lemma 4. *If $(j, s) \in \mathcal{D}$ is assigned to $p(i_k, s)$ in Phase 2 of Algorithm 1, then we have $c^{ss}_{jp(i_k,s)} \leq 5\alpha^s_j/q_s$.*

Finally, we are ready to present the approximation ratio of Algorithm 1.

Theorem 5. *Algorithm 1 is a 7-approximation combinatorial algorithm for the k-SFLP.*

Proof. Denote SOL as the solution of Algorithm 1, whose cost consists of facility cost F_{SOL} and connection cost C_{SOL} which involve q_t or q_s. It follows from Lemmas 2, 3, and 4 that the total cost of SOL is at most

$$cost(SOL) = F_{SOL} + C_{SOL} \leq 2 \sum_{(j,s)\in\mathcal{D}} \alpha^s_j + 5 \sum_{(j,s)\in\mathcal{D}} \alpha^s_j = 7 \sum_{(j,s)\in\mathcal{D}} \alpha^s_j.$$

4 Concluding Remarks

Further improvement of the approximation ratio for the k-level stochastic facility location problem will be interesting, particularly for combinatorial algorithms.

Acknowledgments. The authors would like to thank three anonymous referees for their helpful comments. The research of the first author is supported by NSF of China (Grant 60773185). The second author's research is supported by the Natural Sciences and Engineering Research Council of Canada (NSERC) grant 283103. The third author's research is supported by Beijing Natural Science Foundation (No. 1102001) and PHR(IHLB).

References

1. Aardal, K.I., Chudak, F., Shmoys, D.B.: A 3-approximation algorithm for the k-level uncapacitated facility location problem. Information Processing Letters 72, 161–167 (1999)
2. Ageev, A., Ye, Y., Zhang, J.: Improved combinatorial apporximation algorithms for the k-level facility location problem. SIAM Journal on Discrete Mathematics 18(1), 207–217 (2004)
3. Bumb, A., Kern, W.: A simple dual ascent algorithm for the multilevel facility location problem. In: Proceedings of APPROX-RANDOM, pp. 55–63 (2001)
4. Byrka, J., Aardal, K.I.: An optimal bifactor approximation algorithm for the metric uncapacitated facility location problem. SIAM Journal on Computing (to appear)
5. Chen, X., Chen, B.: Approximation algorithms for soft-capacitated facility location in capacitated network design. Algorithmica 53(3), 263–297 (2007)
6. Chudak, F., Shmoys, D.B.: Improved approxiamtion algorithms for the uncapaciteted facility location problem. SIAM Journal on Computing 33(1), 1–25 (2003)
7. Gabor, A.F., van Ommeren, J.-K.C.W.: A new approximation algorithm for the multilevel facility location problem. Discrete Applied Mathematics 158(5-6), 453–460 (2010)
8. Guha, S., Khuller, S.: Greedy strike back: Improved facility location algorithms. Journal of Algorithms 31, 228–248 (1999)
9. Jain, K., Mahdian, M., Markakis, E., Saberi, A., Vazirani, V.V.: Greedy facility location algorithm analyzed using dual fitting with factor-revealing LP. Jounal of the ACM 50, 795–824 (2003)
10. Jain, K., Vazirani, V.V.: Approximation algorithms for metric facility location and k-median problems using the primal-dual schema and Lagrangian relaxation. Journal of the ACM 48, 274–296 (2001)
11. Mahdian, M.: Facility location and the analysis of algorithms through factor-revealing programs, Ph. D. thesis, MIT, Cambridge, MA (2004)
12. Mahdian, M., Ye, Y., Zhang, J.: Approximation algorithms for metric facility location problems. SIAM Journal on Computing 36(2), 411–432 (2006)
13. Ravi, R., Sinha, A.: Hedging uncertainty: approximation algorithhms for stochastic optimization problems. Mathematical Programming 108, 97–114 (2006)
14. Schrijver, A.: Combinatorial optimization: Polyhedra and Efficiency. In: Algorithms and Combinatorics, vol. 24(A). Springer, New York (2003)
15. Shmoys, D.B., Swamy, C.: An approximation scheme for stochastic linear programming and its application to stochastic integer programs. Journal of the ACM 53, 978–1012 (2006)
16. Shmoys, D.B., Tardös, E., Aardal, K.I.: Approximation algorithms for facility location problems. In: Proceedings of STOC, pp. 265–274 (1997)
17. Shu, J.: An efficient greedy heuristic for warehouse-retailer network design optimization. Transportation Science, doi:10.1287/trsc.1090.0302
18. Shu, J., Ma, Q., Li, S.: Integrated location and two-echelon inventory network design under uncertainty. Annals of Operations Research, doi:10.1007/s10479-010-0732-z
19. Shu, J., Teo, C.P., Max Shen, Z.J.: Stochastic transportation-inventory network design problem. Operations Research 53, 48–60 (2005)
20. Srinivasan, A.: Approximation algorithms for stochastic and risk-averse optimization. In: Proceedings of SODA, pp. 1305–1313 (2007)

21. Wang, Z., Du, D., Gabor, A., Xu, D.: An approximation algorithm for the k-level stochastic facility location problem. Operations Research Letters, doi:10.1016/j.orl.2010.04.010
22. Ye, Y., Zhang, J.: An approximation algorithm for the dynamic facility location problem. In: Combinatorial Optimization in Communication Networks, pp. 623–637. Kluwer Academic Publishers, Dordrecht (2005)
23. Zhang, J.: Approximating the two-level facility location problem via a quasi-greedy approach. Mathematical Programming 108, 159–176 (2006)
24. Zhang, J., Chen, B., Ye, Y.: A multiexchange local search algorithm for the capacitated facility location problem. Mathematics of Operations Research 30(2), 389–403 (2005)
25. Zhang, P.: A new approximation algorithm for the k-facility location problem. Theoretical Computer Science 384(1), 126–135 (2007)

Optimal Semi-online Scheduling Algorithms on Two Parallel Identical Machines under a Grade of Service Provision

Yong Wu[1,*] and Qifan Yang[2]

[1] Department of Fundamental Education, Ningbo Institute of Technology, Zhejiang University, Ningbo 315100, PR China
[2] Department of Mathematics, Zhejiang University, Hangzhou 310027, PR China
wuyong0417@yahoo.com.cn

Abstract. This paper investigates semi-online scheduling problems on two parallel identical machines under a grade of service (GoS) provision. We consider two different semi-online versions where the optimal offline value of the instance is known in advance or the largest processing time of all jobs is known in advance. Respectively for two semi-online problems, we develop algorithms with competitive ratios of 3/2 and $(\sqrt{5}+1)/2$, which are shown to be optimal.

Keywords: Scheduling; Semi-online; Grade of service; Two machines; Competitive ratio.

1 Introduction

In this paper, we investigate semi-online variants of scheduling problem under a grade of service (GoS) provision. The goal is to minimize the makespan under the constraint that all requests are satisfied. This problem was first proposed by Hwang et al. [1]. It is a common practice in any service industry to provide differentiated services to the customers based on their entitled privileges assigned according to their promised GoS levels. GoS is certainly a highly qualitative concept, yet it is often translated into the level of access privilege to service capacity. Scheduling under a GoS provision has many applications coming from the service industry, computer systems, hierarchical databases, etc.

A scheduling problem is called *online* if jobs arrive one by one, and we are required scheduling jobs irrevocably on machines as soon as they are given, without any knowledge of the successive jobs. If we have full information of job sequence before constructing a schedule, the problem is called *offline*. We call a problem *semi-online* if we know some partial information about the jobs in advance. Algorithms for online (semi-online) problems are called online (semi-online) algorithms.

---

* Corresponding author. Supported by Natural Science Foundation of China (10801121).

B. Chen (Ed.): AAIM 2010, LNCS 6124, pp. 261–270, 2010.

In the study of online and semi-online scheduling, the performance of an algorithm is often measured by its *competitive ratio*. For a job sequence $\mathcal{J}$ and an algorithm A, let $C^A(\mathcal{J})$ (or shortly C^A) be the makespan produced by A and let $C^*(\mathcal{J})$ (or shortly C^*) be the optimal makespan in an offline version (optimal offline value). Then the competitive ratio of A is the smallest number c such that for any instance $\mathcal{J}$, $C^A(\mathcal{J}) \leq cC^*(\mathcal{J})$. An online (semi-online) scheduling problem has a *lower bound* ρ if no online (semi-online) algorithm has a competitive ratio smaller than ρ. An online (semi-online) algorithm A is called *optimal* if its competitive ratio matches the lower bound of the problem.

Hwang et al. [1] first studied the (offline) problem of parallel machine scheduling with GoS eligibility. They proposed an approximation algorithm LG-LPT, and proved that its makespan is not greater than 5/4 times the optimal makespan for $m = 2$ and not greater than $2 - 1/(m-1)$ times the optimal makespan for $m \geq 3$. Glass and Kellerer [2] gave an improved algorithm with a worst-case ratio at most 3/2 for m machines. The hierarchical model considered in [3] is exactly identical to our model, they presented an $e + 1$ for general m machines (also in [4]).

Online scheduling under GoS eligibility was first studied by Park et al. [5] and Jiang et al. [6]. For the problem of online scheduling on two machines with GoS constraint, they independently proposed an optimal algorithm with competitive ratio of 5/3. Afterwards, Jiang [7] extended the result to the general case that there are exactly two GoS levels on m machines. He proved that 2 is a lower bound of online algorithms and proposed an online algorithm with competitive ratio of $(12+4\sqrt{2})/7$. The result was improved to $1+\frac{m^2-m}{m^2-mk+k^2} \leq 7/3$ by Zhang et al. [8], where k is the number of machines with high hierarchy.

The semi-online scheduling under GoS eligibility was also first studied by Park et al. [5]. They considered the semi-online version where the total processing time of all jobs is known in advance, and proposed an optimal algorithm with competitive ratio 3/2. Recently, Liu et al. [9] studied two semi-online versions with bounded jobs, i.e. the processing time of each job is bounded by an interval $[a, \alpha a]$. For both problems, they showed lower bounds and proposed semi-online algorithms. Optimal algorithms were given for some situations. In [10], Chassid and Epstein extended the hierarchial scheduling model to two uniform machines, online and semi-online problems were studied and optimal algorithms were proposed.

In this paper, we analyze the semi-online scheduling problem on two parallel identical machines under GoS provision. We consider two different semi-online versions where the optimal offline value of the instance (denoted by C^*) is known in advance or the largest processing time of all jobs (denoted by p_{max}) is known in advance. C^* and p_{max} are often assumed to be known in advance in semi-online scheduling literature for various reasons as stated in [11,12] and [13], respectively. When we know C^* in advance, we show an optimal algorithm $Gos - Opt$ with competitive ratio of 3/2. The competitive ratio is the same as the optimal algorithm [5] designed for the semi-online problem where the total processing time of all jobs is known in advance. For the second problem, know p_{max} in advance, we also design an optimal algorithm $Gos - Max$. The competitive ratio of $Gos - Max$ is $(\sqrt{5}+1)/2$. Competitive ratios of both algorithms are all better than 5/3 of the online version. These results indicate that knowing C^* is much more useful than p_{max} for designing algorithms for semi-online scheduling problems under GoS provision.

The rest of the paper is organized as follows. In Section 2, we develop formal notations and definitions of our problems. Sections 3 and 4 propose lower bounds and optimal algorithms for two semi-online problems respectively. Finally, some concluding remarks are made in Section 5.

2 Definitions

We are given two parallel identical machines M_1, M_2 and a set $\mathcal{J}$ of n independent jobs $J_1, J_2, \ldots, J_n$. We denote each job by $J_i = (p_i,\ g_i)$, where p_i is the processing time of J_i and $g_i \in \{1, 2\}$ is the GoS level of J_i. $g_i = 1$ if the job J_i must be processed by the first machine M_1, and $g_i = 2$ if it can be processed by either of the two machines. p_i and g_i are not known until the arrival of job J_i. Each job J_i is presented immediately after J_{i-1} is scheduled. Let $G_1 = \{J_i | g_i = 1\}$ and $G_2 = \{J_i | g_i = 2\}$, thus $\mathcal{J} = G_1 \cup G_2$. The schedule can be seen as a partition of $\mathcal{J}$ into two subsequences, denoted by S_1 and S_2, where S_1 and S_2 consist of jobs assigned to machines M_1 and M_2, respectively. Let $L_1 = t(S_1) = \sum\limits_{J_i \in S_1} p_i$ and $L_2 = t(S_2) = \sum\limits_{J_i \in S_2} p_i$ denote the loads (or total processing times) of machines M_1 and M_2, respectively. Hence, the makespan of one schedule is $\max\{L_1,\ L_2\}$. The online problem can be written as:

$$\text{Given } \mathcal{J}, \text{ find } S_1 \text{ and } S_2 \text{ to minimize } \max\{L_1,\ L_2\}.$$

With C^* (or p_{max}) known in advance, the semi-online variant can be stated as:

$$\text{Given } \mathcal{J} \text{ and } C^* \text{ (or } p_{max}\text{), find } S_1 \text{ and } S_2 \text{ to minimize } \max\{L_1,\ L_2\}.$$

To simplify the presentation, the following notations and definitions are required in the remainder of the paper.

- T^k is half of the total processing time of the first k jobs.
- G_i^k is the set of jobs with the GoS level of i, $i = 1,\ 2$ immediately after job J_k is assigned.
- S_i^k is the set of jobs assigned to machine M_i, $i = 1,\ 2$ immediately after job J_k is assigned.
- $t(\delta)$ is the total processing time of jobs in any job set δ.
- $t(G_i^k)$ is the total processing time of jobs in the job set G_i^k, $i = 1,\ 2$. It clearly follows that $t(G_i^n) = t(G_i)$, $i = 1,\ 2$.
- $t(S_i^k)$ is the total processing time of jobs in the job set S_i^k, $i = 1,\ 2$. It clearly follows that $C^A = \max\{L_1,\ L_2\} = \max\{t(S_1^n),\ t(S_2^n)\}$.
- p_{max} is the largest processing time of all jobs.

Let $L^k = \max\{p_{max},\ T^k,\ t(G_1^k)\}$, then we have a lower bound of optimal makespan as described as the following lemma.

Lemma 1. *The optimal makespan of the problem* $C^* \geq L^n \geq L^k$, $1 \leq k \leq n$.

Proof. It is clear that the optimal makespan satisfies $C^* \geq \max\{p_{max},\ T^n\}$ since $T^n = \sum_{1\leq i\leq n} p_i/2$. By the definition of the problem, all the jobs in G_1 only can be processed on the machine M_1, which implies that the optimal makespan $C^* \geq t(G_1) = t(G_1^n)$ according to the definition of $t(G_1^n)$. Thus we have $C^* \geq \max\{p_{max},\ T^n,\ t(G_1^n)\} = L^n \geq L^k$, $1 \leq k \leq n$.

3 Known Optimal Value

In this section, we will show an optimal algorithm for the semi-online variant where the optimal offline value C^* is known in advance. At first, a lower bound of competitive ratio will be shown in the following subsection.

3.1 Lower Bound of Competitive Ratio

Theorem 1. *Any semi-online algorithm A for the problem has a competitive ratio of at least* 3/2.

Proof. First, we declare that the optimal offline value is 2, i.e. $C^* = 2$, and begin with job $J_1 = (1,\ 2)$. If job J_1 is scheduled on the first machine M_1, we generate jobs $J_2 = (2,\ 1)$ and $J_3 = (1,\ 2)$. At this point $C^* = 2$, and we have $C^A \geq 3$ since job J_2 must be scheduled on the first machine, thus $C^A/C^* \geq 3/2$. If job J_1 is scheduled on the second machine M_2, we generate job $J_2 = (2,\ 2)$. If job J_2 is scheduled on the first machine, we generate job $J_3 = (1,\ 1)$ which yields $C^A/C^* = 3/2$. Otherwise, if job J_2 is scheduled on the second machine, we generate job $J_3 = (1,\ 2)$. We have $C^A \geq 3$ no matter which machine job J_3 is scheduled. Thus $C^A/C^* \geq 3/2$ while $C^* = 2$. □

3.2 Optimal Semi-online Algorithm *Gos − Opt*

Next we will design an optimal algorithm with competitive ratio of 3/2. Since we known C^* in advance, we only need to make that the load of each machine is not exceeds $3C^*/2$. Then, when job J_i with $g_i = 2$ arrives, the algorithm assigns it to machine M_2 as far as $t(S_2) + p_i \leq 3C^*/2$, and otherwise to machine M_1. For the analysis of the competitive ratio of the algorithm, we define S_1^{k-1} and S_2^{k-1} to be S_1 and S_2 that we have immediately before we schedule job J_k.

Algorithm *Gos − Opt*

1. Let $S_1 = \emptyset$, $S_2 = \emptyset$;
2. Receive arriving job $J_i = (p_i,\ g_i)$;
3. If $g_i = 1$, let $S_1 = S_1 \cup J_i$. Go to Step 5;

4. If $t(S_2) + p_i \leq 3C^*/2$, let $S_2 = S_2 \cup J_i$; Else, let $S_1 = S_1 \cup J_i$;
5. If no more jobs arrive, stop and output S_1 and S_2; Else, let $i = i + 1$ and go to Step 2.

Theorem 2. *The competitive ratio of the algorithm Gos – Opt for the problem is at most* $3/2$.

Proof. We will show that when the algorithm terminates, $t(S_i) \leq 3C^*/2$, $i = 1, 2$. Which implies that the competitive ratio of *Gos – Opt* is at most $3/2$. According to the algorithm we have $t(S_2) \leq 3C^*/2$. If $t(S_1) \leq 3C^*/2$, then we are done.

Assume $t(S_1) > 3C^*/2$, which yields $t(S_2) < C^*/2$. At this point, we have $t(S_1) > 3C^*/2 \geq 3t(G_1)/2$ since $C^* \geq t(G_1)$. Obviously, the first machine M_1 must process at least one job in G_2. Let $J_k = (p_k,\ g_k)$ be the first job scheduled on the first machine with $g_k = 2$. When job J_k arrives, $t(S_2^{k-1}) \leq t(S_2) < C^*/2$. Together with $p_k \leq C^*$, we have

$$t(S_2^{k-1}) + p_k < 3C^*/2.$$

According to the Step 4 of algorithm *Gos – Opt*, job J_k must be scheduled on the second machine M_2, which is contradicts to the definition of J_k. Thus $t(S_1) \leq 3C^*/2$, and

$$C^{Gos-Opt} = \max\{t(S_1),\ t(S_2)\} \leq 3C^*/2.$$

The competitive ratio of the algorithm *Gos – Opt* is at most $\frac{3}{2}$. □

From Theorems 1 and 2, we know that *Gos – Opt* is the optimal algorithm for the semi-online variant where the optimal offline value is known in advance. And its competitive ratio is $3/2$.

4 Known Largest Processing Time

In this section, we will show an optimal algorithm for the semi-online variant where the largest processing time of all jobs p_{max} is known in advance. The lower bound of competitive ratio will be shown at first.

4.1 Lower Bound of Competitive Ratio

Theorem 3. *Any semi-online algorithm A for the problem has a competitive ratio of at least* $(\sqrt{5} + 1)/2$.

Proof. Without loss of generality, let $p_{max} = 2$. And we begin with job $J_1 = (x,\ 2)$ for $0 < x < 2$, whose exact value we will choose later. If job J_1 is scheduled on the first machine, we generate jobs $J_2 = (2,\ 1)$ and $J_3 = (2 - x,\ 2)$. At this point $C^* = 2$, we have $C^A \geq 2 + x$ since job J_2 must be scheduled on the first machine, thus $C^A/C^* \geq (2 + x)/2$. If job J_1 is scheduled on the second machine, we generate job $J_2 = (2,\ 2)$. If job J_2 is scheduled on the second machine, we generate job $J_3 = (2-x,\ 1)$ which yields $C^A/C^* = (2+x)/2$. Otherwise, if job J_2 is scheduled on the first machine, we generate jobs $J_3 = (2,\ 1)$ and $J_4 = (x,\ 1)$. At this point $C^* = 2+x$ and $C^A = 4+x$, which yields $C^A/C^* \geq (4+x)/(2+x)$.

What remains is to find exact value of x so as to make the competitive ratio of A as large as possible. Based on the above analysis, we have $C^A/C^* \geq \min\{(2+x)/2,\ (4+x)/(2+x)\}$ for any $x:\ 0 < x < 2$. Let $(2+x)/2 = (4+x)/(2+x)$, we have $x = \sqrt{5}-1$. Thus we get $C^A/C^* \geq (\sqrt{5}+1)/2$. □

4.2 Optimal Semi-online Algorithm *Gos − Max*

In this section, we will design an optimal semi-online algorithm $Gos - Max$ with competitive ratio of $\alpha = (\sqrt{5}+1)/2$. Before describing the algorithm, we give some notations. At the arrival of each job, T is updated to become a half of the total processing times of all jobs arrived. Also, $t(G_1)$ is updated to be the total processing time of all arrived jobs with $g_i = 1$. Thus, according to Lemma 1, we have $C^* \geq L = \max\{p_{max},\ T,\ t(G_1)\}$. Then, when job J_i with $g_i = 2$ arrives, the algorithm assigns it to machine M_2 as far as $t(S_2) + p_i \leq \alpha L$, and otherwise to machine M_1.

For the analysis of the competitive ratio of the algorithm, we define T^k, L^k, S_1^k and S_2^k to be T, L, S_1 and S_2 that we have immediately after we schedule job J_k. And we set $T^0 = L^0 = 0$.

Algorithm $Gos - Max$

1. Let $S_1 = \emptyset$, $S_2 = \emptyset$, $T = 0$ and $t(G_1) = 0$;
2. Receive arriving job $J_i = (p_i,\ g_i)$. Let $T = T + \frac{p_i}{2}$;
3. If $g_i = 1$, let $S_1 = S_1 \cup J_i$ and $t(G_1) = t(G_1) + p_i$. Go to Step 5;
4. Let $L = \max\{p_{max},\ T,\ t(G_1)\}$. If $t(S_2) + p_i \leq \alpha L$, let $S_2 = S_2 \cup J_i$; Else, let $S_1 = S_1 \cup J_i$;
5. If no more jobs arrive, stop and output S_1 and S_2; Else, let $i = i + 1$ and go to Step 2.

The proof for the competitive ratio of the proposed semi-online algorithm $Gos - Max$ is by contradiction. Hence, we suppose that there exists a problem instance that we call, a counter example, for which the semi-online algorithm yields a schedule with makespan bigger than 3/2 of the optimal value C^*. Then, the counter example with the least number of jobs is defined to be the minimal counter example. For notational ease in the remainder of this paper, we let $\mathcal{J} = \{J_1, J_2, \cdots, J_n\}$ be the minimal counter example. A lemma about the minimal counter example will be shown at first.

Lemma 2. *For a minimal counter example $\mathcal{J} = \{J_1, J_2, \cdots, J_n\}$, $t(G_2) > \alpha L$ must hold.*

Proof. Suppose $t(G_2) \leq \alpha L$. For any job $J_k = (p_k,\ 2)$, we have $t(S_2^{k-1}) + p_k \leq t(G_2) \leq \alpha L$. According to the Step 4 of the semi-online algorithm $Gos - Max$, job J_k should be assigned to machine M_2. Thus we get $S_1 = G_1$ and $S_2 = G_2$. At this point, we have $C^A = \max\{t(S_1^n),\ t(S_2^n)\} = \max\{t(S_1),\ t(S_2)\} \leq \alpha C^*$ since $t(G_2) \leq \alpha L \leq \alpha C^*$ and $t(G_1) \leq C^* \leq \alpha C^*$. Which is contradicts to the definition of instance $\mathcal{J}$. □

Theorem 4. *The competitive ratio of the algorithm Gos$-$Max for the problem is at most* $\alpha = (\sqrt{5}+1)/2$.

Proof. Suppose that the theorem is false. There must exist a minimal counter example $\mathcal{J} = \{J_1, J_2, \cdots, J_n\}$. Then, due to the minimality, the makespan is not determined until the arrival of job J_n. Therefore, we have

$$C^A = \max\{t(S_1^n),\ t(S_2^n)\} > \alpha C^*, \tag{1}$$

but

$$\max\{t(S_1^{n-1}),\ t(S_2^{n-1})\} \leq \alpha C^*. \tag{2}$$

Next, we distinguish two possible cases according to the GoS level of job J_n.

Case 1 $g_n = 2$.

If J_n is assigned to machine M_2, we have $t(S_2^{n-1}) + p_n \leq \alpha L^n \leq \alpha C^*$ and $t(S_1^n) = t(S_1^{n-1})$. By inequality (1), it follows that $t(S_1^{n-1}) = t(S_1^n) > \alpha C^*$. This contradicts inequality (2).

Hence, job J_n must be assigned to machine M_1. At this point, we have $t(S_2^{n-1}) + p_n > \alpha L^n$, $t(S_2^n) = t(S_2^{n-1})$ and

$$T^n = \frac{t(\mathcal{J})}{2} = \frac{t(S_1^{n-1}) + t(S_2^{n-1}) + p_n}{2} \leq L^n \leq C^*.$$

Hence $t(S_1^{n-1}) < (2-\alpha)L^n \leq (2-\alpha)C^*$, together with $p_n \leq C^*$ we have

$$t(S_1^n) = t(S_1^{n-1}) + p_n < (2-\alpha)L^n + p_n \leq (2-\alpha)C^* + C^* = (3-\alpha)C^*. \tag{3}$$

Note that $(3-\alpha) \leq \alpha$ since $\alpha = (\sqrt{5}+1)/2$, and $t(S_2^n) = t(S_2^{n-1}) \leq \alpha C^*$ implied by (2). Thus the two inequalities (1) and (3) contradict each other.

Case 2 $g_n = 1$.

Due to the minimality, we have

$$C^A = \max\{t(S_1^n),\ t(S_2^n)\} = t(S_1^n) = t(S_1) > \alpha C^*. \tag{4}$$

By Lemma 2, we know that there exists at least one job in G_2 scheduled on machine M_1. Otherwise, we have $C^A/C^* \leq \alpha$ and there exists a contradiction. Hence, we let J_k be the last job with $g_k = 2$ assigned to machine M_1. Also let δ be the job set of jobs assigned to machine M_1 after job J_k assigned to machine M_1. Then $t(\delta) = \sum_{g_i=1,\ i>k} p_i$ and $t(S_1) = t(S_1^n) = t(S_1^k) + t(\delta)$.

In the rest of the proof, we will show that the minimal counter example $\mathcal{J} = \{J_1, J_2, \cdots, J_n\}$ with $g_n = 1$ does not exist exactly. And at first we will give some lemmas about the loads of two machines at the arrival time of job J_k.

Lemma 3. *If job J_k is the last job with $g_k = 2$ assigned to the first machine by the proposed semi-online algorithm $Gos - Max$, $t(S_1^k) > \frac{\alpha-1}{2-\alpha} t(S_2^k)$ must hold.*

Proof. According to (4) and lemma 1, we have

$$t(S_1^k) + t(\delta) > \alpha C^* \geq \alpha t(G_1) \geq \alpha t(\delta),$$

it follows that

$$t(S_1^k) > (\alpha - 1)t(\delta),\ t(\delta) < \frac{1}{(\alpha-1)} t(S_1^k). \tag{5}$$

We also have

$$t(S_1^k) + t(\delta) > \alpha C^* \geq \alpha L^n \geq \alpha \frac{t(S_1^k) + t(S_2^k) + t(\delta)}{2},$$

which implies that

$$t(S_1^k) + t(\delta) > \frac{\alpha}{2-\alpha} t(S_2^k). \tag{6}$$

Combining with (5) and (6), we obtain

$$t(S_1^k) > \frac{\alpha - 1}{2-\alpha} t(S_2^k).$$

□

Lemma 4. *If job J_k is the last job with $g_k = 2$ assigned to the first machine by the proposed semi-online algorithm $Gos - Max$, $t(S_2^k) < \frac{4-2\alpha}{3\alpha-4} p_k$ must hold.*

Proof. According to the definition of job J_k, we have $t(S_2^k) = t(S_2^{k-1})$, thus

$$t(S_2^k) + p_k = t(S_2^{k-1}) + p_k > \alpha L^k. \tag{7}$$

According to the algorithm $Gos - Max$, the following inequality is hold.

$$L^k \geq \frac{t(S_1^k) + t(S_2^k)}{2} = \frac{t(S_1^{k-1}) + p_k + t(S_2^k)}{2}. \tag{8}$$

By (7), (8) and Lemma 3, we have

$$t(S_2^k) + p_k > \alpha \frac{t(S_1^k) + t(S_2^k)}{2} > \frac{\alpha}{2}(\frac{\alpha-1}{2-\alpha} + 1) t(S_2^k) = \frac{\alpha}{2(2-\alpha)} t(S_2^k)$$

$$t(S_2^k) < \frac{4-2\alpha}{3\alpha-4} p_k.$$

□

Lemma 5. *If job J_k is the last job with $g_k = 2$ assigned to the first machine by the proposed semi-online algorithm $Gos - Max$, $t(S_1^{k-1}) < \frac{2-\alpha}{3\alpha-4} p_k$ must hold.*

Proof. According to (7) and (8), we can obtain

$$t(S_1^{k-1}) < \frac{2-\alpha}{\alpha}(p_k + t(S_2^k)). \tag{9}$$

Combine with (9) and Lemma 4, $t(S_1^{k-1}) < \frac{2-\alpha}{3\alpha-4} p_k$ is hold. □

Now we are going to proof that there exists contradiction for the minimal counter example $\mathcal{J}$.

By Lemmas 4 and 5, we have the following inequality since $\alpha = (\sqrt{5}+1)/2$.

$$t(S_2^k) + t(S_1^{k-1}) = t(S_2^{k-1}) + t(S_1^{k-1}) < \frac{6-3\alpha}{3\alpha-4} p_k \leq \alpha p_{max}. \quad (10)$$

The equality (10) implying that job J_k is the only job in G_2 assigned to machine M_1. Otherwise, suppose there is another job $J_b = (p_b,\ 2) \in G_2$ assigned to machine M_1. Note that job J_k is the last job with $g_k = 2$ assigned to machine M_1. Hence job J_b is arrive before job J_k, i.e. $b \leq k-1$, implying that

$$t(S_2^b) + p_b \leq t(S_2^k) + t(S_1^{k-1}) \leq \alpha p_{max} \leq \alpha L^b. \quad (11)$$

Inequality (11) shows that job J_b must be assigned to machine M_2 by Step 4 of algorithm $Gos-Max$, which contradicts the definition of job J_b. Thus there is only one job in $G_2 \cap S_1$, i.e. $G_2 \cap S_1 = \{J_k\}$. Namely, $S_1 = G_1 \cup \{J_k\}$ and $t(S_1) = t(G_1) + p_k$. Together with inequality (4), we have

$$t(S_1) = t(G_1) + p_k > \alpha C^* > \alpha L^n. \quad (12)$$

Note that

$$L^n \geq (t(S_1) + t(S_2))/2 = (t(G_1) + p_k + t(S_2))/2 \geq (t(G_1) + p_k + t(S_2^k))/2,$$

together with (12) we obtain $t(S_2^k) \leq t(S_2) \leq \frac{2-\alpha}{\alpha}(t(G_1) + p_k)$. We also have $t(G_1) < \frac{1}{(\alpha-1)} p_k$ according to (12) since $L^n \geq t(G_1)$. Thus, we have

$$t(S_2^k) \leq \frac{2-\alpha}{\alpha}(t(G_1) + p_k) \leq \frac{2-\alpha}{\alpha-1} p_k. \quad (13)$$

From the definitions of L^k and job J_k, we know that $t(S_2^k) + p_k = t(S_2^{k-1}) + p_k > \alpha L^k \geq \alpha p_{max}$. At this point, we have

$$t(S_2^k) > (\alpha - 1) p_{max}. \quad (14)$$

Combined inequalities (13) and (14), we get $p_k > \frac{(\alpha-1)^2}{2-\alpha} p_{max} \geq p_{max}$ since $\alpha = (\sqrt{5}+1)/2$. That's a flat contradiction.

According to the above discussion of two possible cases of the minimal counter example, we can make sure that there exists no such minimal counter example exactly. Therefore, the competitive ratio of the algorithm $Gos-Max$ is at most $\alpha = (\sqrt{5}+1)/2$. □

From Theorems 3 and 4, we know that $Gos-Max$ is the optimal algorithm for the semi-online variant where the largest processing time of all jobs is known in advance. And its competitive ratio is $(\sqrt{5}+1)/2$.

5 Conclusions

In this paper, we have studied two semi-online scheduling problems on parallel identical machines under a grade of service provision. We considered two semi-online versions where we known C^* or p_{max} in advance. Optimal algorithms were proposed for both scheduling problems on two machines. It is left as open problems to design (optimal) algorithms on $m > 2$ parallel identical machines.

References

1. Hwang, H., Chang, S., Lee, K.: Parallel machine scheduling under a grade of service provision. Comput. Oper. Res. 31, 2055–2061 (2004)
2. Glass, C., Kellerer, H.: Parallel machine scheduling with job assignment restrictions. Naval Res. Logis. 54, 250–257 (2007)
3. Bar-Noy, A., Freund, A., Naor, J.: On-line load balancing in a hierarchical server topology. SIAM J. Comput. 31, 527–549 (2001)
4. Crescenzi, P., Gambosi, G., Penna, P.: On-line algorithms for the channel assignment problem in cellular networks. Discret. Appl. Math. 137(3), 237–266 (2004)
5. Park, J., Chang, S.Y., Lee, K.: Online and semi-online scheduling of two machines under a grade of service provision. Oper. Res. Lett. 34, 692–696 (2006)
6. Jiang, Y.W., He, Y., Tang, C.M.: Optimal online algorithms for scheduling on two identical machines under a grade of service. J. Zhejiang Univ. Sci. A 7(3), 309–314 (2006)
7. Jiang, Y.W.: Online scheduling on parallel machines with two GoS levels. J. Comb. Optim. 16, 28–38 (2008)
8. Zhang, A., Jiang, Y.W., Tan, Z.Y.: Online parallel machines scheduling with two hierarchies. Theoret. Comput. Sci. 410, 3597–3605 (2009)
9. Liu, M., Chu, C.B., Xu, Y.F., Zheng, F.F.: Semi-online scheduling on 2 machines under a grade of service provision with bounded processing times. J. Comb. Optim. (2009), doi:10.1007/s10878-009-9231-z
10. Chassid, O., Epstein, L.: The hierarchical model for load balancing on two machines. J. Comb. Optim. 15, 305–314 (2008)
11. Azar, Y., Epstein, L.: On-line machine covering. J. Sched. 1, 67–77 (1998)
12. Azar, Y., Regev, O.: On-line bin-stretching. Theoret. Comput. Sci. 168, 17–41 (2001)
13. He, Y., Zhang, G.C.: Semi on-line scheduling on two identical machines. Computing 62(3), 179–187 (1999)

Varieties of Regularities in Weighted Sequences

Hui Zhang[1], Qing Guo[2,*], and Costas S. Iliopoulos[3]

[1] College of Computer Science and Technology,
Zhejiang University of Technology, Hangzhou, Zhejiang 310023, China
zhangh@zjut.edu.cn
[2] College of Computer Science and Engineering,
Zhejiang University, Hangzhou, Zhejiang 310027, China
Tel.: 0086-571-88939701, Fax: 0086-571-88867185
guoqing@tiansign.com
[3] Department of Computer Science, King's College London Strand,
London WC2R 2LS, England
csi@dcs.kcl.ac.uk

Abstract. A weighted sequence is a string in which a set of characters may appear at each position with respective probabilities of occurrence. A common task is to identify repetitive motifs in weighted sequences, with presence probability not less than a given threshold. We consider the problems of finding varieties of regularities in a weighted sequence. Based on the algorithms for computing all the repeats of every length by using an iterative partitioning technique, we also tackle the all-covers problem and all-seeds problem. Both problems can be solved in $O(n^2)$ time.

1 Introduction

A weighted biological sequence, called for short a *weighted sequence*, can be viewed as a compressed version of multiple alignment, in the sense that at each position, a set of characters appear with respective probability, instead of a fixed single character occurring in a normal string.

Weighted sequences are apt at summarizing poorly defined short sequences, e.g. transcription factor binding sites and the profiles of protein families and complete chromosome sequences [7]. With this model, one can attempt to locate the biological important motifs, to estimate the binding energy of the proteins, even to infer the evolutionary homology. It thus exhibits theoretical and practical significance to design powerful algorithms on weighted sequences.

This paper concentrates on those repetitive motifs, specially *regularities*, in a weighted sequence. It has been an effort for a long time to identify special areas in a biological sequence by their structure. Examples are repetitive genomic segments such as tandem repeats, long interspersed nuclear sequences and short interspersed nuclear sequences. The motivation comes from the striking feature of DNA that vast quantities of repetitive structures occur in the genome.

* Corresponding author.

B. Chen (Ed.): AAIM 2010, LNCS 6124, pp. 271–280, 2010.

The most simple repetitive motifs are repeatedly occurred segments, called *repeats*. Formally speaking, a repeat of a string x is a substring that repeatedly occurs in x. As the extensions to repeats, the most common regularities in strings have been found to be those that are periodically repetitive. We focus on two typical regularities, notably the *covers* and the*seeds*. A cover is a substring w of x such that x is structured by concatenations and superpositions of w. A seed is an extended cover in the sense of a cover of a superstring of x. For instance, The substring *aba* is a cover of $x = abababa$, while *bab* is a seed of x.

It is of biological interest to locate regularities in biological sequences. As early as in 1970's, Ohno proposed that primordial proteins might evolve from periodic amplifications of oligopeptides [13]. Thus internal repeating segments in proteins may serve important roles in functional evolution of proteins.

It turns out that locating all the repeats forms a basis for further discerning covers and seeds from them. Since we have designed efficient algorithms for computing all the repeats in a weighted sequence [15], we will rely on these efforts and apply the results about the repeats to the computation of covers and seeds in the paper.

Large amount of work has been done to locate repeats and regularities in non-weighted strings [1,5,9,12], but relatively small in weighted sequences. Iliopoulos et al. [8] were the first to touch this field, and extract repeats and other types of repetitive motifs in weighted sequences by constructing weighted suffix tree. They also apply Crochemore's partitioning technique [4] into weighted sequences, and presented an $O(n^2)$-time algorithm for finding all tandem repeats [10,11]. Another solution [2,3] finds all the repeats as well as covers of length d in $O(n \log d)$ time.

The paper is organized as follows. In the next section we give the necessary theoretical preliminaries used, and make a simple description of our algorithms for computing all the repeats of weighted sequences. Based on these outcome, we tackle the all-covers problem and all-seeds problem respectively in Section 3 and Section 4. Finally in Section 5 we conclude and discuss our research interest.

2 Preliminaries

A biological sequence used throughout the paper is a string either over the 4-character DNA alphabet Σ ={A,C,G,T} of nucleotides or the 20-character alphabet of amino acids. Assume that readers have essential knowledge of the basic concepts of strings, now we extend parts of it to weighted sequences. Formally speaking:

Definition 1. *Let an alphabet be $\Sigma = \{\sigma_1, \sigma_2, \ldots, \sigma_l\}$. A weighted sequence X over Σ, denoted by $X[1,n] = X[1]X[2]\ldots X[n]$, is a sequence of n sets $X[i]$ for $1 \leq i \leq n$, such that:*

$$X[i] = \left\{(\sigma_j, \pi_i(\sigma_j)) | \, 1 \leq j \leq l, \pi_i(\sigma_j) \geq 0, \text{ and} \sum_{j=1}^{l} \pi_i(\sigma_j) = 1\right\}$$

Each $X[i]$ *is a set of couples* $(\sigma_j, \pi_i(\sigma_j))$*, where* $\pi_i(\sigma_j)$ *is the non-negative* weight *of* σ_j *at position* i*, representing the probability of having character* σ_j *at position* i *of* X*.*

Let X be a weighted sequence of length n, σ be a character in Σ. We say that σ *occurs* at position i of X if and only if $\pi_i(\sigma) > 0$, written as $\sigma \in X[i]$. A nonempty string $f[1,m]$ ($m \in [1,n]$) *occurs* at position i of X if and only if position $i+j-1$ is an occurrence of the character $f[j]$ in X, for all $1 \leq j \leq m$. Then f is said to be a *factor* of X, and i is an *occurrence* of f in X. The probability of the presence of f at position i of X is called the *weight* of f at i, written as $\pi_i(f)$, which can be obtained by using different weight measures. We exploit the one in common use, called the *cumulative weight*, defined as the product of the weight of the character at every position of f: $\pi_i(f) = \prod_{j=1}^{m} \pi_{i+j-1}(f[j])$.

Considering a weighted sequence:

$$X = \left\{ \begin{array}{c} (A, 0.5) \\ (C, 0.25) \\ (G, 0.25) \end{array} \right\} G \left\{ \begin{array}{c} (A, 0.6) \\ (C, 0.4) \end{array} \right\} \left\{ \begin{array}{c} (A, 0.25) \\ (C, 0.25) \\ (G, 0.25) \\ (T, 0.25) \end{array} \right\} C \quad (1)$$

the weight of f=CGAT at position 1 of X is: $\pi_1(f) = 0.25 \times 1 \times 0.6 \times 0.25 = 0.0375$. That is, CGAT occurs at position 1 of X with probability 0.0375.

Repeats are those repeated factors in weighted sequences, however, the following remarks draws a distinction due to the feature of weighted sequences.

Remark 1. The weight of each appearance of a repeat can be highly different.

Remark 2. Observe the above example (1), the factor AGC has two occurrences at position 1 and 3 in X, can we thereby approve it to be an overlapping repeat? Undoubtedly, the two appearances of AGC do have one common area, i.e. position 3, but simply in structure rather than in symbol. In other words, position 3 simultaneously "contributes" two different characters to each presence of AGC respectively, C for the first and A for the second. This is unique in weighted sequences, which comes from the uncertainty at each position of weighted sequences.

Definition 2. *Let* $f_1[1, m_1]$ *and* $f_2[1, m_2]$ *be two factors of a weighted sequence* $X[1, n]$*. We say that there exist structural overlaps between* f_1 *and* f_2 *at position* i *of* X *if for some* $j \in [1, \min(m_1, m_2)]$*:*

1. $f_1[m_1 - j + l] \in X[i + l - 1]$ *and* $f_2[l] \in X[i + l - 1]$ *for all* $l \in [1, j]$
2. *there exists at least a* j *such that* $f_1[m_1 - j + l] \neq f_2[l]$

Depending on whether structural overlaps are acceptable or not, we reinterpret the repeats, further the covers and the seeds, in weighted sequences as two types:

Definition 3. *A repeat in a weighted sequence is a loose repeat if structural overlaps are allowed, otherwise a strict repeat. A factor* f *of a weighted sequence*

$X[1,n]$ is a strict cover of X if concatenations and overlaps of copies of f form a factor of X of length n, and a loose cover if structural overlaps are permitted as well. Moreover, f is a strict (resp. loose) seed of X if f is a strict (resp. loose) cover of a superstring of a factor of X of length n.

As we have mentioned above, we have presented efficient solutions to the all-repeats problem [15] defined as below:

Problem 1. Given a weighted sequence $X[1,n]$ and a real number $k \geq 1$, the *all-loose(strict)-repeats* problem seeks in X all the loose (strict) repeats of every possible length having the probability of appearance at least $1/k$.

The algorithm for picking all the loose repeats is based on the following idea of equivalence relation on positions of the string and the partitioning lemma:

Definition 4. *Given a string $x[1,n] \in \Sigma^*$ and two positions $i, j \in \{1, \ldots, n-p+1\}$ of x, then $(i,j) \in E_p$ iff $x[i, i+p-1] = x[j, j+p-1]$, denoted iE_pj.*

Lemma 1. *Let $p \in \{1, 2, \ldots, n\}$, $i, j \in \{1, 2, \cdots, n-p\}$.*

$$(i,j) \in E_p \text{ iff } (i,j) \in E_{p-1} \text{ and } (i+p-1, j+p-1) \in E_1$$

Computing the strict repeats is a bit complicated. The difficulty arises when adjacent appearances of a repeat in X are overlapping. By introducing the notion of border check array, we can also solve the all-strict-repeats problem in the same $O(n^2)$ time with the loose counterpart. Readers can refer to [15] for more details about the algorithms.

3 Computing the Covers

Problem 2. Given a weighted sequence $X[1,n]$ and a real number $k \geq 1$, the *all-loose(strict)-covers* problem is to find all possible proper loose (strict) covers of X with presence probability at least $1/k$.

The fact that a cover is by all means a repeat of X suggests an immediate solution to the all-covers problem: first locate all the repeats of X, then check each if it is a cover. A cover $f[1,p]$ of X complies with the following two basic facts:

1. The first occurrence of f in X is always 1, and the last occurrence of f in X is $n-p+1$.
2. Any distance between adjacent occurrences of f should not exceed $|f|$.

Testing a repeat of length p if it is a cover is at most the cardinality of the input, and at stage p all the covers of length p are reported in $O(n)$ time. Combining this function with our algorithms for computing loose repeats and strict repeats separately, we succeed in identifying all the loose covers and strict covers, respectively. It is clear that the all-covers problem can be answered in $O(n^2)$ time, the same with the corresponding all-repeats problem.

4 Computing the Seeds

Problem 3. Given a weighted sequence $X[1, n]$ and a real number $k \geq 1$, the *all-loose(strict)-seeds* problem is to find all possible proper loose (strict) seeds of X with presence probability at least $1/k$.

Commonly, the first (resp. last) appearance of a seed in X might be incomplete, shown as the structure of a suffix (resp. prefix) of this seed.

Definition 5. *Given a weighted sequence $X[1, n]$, a real factor $f[1, p]$ is called a* candidate seed *of X if there is a factor x' of $X = Hx'T$ such that f covers x' and $|H|, |T| < p$. For maximal such x', we call H (resp. T) the* head *(resp.* tail*) of X with respect to f.*

Note that both H and T could be weighted or normal strings. In order for a candidate seed to be a true one, it must suffice to cover a left extension of the sequence Hf as well as a right extension of the sequence fT. If it does, a seed of X can be reported.

4.1 The ECT and the RECT

To help locating the seeds, we reinterpret the idea of the Equivalence Class Tree (ECT) and the Reversed Equivalence Class Tree (RECT) for weighted sequences that was first introduced for non-weighted strings [6,14].

Let $\{f_1, \ldots, f_r\}$ be the real factors of length $p-1$ of X, denote $\{C_{f_1}, \ldots, C_{f_r}\}$ to be the E_{p-1}-classes associated with these factors. The ECT is created as follows: The root has label 0. There are r nodes of depth $p-1$, each of which is a pair $(C_{f_i}, f_i)(i \in [1, r])$. For the convenience of explanation, we label each node by f_i instead of the pair in the ECT. The children of f_i are the E_p-classes partitioned by C_{f_i} according to Lemma 1, corresponding to those real factors of length p produced by each f_i reading one character to the right. The construction of the ECT proceeds along with the computation of equivalence classes, until at stage L all the nodes are not repeats of X.

Constructing the RECT is similar, except that the refinement of E_p from E_{p-1} counts on the following corollary of Lemma 1:

Corollary 1. *Let $p \in \{1, 2, \ldots, n\}$, $i, j \in \{2, \cdots, n-p+1\}$. Then:*

$$(i-1, j-1) \in E_p \text{ iff } (i, j) \in E_{p-1} \text{ and } (i-1, j-1) \in E_1$$

Let $C_f = \{i_1, i_2, \ldots, i_q\}$ be an E_{p-1}-class associated with a factor f of X. This corollary indicates another partitioning technique, which partitions those positions $i_1 - 1, i_2 - 1, \ldots, i_q - 1$ to generate a set of E_p-classes, corresponding to those factors extended by each presence of f in X one character to the left. We add these factors into the RECT as the children of f instead.

For example, Figure 1 shows the ECT and the RECT of the following weighted sequence, when $k = 5$.

$$X = \text{TAT[(A,0.5),(C,0.3),(T,0.2)]AT[(A,0.5),(C,0.5)] A[(C,0.5),(T,0.5)]A}$$

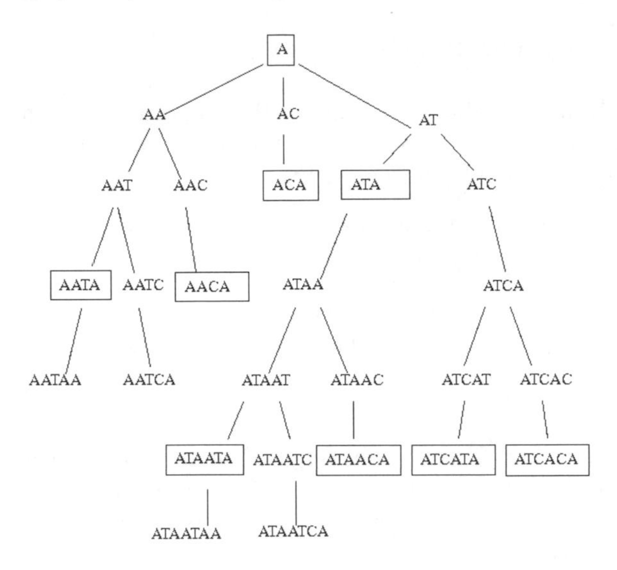

Fig. 1. A subtree of the ECT of X rooted at A

We label each factor that ends to position n in the ECT, called an *end-aligned* factor in the following context. Correspondingly, each factor that starts at position 1, called a *start-aligned* factor is marked in the RECT. The following facts about the ECT and the RECT are easily observed:

Fact 1. *In any branch of the ECT, every internal node represents a proper prefix of the leaf, of decreasing length from bottom to top.*

Fact 2. *In any path from the leaf to the root of the RECT, each internal node represents a proper suffix of the leaf in order of decreasing length.*

Both the ECT and the RECT are rooted trees built upon the partitioning of equivalence classes, each of which expresses the relationship between each E_{p-1}-class and its corresponding E_p-classes. The declaration that each tree is constructed along with the partitioning of equivalence classes implies that the construction takes time no more than the partitioning, that is $O(n^2)$ for either the ECT or the RECT.

4.2 Loose Seeds

The solution to the all-loose-seeds problem is straightforward: first locate all the loose repeats with probability at least $1/k$, then determine the candidate seeds from them, and check which are true seeds. We simply discuss the latter step.

Consider a E_p-class C_f that corresponds to a loose repeat $f[1, p]$ with probability not below $1/k$. Denote e_1 and e_t to be the first and the last occurrence of f in X respectively, then the head $|H| = e_1 - 1$, the tail $|T| = n - e_t + 1 - p$. During the construction of C_f, We maintain the maximum difference between adjacent occurrences, denoted by max_gap. Hence, in order for f to be a candidate loose seed, its length must suffice to: $p \geq max(e_1, \lceil (n - e_t + 2)/2 \rceil, max\_gap)$.

There are two steps involved to verify a candidate seed f:

1. Check the tail T to test if f covers a right extension of fT.

If f covers a right extension of fT, we say that f is *end covered.* This work can be done depending on the ECT. Trivially, an end-aligned candidate is end covered since the tail T is an empty string in this case. If f is not marked in the ECT, we turn to consider its nearest marked ancestor $anc(f)$. By Fact 1, $anc(f)$ is a prefix of f. The definition of end-aligned factors says that:

- if T includes branching positions: there is a factor f' of fT of the same length $n - e_t + 1$ such that $anc(f)$ is a suffix of f'.
- if T is a normal string: $anc(f)$ is a suffix of fT.

Thus in each case, $anc(f)$ is the longest substring that is both a suffix and a prefix, i.e. the border of f' or fT. Then:

- if $|anc(f)| \geq |T| = n - e_t + 1 - p$: f' or fT is a concatenation or superposition of f and $anc(f)$. Hence, f is end covered.
- if $|anc(f)| < |T|$: f cannot cover any right extension of f' or fT.

Consider the above example weighted sequence, set $k = 5$. A candidate seed ATAA is not marked in the ECT. As $C_{ATAA} = \{2, 5\}$, $e_t = 5$, we can compute the tail $|T| = 2$. $anc(ATAA) = ATA$. Thus we conclude that ATAA is also end covered since $|anc(ATAA)| = 3 > |T|$.

To efficiently implement the above tail testing, our algorithm adds two elements into the node pair, then each node in the ECT is redefined as below:

Definition 6. *A node f in the ECT is a quadruple: $Node(f)$=(C_f, f, P-Align, E-Align), where C_f stands for the equivalence class corresponding to a factor f of X. P-Align points to the nearest ancestor of f that is marked in the ECT. E-Align is a boolean value for f, where E-Align is TRUE if f itself is end-aligned, and FALSE otherwise.*

For instance, a node ATAA in the ECT can be denoted to be: $Node(ATAA) = (\{2, 5\}, ATAA, ATA,$ FALSE). All the values of each node is updated along with the construction of the ECT, which yields a direct one-step checking for a candidate loose seed, to be end covered or notas shown in Algorithm 1.

2. Check the head H to test if f covers a left extension of Hf.

This step is symmetric to step (1). If f covers a left extension of Hf, we say that f is *start covered.* We utilize the RECT to help checking the head.

Trivially, a start-aligned candidate is start covered since it is equivalent to an empty head H. If f is not marked in the RECT, we turn to consider its nearest start-aligned ancestor $anc'(f)$. By Fact 2, $anc'(f)$ is a suffix of f. The definition

Algorithm 1. Test if a candidate loose seed of length p is end covered

Input: An E_p node: $Node(f_p)=(C_{f_p}, f_p, P\text{-}Align, E\text{-}Align)$
Output: A boolean variable

1: **Function** Test-Tail-Loose($Node(f_p)$)
2: **if** $p = 0$ **then**
3: $f_p \leftarrow$ NULL
4: $par(f_p) \leftarrow$ parent of f_p in the ECT
5: **if** $e_t + p - 1 = n$ **then**
6: $Node(f_p).E\text{-}Align \leftarrow$ TRUE
7: **if** $Node(par(f_p)).E\text{-}Align$=TRUE **then**
8: $Node(f_p).P\text{-}Align \leftarrow par(f_p)$
9: **else**
10: $Node(f_p).P\text{-}Align \leftarrow Node(par(f_p)).P\text{-}Align$
11: TailTag $\leftarrow$ TRUE
12: **else**
13: $Node(f_p).E\text{-}Align \leftarrow$ FALSE
14: **if** $Node(par(f_p)).E\text{-}Align$=TRUE **then**
15: $Node(f_p).P\text{-}Align \leftarrow par(f_p)$
16: **else**
17: $Node(f_p).P\text{-}Align \leftarrow Node(par(f_p)).P\text{-}Align$
18: **if** $Node(f_p).P\text{-}Align$=NULL **or** $|Node(f_p).P\text{-}Align| < n - e_t - p + 1$ **then**
19: TailTag $\leftarrow$ FALSE
20: **else**
21: TailTag $\leftarrow$ TRUE
22: **return** Tailtag

of start-aligned factors says that $anc'(f)$ occurs at position 1, thus $anc'(f)$ is the border of either Hf if H is a normal string, or a factor of Hf of length exactly $e_1 + p - 1$ otherwise. If $|anc'(f)| \geq |H| = e_1 - 1$, f is verified to be start covered, otherwise not.

Every node in the RECT is also denoted by a quadruple in our algorithm: $Node(f)=(C_f, f, P\text{-}Align, E\text{-}Align)$. The difference is that *E-Align* is a boolean symbol identifying if f itself is start-aligned or not, and *P-Align* points to the nearest marked ancestor of f in the RECT. Therefore, with a slight modification to Algorithm 1, we can obtain the function for testing if a candidate loose seed of length p is start covered, with the same time complexity.

As we mentioned in Section 4.1, both the ECT and the RECT can be constructed in $O(n^2)$ time. During an refinement of E_p from E_{p-1}, all the values of the corresponding node quadruple $Node(f)$ of length p in the ECT and the RECT are updated in constant time. Then checking f if it is end covered or start covered simply takes one step by checking the length of the border of f. Thus it needs $O(1)$ time to test a loose repeat of length p if it is a seed.

As a matter of fact, our algorithm proceeds along with the construction of the two trees. Once an equivalence class of length p is computed, we immediately report if it is a seed or not. Therefore, the overall running time of our solution to the all-loose-seeds problem is exactly the same with that of all-loose-repeats problem, i.e. $O(n^2)$.

4.3 Strict Seeds

Answering the all-strict-seeds problem is similar, that is, first compute all the strict repeats with probability at least $1/k$, then recognize the strict seeds from them. The method for determining a candidate strict seed is the same as what have done for loose counterpart. However, the distinction arises when we test if a candidate strict seed is end covered.

As we discussed above, if the nearest marked ancestor $anc(f)$ of a candidate loose strict f in the ECT is longer than the length of the tail, it is definitely correct that f is end covered. But this argument might not infer a candidate strict seed f. As $anc(f) > |T|$ implies, fT (if T is a normal string) or f' (if T is a weighted sequence) is a superposition of f and $anc(f)$. Thus the reason we hesitate is that, the possibility of branching positions to exist in these overlapping positions might result in structural overlaps between f and $anc(f)$, which is not allowed for strict seeds.

Our solution to this uncertainty is that for each such overlapping position that is a branching position, we execute a character comparison between $anc(f)$ and f. If each comparison comes out a match, f is verified to be end covered. Otherwise, we climb up to test the next end-aligned ancestor of f in the ECT following the above process, until we reach the root or the length of the end aligned ancestor to be touched is less than $|T|$. The candidate f is rejected if no eligible ancestor is available.

Back to the example given in Section 4.1, ATAAT is a candidate strict seed that is not marked in the ECT, as $C_{ATAAT} = \{2, 5\}$. The tail $|T| = n - (e_t + p - 1) = 1$, $anc(ATAAT) = ATA$. There are two overlapping positions between ATAAT and ATA, specially, the second one is a weighted position. Thus we need to check if $anc(ATAAT)[2] = ATAAT[5]$. Clearly it is a match telling us ATAAT is end covered. However, although $anc(ATAAC) = ATA$, $anc(ATAAC)[2] \neq ATAAC[5]$, we have to test a shorter border A that also guarantees ATAAC to be end covered.

We simply modify Algorithm 1 to implement testing the tail for strict seeds. The only difference between the algorithms for strict seeds and those for loose ones, is that the former might trace back to more than one marked ancestor, however, it runs constant times as well. Testing a candidate if it is start covered is similar performed. Therefore, the all-strict-seeds problem can also be answered in $O(n^2)$ time.

5 Conclusions

The paper investigated a series of problems on the regularities arisen in weighted sequences, including finding all the covers and seeds, in both loose and strict sense. As opposed to the loose versions, identifying strict regularities needed more skills when structural overlaps are not permitted. However, we devised efficient algorithms for all these problem, each of which operates in $O(n^2)$ time.

Nevertheless, it still leaves space to improve the time complexity. Thus we are tempting to save the space and implement these algorithms in a more efficient way in the future research.

References

1. Brodal, G.S., Lyngsø, R.B., Pedersen, C.N.S., Stoye, J.: Finding Maximal Pairs with Bounded Gap. Journal of Discrete Algorithms, Special Issue of Matching Patterns 1(1), 77–104 (2000)
2. Christodoulakis, M., Iliopoulos, C.S., Mouchard, L., Perdikuri, K., Tsakalidis, A., Tsichlas, K.: Computation of repetitions and regularities on biological weighted sequences. Journal of Computational Biology 13(6), 1214–1231 (2006)
3. Christodoulakis, M., Iliopoulos, C.S., Perdikuri, K., Tsichlas, K.: Searching the regularities in weighted sequences. In: Proc. of the International Conference of Computational Methods in Science and Engineering. Lecture Series on Computer and Computational Sciences, pp. 701–704. Springer, Heidelberg (2004)
4. Crochemore, M.: An Optimal Algorithm for Computing the Repetitions in a Word. Information Processing Letter 12(5), 244–250 (1981)
5. Franêk, F., Smyth, W.F., Tang, Y.: Computing All Repeats Using Suffix Arrays. Journal of Automata, Languages and Combinatorics 8(4), 579–591 (2003)
6. Guo, Q., Zhang, H., Iliopoulos, C.S.: Computing the λ-covers of a string. Information Sciences 177(19), 3957–3967 (2007)
7. Gusfield, D.: Algorithms on Strings, Trees and Sequences: Computer Science and Computational Biology. Cambridge University Press, Cambridge (1997)
8. Iliopoulos, C.S., Makris, C., Panagis, Y., Perdikuri, K., Theodoridis, E., Tsakalidis, A.: Efficient Algorithms for Handling Molecular Weighted Sequences. IFIP Theoretical Computer Science 147, 265–278 (2004)
9. Iliopoulos, C.S., Moore, D.W.G., Park, K.: Covering a String. Algorithmica 16, 288–297 (1996)
10. Iliopoulos, C.S., Mouchard, L., Perdikuri, K., Tsakalidis, A.: Computing the repetitions in a weighted sequence. In: Proc. of the 8th Prague Stringology Conference (PSC 2003), pp. 91–98 (2003)
11. Iliopoulos, C.S., Perdikuri, K., Zhang, H.: Computing the regularities in biological weighted sequence. In: String Algorithmics. NATO Book series, pp. 109–128 (2004)
12. Li, Y., Smyth, W.F.: Computing the Cover Array in Linear Time. Algorithmica 32(1), 95–106 (2002)
13. Ohno, S.: Repeats of base oligomers as the primordial coding sequences of the primeval earth and their vestiges in modern genes. Journal of Molecular Evolution 20, 313–321 (1984)
14. Zhang, H., Guo, Q., Iliopoulos, C.S.: Algorithms for Computing the λ-regularities in Strings. Fundamenta Informaticae 84, 33–49 (2008)
15. Zhang, H., Guo, Q., Iliopoulos, C.S.: Loose and strict repeats in weighted sequences. In: Proc. of the International Conference on Intelligent Computing, ICIC 2009 (accepted 2009)

Online Uniformly Inserting Points on Grid

Yong Zhang[1,2], Zhuo Chang[1], Francis Y.L. Chin[2,⋆],
Hing-Fung Ting[2,⋆⋆], and Yung H. Tsin[3,⋆⋆⋆]

[1] College of Mathematics and Computer Science, Hebei University, China
changzhuo@cmc.hbu.cn
[2] Department of Computer Science, The University of Hong Kong, Hong Kong
{yzhang,chin,hfting}@cs.hku.hk
[3] School of Computer Science, University of Windsor, Canada
peter@uwindsor.ca

Abstract. In this paper, we consider the problem of inserting points in a square grid, which has many background applications, including halftone in reprographic and image processing. We consider an online version of this problem, i.e., the points are inserted one at a time. The objective is to distribute the points as uniformly as possible. Precisely speaking, after each insertion, the gap ratio should be as small as possible. In this paper, we give an insertion strategy with a maximal gap ratio no more than $2\sqrt{2} \approx 2.828$, which is the first result on uniformly inserting point in a grid. Moreover, we show that no online algorithm can achieve the maximal gap ratio strictly less than 2.5 for a 3×3 grid.

1 Introduction

In this paper, we consider the problem of online inserting points in a square grid such that the distribution of the inserted points is as uniform as possible. In the real world, there are many applications needing an uniform distribution of some values in a given area, e.g., halftone, distribution of chain stores in an area.

Halftone is a very important technique in image processing, which simulates the actual continuous image by discrete dots so that in the view of human's eyes, the simulation is almost identical to the original image. To achieve better performances, e.g., higher resolutions, *dithering* method [3] is often applied in halftone. One of the most important tasks in dithering is how to generate the dither matrix, on which the quality of the simulation heavily depends. Each element in the dither matrix represents a threshold value of the grey level between black and white. For example, consider the dither matrix in Figure 1. An absolutely dark spot with the grey level of 32 will be able to meet all the threshold values of the dither matrix, thus, all the 64 elements (pixels) will be black; similarly, a grey level of 0 will have all white pixels. As for any of the remaining grey level x, only

⋆ Research supported by HK RGC grant HKU-7113/07E and the William M.W. Mong Engineering Research Fund.
⋆⋆ Research supported by HK RGC grant HKU-7171/08E.
⋆⋆⋆ Research supported by NSERC under grant NSERC 7811-2009.

B. Chen (Ed.): AAIM 2010, LNCS 6124, pp. 281–292, 2010.

those matrix elements (pixels) whose values (thresholds) are equal to or below x will turn black, e.g., for grey level = 10, only 10 elements (pixels) are black and these elements have to be distributed uniformly inside the matrix. Since the uniformity has to be applied to all grey levels, this reduces to our problem which is online inserting points uniformly in a square grid. Figure 1 gives an example of a dither matrix in [3] and the simulation based on this matrix. Formally, the dither matrix is an $n \times n$ matrix in which the value of each element is in the range from 0 to $n^2 - 1$, and the values up to each i ($0 \leq i \leq n^2 - 1$) are uniformly distributed.

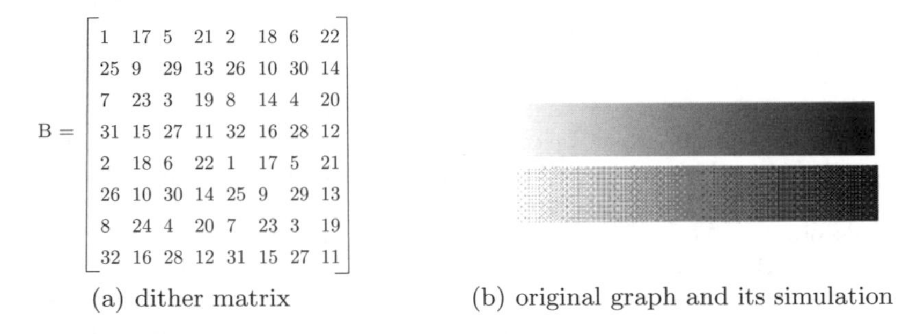

(a) dither matrix (b) original graph and its simulation

Fig. 1. A simulation based on a dither matrix

Another motivation is the distribution of chain stores in an area. A famous chain store has planned to establish its business in a district by establishing a number of stores at the road junctions in a city with Manhattan-like road network one at a time. Assuming that the clients are distributed uniformly, and each client will be served by its nearest store. In order to minimize the unnecessary competition among its own stores, the established stores at any time should be distributed as uniformly as possible while the stores once established cannot be dismantled or relocated.

In this problem, we consider the insertion of the points in an online manner, i.e., the points are inserted one by one, and the algorithm does not know the number of inserted points in advance. After the insertion of each point, the uniformity is guaranteed. The uniformity is a measurement of how uniform the inserted points are distributed. There are several ways to define the uniformity of a set of points. Some studies define the uniformity by the minimal pairwise distance [5,7]. In discrepancy theory [4,6], uniformity is defined by the ratio between the maximal and minimal number of points in a fixed shape within the area. In this paper, uniformity is defined by the *gap ratio*, i.e., the ratio between the maximal gap (the diameter of the largest empty circle) and the minimal gap (the minimal pairwise distance).

Problem Statement
Let $\mathcal{S}^2$ be an $m \times m$ unit square grid in the 2-dimensional square $\mathcal{R}^2$ such that the four corners of $\mathcal{S}^2$ and $\mathcal{R}^2$ are located at the same position. Consider any

point request sequence with n requests. In the initial state, each of the four corner grid positions in $\mathcal{S}^2$ are assigned a point. Each following request must be assigned on some grid position in $\mathcal{S}^2$, and each grid position can satisfy at most one request, thus $n \leq (m+1)^2 - 4$. Let p_i be the grid position used in satisfying the i-th request, and $\mathcal{S}_i = \{p_1, ..., p_i\} \bigcup \mathcal{S}_0$ be the configuration in $\mathcal{S}^2$ after inserting the i-th point, where $\mathcal{S}_0$ consists of the four corner points of $\mathcal{S}^2$.

Define the maximal gap at step i to be $G_i = \max_{p \in \mathcal{R}^2} \min_{q \in \mathcal{S}_i} 2d(p,q)$, the minimal gap at step i to be $g_i = \min_{p,q \in \mathcal{S}_i, p \neq q} d(p,q)$, where $d(\cdot,\cdot)$ is the Euclidean distance, and define the i-th gap ratio as $r_i = G_i/g_i$. The maximum gap and the minimum gap imply the diameter of the largest empty circle[1] and the minimum pairwise distance, respectively.

The objective of this problem is assigning points into the grid as uniformly as possible, i.e., minimize the maximal gap ratio ($\min \max_i r_i$) for each insertion.

For the $m \times m$ square grid, let $(0,0)$ represent its upper-left-most point a. Each grid point p is represented by (i,j), where i is the difference between the x-coordinate of a and p, j is the difference between the y-coordinate of a and p. We say that a square or rectangle is of size $i \times j$ if the lengths of two adjacent edges of the square or rectangle are i and j respectively. Let R be the circumradius of a triangle UVW, we have $|R| = \frac{uvw}{4\Delta}$, where u, v and w are the length of edges of the triangle and Δ is the area of the triangle.

Now we give an example to illustrate the maximal gap, the minimal gap, and the gap ratio. At the initial state, there are only four assigned points at the four corners a, b, c and d as shown in Fig. 2, the maximal gap $G_0 = \sqrt{2} \cdot m$ while the minimal gap $g_0 = m$, the gap ratio $r_0 = \sqrt{2}$. If the first point p_1 is inserted at the center of the square, the current maximal gap $G_1 = m$ while the minimal gap $g_1 = \sqrt{2} \cdot m/2$ and the gap ratio $r_1 = \sqrt{2}$.

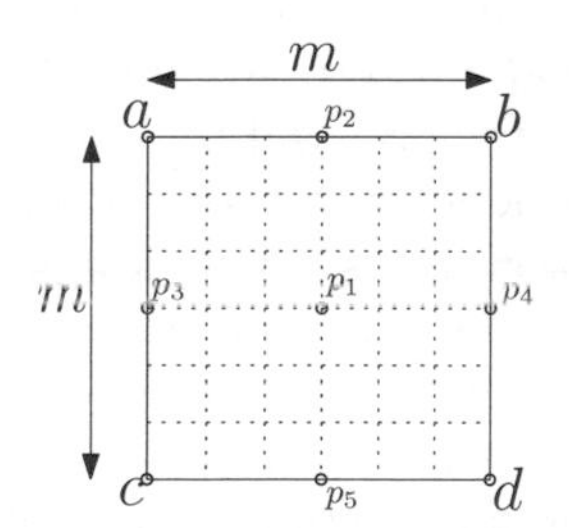

Fig. 2. An insertion of five points in a grid

Related Works:

Uniformly inserting points in a given area had been studied before. If the points can be inserted at any position in the given area, Teramoto et al [8] and Asano et al [2] showed that the greedy algorithm (voronoi insertion) has uniformity 2.

[1] Since we only focus on the area $\mathcal{R}^2$, the center of the largest empty circle must be within $\mathcal{R}^2$.

In one dimensional case, if the algorithm knows the number n of the inserted points, an insertion strategy with maximal gap ratio $2^{\lfloor n/2 \rfloor / (\lfloor n/2 \rfloor + 1)}$, which is slightly less than 2, can be achieved. Moreover, they gave a local search heuristic for uniformly inserting points on two dimensional square. Experimental results showed that the maximal gap ratio is less than 2 if the number of inserted points is small. If the points must be inserted at the grid points, Asano [1] gave an insertion strategy with uniformity 2 for one dimensional case.

Our Contributions:
To uniformly insert the points into a square, an intuitive idea is to insert each point at the center of the largest empty circumcircle of a triangle within the square. But this idea is not a good strategy when implementing on the square grid. For example, consider a 6×6 square grid as shown in Figure 2. If each point is inserted at the center of the largest empty circumcircle, the first five points must be inserted as shown in Figure 2. No matter where the next point is inserted, the gap ratio will be no less than 3.

In the following part of this paper, we give an insertion strategy for the problem of inserting points in a square grid with the maximal gap ratio no more than $2\sqrt{2} \approx 2.828$. For the problem of uniformly inserting points, this is the first result on inserting points at grid position. Moreover, we show that no online inserting strategy can hold the maximal gap ratio to strictly less than 2.5 for 3×3 grids.

2 Inserting Method

Inserting each point at the center of the largest empty circle is a good strategy if the size of the grid is some power of 2, i.e., $m = 2^k$. In this case, insertion at the center can always hold the gap ratio to no more than 2. Another observation is that once a point is inserted, the grid will be somewhat partitioned into regions which can be handled independently and locally. In the following, we devise our heuristic based on these observations, to achieve good performance. Instead of inserting each point at the center of the largest empty circle, we choose a proper position which partitions the grid into several parts: some of them are square grids whose sizes are some power of 2; some are square grids with sizes similar to the above ones; the others are rectangles with sizes between the above two types of square grids. Assigning the points at such positions can guarantee that the maximal gap ratio is not large.

Our strategy is carried out phase by phase. The position at which each point is inserted depends on the current configuration of the grids. At the beginning and end of each phase, the square grid is partitioned into four disjoint parts: the up-left part is a combination of small square grids of the same size $2^{k'} \times 2^{k'}$; the down-right part is a single square grid of size $m' \times m'$; the up-right and down-left parts are combinations of rectangle grids of size $2^{k'} \times m'$, as shown in Figure 3(a).

When starting to insert points in a phase, if the small square grids in the up-left part are larger than the down-right square, we insert a point into the

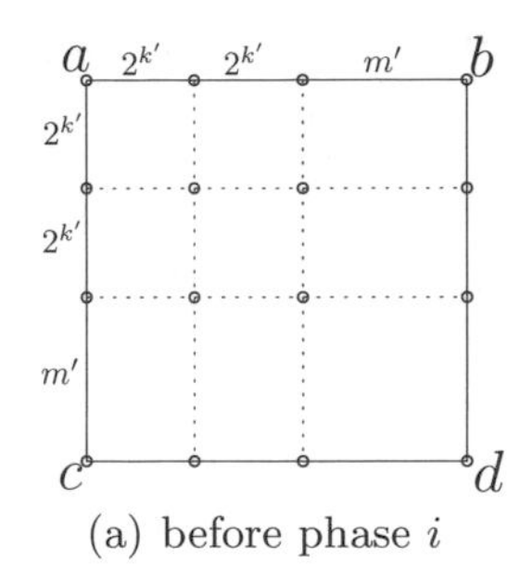

(a) before phase i

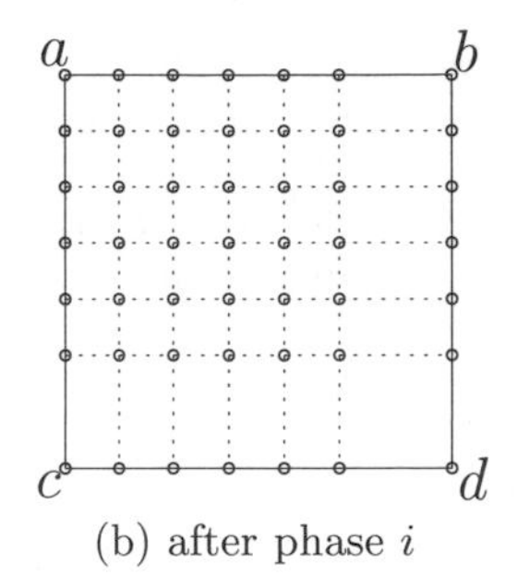

(b) after phase i

Fig. 3. The configuration of the square grid before and after a phase of insertion

center of each small square grids; otherwise, the down-right part is larger, then we insert points into the down-right square. According to the following insertion strategy, we can insert some points such that either the size of the small square grids in the up-left part is decreased, or the up-left part is enlarged. From the analysis of the strategy, the gap ratio is bounded by $2\sqrt{2}$ after each insertion. The configuration after a phase of insertion from the configuration in Figure 3(a) is shown in Figure 3(b).

2.1 The First Phase

In the initial state, there are four assigned points located at the four corners of the square grid. To assign the first point, we must determine the (x, y)-coordinate for p_1. Find the integer k such that $3 \cdot 2^{k-1} \leq m < 3 \cdot 2^k$. Insert the first point p_1 at $(2^k, 2^k)$.

Case 1: $3 \cdot 2^{k-1} \leq m < (2 + \sqrt{2}) \cdot 2^{k-1}$

In this case, we insert p_2 at $(2^{k-1}, 2^{k-1})$, p_3 at $(2^k, 0)$, p_4 at $(0, 2^k)$, and so on until p_{12} is assigned as shown in Figure 4.

Let $m = (1 + x) \times 2^k$, we have $1/2 \leq x < \sqrt{2}/2$. Now we analyze the gap ratio after each insertion.

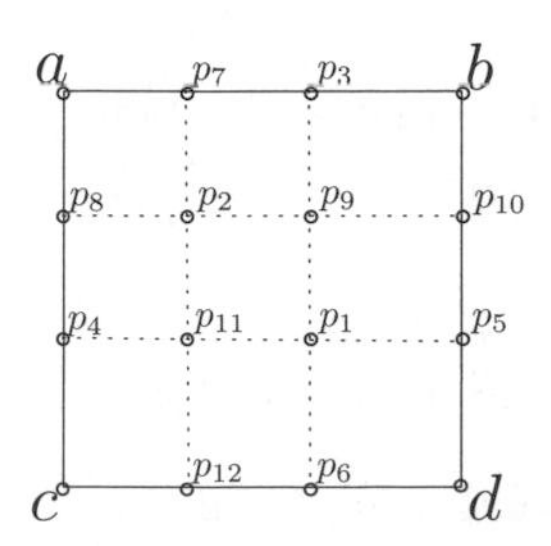

Fig. 4. Case 1 of the first phase

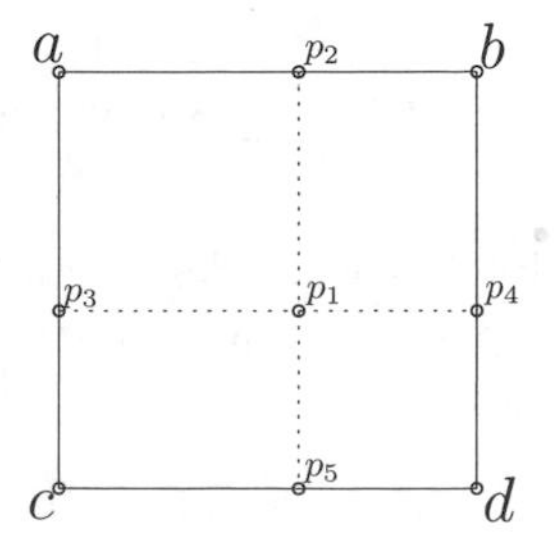

Fig. 5. Case 2 of the first phase

Lemma 1. *In case 1: $3 \cdot 2^{k-1} \leq m < (2+\sqrt{2}) \cdot 2^{k-1}$, the gap ratio is no more than $2\sqrt{2}$ after each insertion according to the strategy.*

Proof. After the insertion of p_1, the maximal gap is twice the length of the circumradius of triangle abp_1, the minimal gap is the length of $p_1 d$. Thus,

$$G_1 = \frac{2(1+x)2^k \times \sqrt{2}2^k \times \sqrt{1+x^2}2^k}{2(1+x)2^k \times 2^k} = \frac{\sqrt{1+x^2}2^{k+1}}{\sqrt{2}}$$

$$g_1 = x\sqrt{2}2^k$$

The gap ratio at this step is $\sqrt{1+1/x^2}$, which is at most $\sqrt{5}$ since $1/2 \leq x < \sqrt{2}/2$.

After the insertion of p_2, the minimal gap is the length of ap_2, which is $\sqrt{2}2^{k-1}$. The maximal gap must appear in the triangle abp_2, bp_1p_2, or bp_1d. We shall consider these three triangles separately.

- Suppose the maximal gap appears in triangle abp_2, since $x < \sqrt{2}/2$, the maximal gap must be $(x+y)2^{k+1}$, such that $(1/2-y)^2 + (1/2)^2 = (x+y)^2$. Thus, $y = (1/2-x^2)/(2x+1)$ and $x+y = (x^2+x+1/2)/(2x+1)$. Therefore, the gap ratio is $\frac{\sqrt{2}(2x^2+2x+1)}{2x+1} < 2\sqrt{2}$.
- Suppose the maximal gap appears in triangle bp_1p_2. Since $|bp_1| = \sqrt{1+x^2}2^k$, $|p_1p_2| = \sqrt{2}2^{k-1}$, $|bp_2| = \sqrt{1/4+(1/2+x)^2}2^k$, and $\Delta_{bp_1p_2} = (1+x)2^{2k-2}$, the maximal gap is $\frac{\sqrt{2}\sqrt{1+x^2}\sqrt{1/4+(1/2+x)^2}2^k}{(1+x)}$. Therefore, the gap ratio is $\frac{2\sqrt{1+x^2}\sqrt{x^2+x+1/2}}{1+x}$, this value is strictly less than $2\sqrt{2}$ since $1/2 \leq x < \sqrt{2}/2$.
- Suppose the maximal gap appears in triangle bp_1d. Since $x < \sqrt{2}/2 < 1$, the maximal gap will be $2(1/2+y)2^k$, such that $(1/2-y)^2 + x^2 = (1/2+y)^2$. Thus, we have $y = x^2/2$ and the maximal gap is $(x^2+1)2^k$. The gap ratio is $\sqrt{2}(x^2+1) < 2\sqrt{2}$.

After the insertion of p_3, the maximal gap remains same as the previous step, but the minimal gap is decreased to $x2^k$. Similar to above analysis,

- Suppose the maximal gap appears in triangle acp_2, the maximal gap must be $(x^2+x+1/2)/(2x+1)2^{k+1}$. Therefore, the gap ratio is $\frac{2x^2+2x+1}{2x^2+x} < 2\sqrt{2}$ since $1/2 \leq x < \sqrt{2}/2$.
- Suppose the maximal gap appears in triangle cp_1p_2. The maximal gap is $\frac{\sqrt{2}\sqrt{1+x^2}\sqrt{1/4+(1/2+x)^2}2^k}{(1+x)}$. Therefore, the gap ratio is $\frac{\sqrt{2}\sqrt{1+x^2}\sqrt{x^2+x+1/2}}{x(1+x)}$; this value is strictly less than $2\sqrt{2}$ since $1/2 \leq x < \sqrt{2}/2$.
- Suppose the maximal gap appears in triangle cp_1d. Since $x < \sqrt{2}/2 < 1$, the maximal gap is $(x^2+1)2^k$. The gap ratio is $(x^2+1)/x < 2\sqrt{2}$.

After the insertion of p_4, p_5, and p_6, the minimal gap is still $x2^{k-1}$ but the maximal gap does not increase. Thus, the gap ratio at this step is still no more than $2\sqrt{2}$.

After the insertion of p_7, the minimal gap is decreased to 2^{k-1}. The maximal gap is twice the length of the circumradius of the rectangle $bp_3p_1p_5$, which is $\sqrt{x^2+1}2^k$. Thus, the gap ratio is no more than $2\sqrt{x^2+1} < 2\sqrt{2}$.

After the insertion of p_j ($8 \leq j \leq 12$), the minimal gap remains as 2^{k-1}, but the maximal gap does not increase. Thus, the gap ratio after each of the insertions is no more than $2\sqrt{2}$. □

After this insertion phase, the up-left part is a combination of small square grids of size $2^{k-1} \times 2^{k-1}$, the down-right part is a square grid of size $(m-2^k) \times (m-2^k)$, the up-right and down-left parts are combinations of rectangles of size $2^{k-1} \times (m-2^k)$.

Case 2: $(2+\sqrt{2}) \cdot 2^{k-1} \leq m < (2+1/\sqrt{2}) \cdot 2^k$

In this case, insert p_2 at $(2^k, 0)$, p_3 at $(0, 2^k)$, p_4 at $(m, 2^k)$, and p_5 at $(2^k, m)$, as shown in Figure 5.

Let $m = (1+x)2^k$, we have $\sqrt{2}/2 \leq x \leq 1+\sqrt{2}/2$. Now we analyze the gap ratio after each insertion.

Lemma 2. *In case 2: $(2+\sqrt{2}) \cdot 2^{k-1} \leq m < (2+1/\sqrt{2}) \cdot 2^k$, the gap ratio is no more than $2\sqrt{2}$ after each insertion according to the strategy.*

After this phase, the up-left part is a square grid of size $2^k \times 2^k$, the down-right part is a square grid of size $(m-2^k) \times (m-2^k)$, the up-right and down-left parts are rectangles of size $2^k \times (m-2^k)$.

Case 3: $(2+1/\sqrt{2}) \cdot 2^k \leq m < 3 \cdot 2^k$

In this case, we insert p_2 at $(2^{k+1}, 2^{k+1})$, p_3 at $(2^k, 0)$, p_4 at $(0, 2^k)$, and so on until p_{12} is assigned as shown in Figure 6.

Let $m = (1+x) \times 2^k$, we have $1+\sqrt{2}/2 \leq x < 2$. Now we analyze the gap ratio after each insertion.

Lemma 3. *In case 3: $(2+1/\sqrt{2}) \cdot 2^k \leq m < 3 \cdot 2^k$, the gap ratio is no more than $2\sqrt{2}$ after each insertion according to the strategy.*

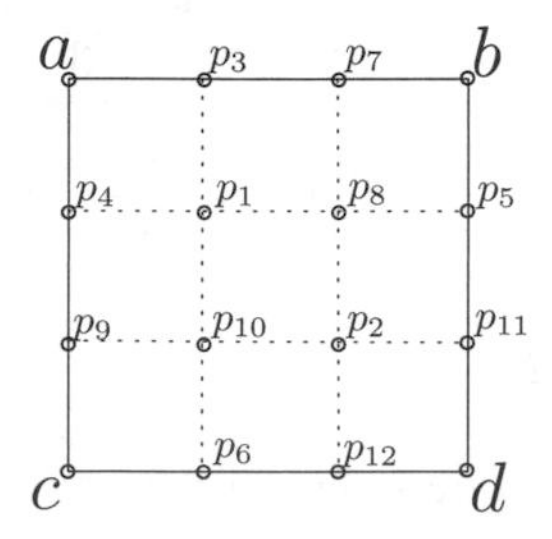

Fig. 6. Case 3 of the first phase

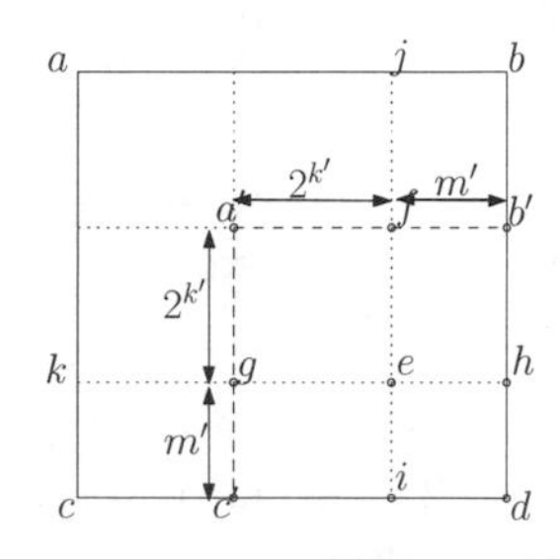

Fig. 7. Before inserting points in next phase

After this phase, the up-left part is a combination of small square grids of size $2^k \times 2^k$, the down-right part is a square grid of size $(m-2^{k+1}) \times (m-2^{k+1})$, the up-right and down-left parts are combinations of rectangles of size $2^k \times (m-2^{k+1})$.

2.2 The Following Phases

After the first phase, the square grid is partitioned into squares and rectangles. There are three types of such square or rectangle, i.e., the small square grids in the up-left part, the square grid in the down-right part, and the rectangles in the up-right and down-left parts. The only square grid in the down-right part is adjacent to the other two types of square grid and rectangle.

For clarity, we shall only consider the insertion of points in four of the squares and rectangles; specifically, the square at the down-right part, the rectangle immediately to its left, the rectangle above it and the square in the top-left part that has a common grid-position with it. Consider the configuration shown in Figure 7, we shall only focus on insertions in the square grid $a'b'c'd$. In this configuration, a', b', c', d to i have been assigned a point each; the down-right part is the square $ehid$, which is adjacent to $a'fge$, $fb'eh$ and $gec'i$. If a point is to be inserted into a square grid (rectangles, resp.) at a specified position according to the strategy, then the same action applies to every square grid (rectangles, resp.) within the part of the grid containing that square grid (rectangles, resp.). For example, in Figure 7, there are four small square grids in the up-left part, if the strategy inserts a point in square grid $a'fge$, then a point is inserted into each of the four small square grids in square $ajke$.

Now consider inserting points into the configuration as shown in Figure 7. In this configuration, points a to i are already assigned. The up-left part is a square grid of size $2^{k'} \times 2^{k'}$, the down-right part is a square grid of size $m' \times m'$, the up-right and down-left part are rectangles of size $2^{k'} \times m'$. When inserting the first point in the down-right square, we use the same strategy as in the first phase, i.e., find the value k'' such that $3 \cdot 2^{k''-1} \leq m' < 3 \cdot 2^{k''}$, then insert the point at $(2^{k'} + 2^{k''}, 2^{k'} + 2^{k''})$.

From Case 1 of the first phase, we have $m' = m - 2^k$, $2^{k'} = 2^{k-1}$, and $3 \cdot 2^{k-1} \leq m < (2+\sqrt{2}) \cdot 2^{k-1}$, thus, $2^{k'} \leq m' \leq \sqrt{2} \cdot 2^{k'}$. From case 2 of the first phase, we have $m' = m - 2^k$, $2^{k'} = 2^k$, and $(2+\sqrt{2}) \cdot 2^{k-1} \leq m < (2+1/\sqrt{2}) \cdot 2^k$, thus, $\sqrt{2} \cdot 2^{k'-1} \leq m' \leq (1+\sqrt{2}/2) \cdot 2^{k'}$. From case 3 of the first phase 1, we have $m' = m - 2^{k+1}$, $2^{k'} = 2^k$, and $(2+1/\sqrt{2}) \cdot 2^k \leq m < 3 \cdot 2^k$, thus, $\sqrt{2} \cdot 2^{k'-1} \leq m' \leq 2^{k'}$.

Combine all these cases, we have $\sqrt{2} \cdot 2^{k'-1} \leq m' \leq (1+\sqrt{2}/2) \cdot 2^{k'}$, this constraint can be relaxed to

$$\sqrt{2} \cdot 2^{k'-1} \leq m' < 2 \cdot 2^{k'}. \tag{1}$$

Let $m' = x \cdot 2^{k'}$, we have $\sqrt{2}/2 \leq x < 2$. Now we show how to insert points in such configuration.

Case 1: $\sqrt{2} \cdot 2^{k'-1} \leq m' < 2^{k'}$

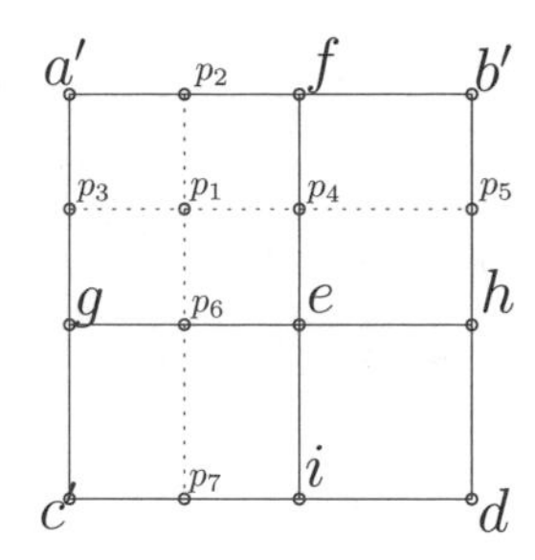

Fig. 8. Case 1 of the following phase

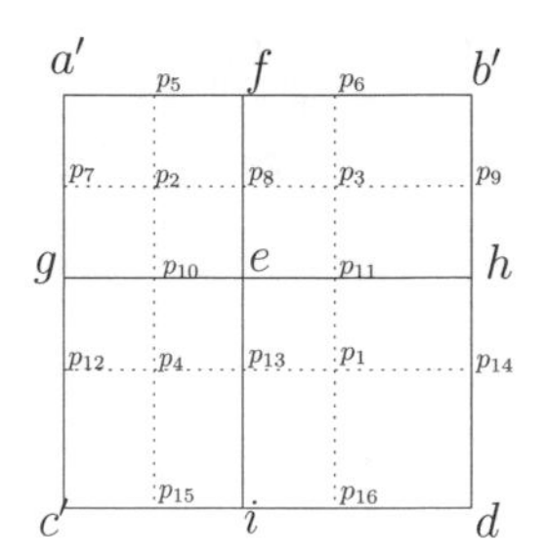

Fig. 9. Case 2 of the following phase

In this case, we insert points p_1 until p_7 in the configuration as shown in Figure 8.

After the insertion of p_1, the minimal gap is decreased to the length of $a'p_1$, which is $\sqrt{2}2^{k'-1}$, the maximal gap is twice the length of the circumradius of the rectangle $b'feh$, which is $\sqrt{1+x^2}2^{k'}$. Thus, the gap ratio is at most $\sqrt{2(x^2+1)}$, which is no more than $2\sqrt{2}$.

After the insertion of p_2, the minimal gap is decreased to $2^{k'-1}$, while the maximal gap is still $\sqrt{x^2+1}2^{k'}$. Thus, the gap ratio is $2\sqrt{x^2+1}$, which is no more than $2\sqrt{2}$ since $x < 1$.

After the insertion of p_3 until p_7, the minimal gap remains as $2^{k'-1}$, and the maximal gap does not increase. Thus, the gap ratio is still no more than $2\sqrt{2}$.

When the insertions of this phase complete, m' remains unchanged but the size of the square grids in up-left part is decreased to $2^{k'-1} \times 2^{k'-1}$. Therefore, in the next phase, the constraint of m' is $\sqrt{2}2^{k'} \leq m' < 2^{k'+1}$.

Case 2: $2^{k'} \leq m' < (1+\sqrt{2}/4)2^{k'}$

In this case, since the down-right square grid is larger, we insert points p_1 until p_{16} in the configuration as shown in Figure 9.

Note that $4 \cdot 2^{k'-2} \leq m' < (2+\sqrt{2}/2)2^{k'-1}$, if we regard $m' = m$ and $2^{k'-1} = 2^k$, this constraint is within the range in case 2 of the first phase, i.e., $(2+\sqrt{2}) \cdot 2^{k-1} \leq m < (2+1/\sqrt{2}) \cdot 2^k$. Thus, similar to the analysis in case 2 of the first phase, we conclude that the ratio after each insertion is no more than $2\sqrt{2}$.

When the insertions complete, m' is decreased by $2^{k'-1}$ and the size of the square grids in up-left part is decreased to $2^{k'-1} \times 2^{k'-1}$, therefore, in the next phase, the constraint of m' is $2^{k'} \leq m' < (1+\sqrt{2}/2)2^{k'}$.

Case 3: $(1+\sqrt{2}/4)2^{k'} \leq m' < 3 \cdot 2^{k'-1}$

In this case, we insert points p_1 until p_{27} in the configuration as shown in Figure 10.

If we regard $m' = m$ and $2^{k'-1} = 2^k$, the constraint in this case is within the range in case 3 of the first phase, i.e., $(2+1/\sqrt{2}) \cdot 2^k \leq m < 3 \cdot 2^k$. Thus, similar to the analysis in case 3 of the first phase, we conclude that the ratio after each insertion is no more than $2\sqrt{2}$.

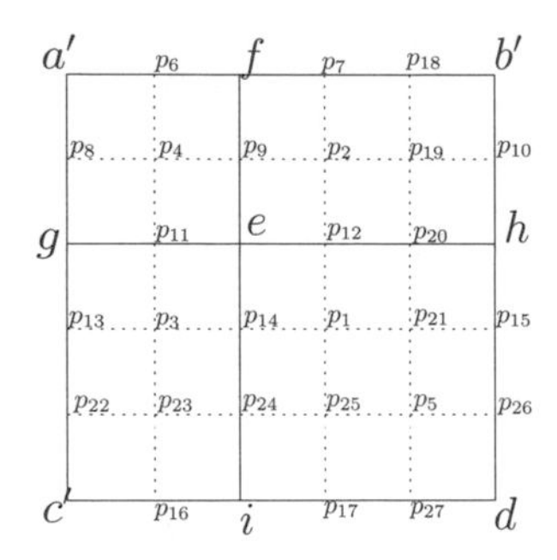

Fig. 10. Case 3 of the following phase

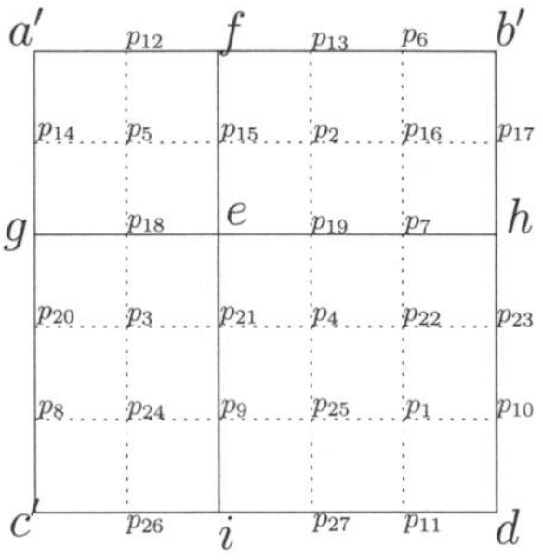

Fig. 11. Case 4 of the following phase

When the insertions complete, m' is decreased by $2^{k'}$ and the size of the square grids in up-left part is decreased to $2^{k'-1} \times 2^{k'-1}$, therefore, in next phase, the constraint of m' is $\sqrt{2}2^{k'-1} \leq m' < 2^{k'}$.

Case 4: $3 \cdot 2^{k'-1} \leq m' < (1 + \sqrt{2}/2) \cdot 2^{k'}$

In this case, we insert points p_1 until p_{27} in the configuration as shown in Figure 11.

If we regard $m' = m$ and $2^{k'-1} = 2^k$, the constraint in this case is within the range in case 1 of the first phase, i.e., $3 \cdot 2^{k-1} \leq m < (2 + \sqrt{2}) \cdot 2^{k-1}$. Thus, similar to the analysis in case 1 of the first phase, we conclude that the ratio after each insertion is no more than $2\sqrt{2}$.

When the insertions complete, m' is decreased by $2^{k'}$ and the size of the small square grids is decreased to $2^{k'-1} \times 2^{k'-1}$, therefore, in next phase, the constraint of m' is $2^{k'} \leq m' < \sqrt{2} \cdot 2^{k'}$.

Case 5: $(1 + \sqrt{2}/2) \cdot 2^{k'} \leq m' < 2^{k'+1}$

In this case, we insert points p_1 until p_7 in the configuration as shown in Figure 12.

If we regard $m' = m$ and $2^{k'} = 2^k$, the constraint in this case is within the range in case 2 of the first phase, i.e., $(2 + \sqrt{2}) \cdot 2^{k-1} \leq m < (2 + 1/\sqrt{2}) \cdot 2^k$.

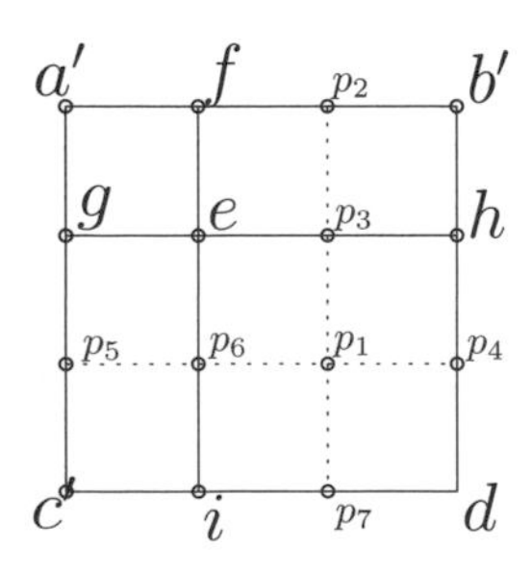

Fig. 12. Case 5 of the following phase

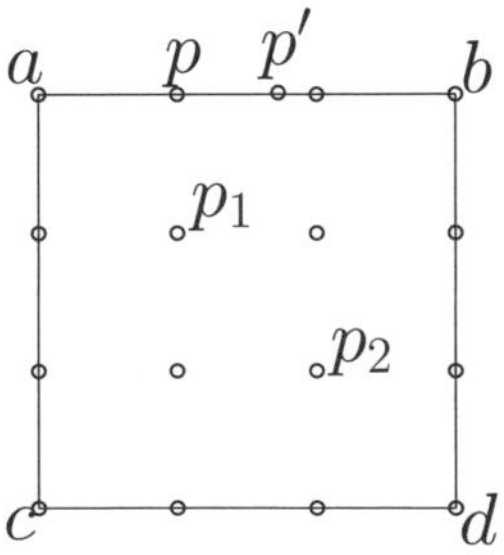

Fig. 13. In a 3×3 grid, the gap ratio is at least 2.5

Thus, similar to the analysis in case 2 of the first phase, we conclude that the ratio after each insertion is no more than $2\sqrt{2}$.

When the insertions complete, m' is decreased to $m' - 2^{k'}$ while the size of the square grids in up-left part remains the same as in previous phase, therefore, in next phase, the constraint of m' is $\sqrt{2} \cdot 2^{k'-1} \le m' < 2^{k'}$.

Combine all these cases, we conclude that the gap ratio is no more than $2\sqrt{2}$ after each insertion. When a phase completes, suppose the size of the down-right square grid is $m' \times m'$, the size of the square grids in up-left part is $2^{k'} \times 2^{k'}$, we have $\sqrt{2}2^{k'-1} \le m' < 2^{k'+1}$, which is consistent with Inequality (1).

Therefore, we have the following conclusion.

Theorem 1. *The maximal gap ratio of the above strategy for inserting points into any square grid is at most* $2\sqrt{2} \approx 2.828$.

3 Lower Bound of the Maximal Gap Ratio

In this part, we prove that for 3×3 grids, the lower bound of the maximal gap ratio is at least 2.5 for any online inserting method. Consider inserting points into a 3×3 grid, as shown in Figure 13.

initial step:

In the initial step, the four corner points are already assigned. The maximal gap $G_0 = 3\sqrt{2}$ and the minimal gap is $g_0 = 3$, thus the gap ratio at this step is $\sqrt{2}$.

inserting the first point:

If the first point is inserted at the boundary line of the square, w.l.o.g., at p, the maximal gap is twice the length of the circumradius of triangle pcd, which is $\sqrt{130}/3$, the minimal gap is the length between a and p, which is 1, so the gap ratio is $\sqrt{130}/3 \approx 3.8$.

If the first point is inserted at some interior point, w.l.o.g., at p_1, the gap ratio will be lower. In this case, the maximal gap is twice the length of the circumradius of triangle bdp_1, which is $\sqrt{10}$, the minimal gap is the length between a and p_1, which is $\sqrt{2}$, so the gap ratio $r_1 = \sqrt{5} \approx 2.236$.

inserting the second point:

We may assume the first point is inserted at p_1 since the gap ratio is larger if the first point is assigned at p.

If the second point is not inserted at p_2, the maximal gap will be $\sqrt{10}$ too, since the maximal gap must appear in triangle bdp_1 or cdp_1, and the minimal gap will be 1, so the gap ratio $r_2 = \sqrt{10} \approx 3.162$.

If the second point is inserted at p_2, the circumradius of triangle bp_1p_2 is $5\sqrt{2}/6$. Consider the point p' on the edge ab such that $|p'b| = 5/4$, the length of $p'p_1$ is also $5/4$, and the length between p' and any other point is larger than $5/4$. Since $5/4 > 5\sqrt{2}/6$, we conclude that the maximal gap is $5/2$. The minimal gap is still $\sqrt{2}$, so the gap ratio $r_2 = 5/(2\sqrt{2}) \approx 1.768$.

inserting the third and following points:

Similarly, we may assume the second point is assigned at p_2, otherwise the gap ratio will be larger. No matter where we assign the third point, the minimal gap will be 1, while the maximal gap remains as $5/2$ owing to the symmetry of the assigned points in this square grid. Thus, the gap ratio $r_3 = 2.5$.

For the insertion of the following points, the minimal gap is 1 and the maximal gap is no larger than $5/2$, so the gap ratio $r_i \leq 2.5$ $(i \geq 3)$.

Combining all the above cases, we conclude that in handling any insertion sequence in a 3×3 square grid, the maximal gap ratio is at least 2.5. Thus, we have the following conclusion.

Theorem 2. *No online inserting method can hold the maximal gap ratio to strictly less than 2.5.*

References

1. Asano, T.: Online uniformity of integer points on a line. Information Processing Letters 109, 57–60 (2008)
2. Asano, T., Teramoto, S.: On-line uniformity of points. In: Book of Abstracts for 8th Hellenic-European Conference on Computer Mathematics and its Applications, Athens, Greece, pp. 21–22 (September 2007)
3. Bayer, B.E.: An Optimum Method for Two-Level Rendition of Continuous-Tone Pictures. In: Proceedings of the IEEE International Conference on Communication, pp. 11–26 (1973)
4. Chazelle, B.: The Discrepancy Method: Randomness and Complexity. Cambridge University Press, Cambridge (2000)
5. Collins, C.R., Stephenson, K.: A circle packing algorithm. Computational Geometry: Theory and Applications 25(3), 233–256 (2003)
6. Matoušek, J.: Geometric Discrepancy. Springer, Heidelberg (1991)
7. Nurmela, K.J., Östergård, P.R.J.: More Optimal Packings of Equal Circles in a Square. Discrete & Computational Geometry 22(3), 439–457 (1999)
8. Teramoto, S., Asano, T., Katoh, N., Doerr, B.: Inserting points uniformly at every instance. IEICE Trans. Inf. Syst. E89-D(8), 2348–2356 (2006)

Kernelization for Cycle Transversal Problems

Ge Xia[1] and Yong Zhang[2]

[1] Department of Computer Science, Lafayette College, Easton, PA 18042
gexia@cs.lafayette.edu

[2] Department of Computer Science, Kutztown University, Kutztown, PA 19530
zhang@kutztown.edu

Abstract. We present new kernelization results for the s-CYCLE TRANSVERSAL problem for $s > 3$. In particular, we show a $6k^2$ kernel for 4-CYCLE TRANSVERSAL and a $O(k^{s-1})$ kernel for s-CYCLE TRANSVERSAL when $s > 4$. We prove the NP-completeness of s-CYCLE TRANSVERSAL on planar graphs and obtain a $74k$ kernel for 4-CYCLE TRANSVERSAL on planar graphs. We also give several kernelization results for a related problem ($\leq s$)-CYCLE TRANSVERSAL.

1 Introduction

Graphs that are free of cycles of a given length s are extensively studied in extremal graph theory, including cases when s is small [2,10,11], or an odd number [14], or an even number [21]. When s is small ($s \leq 5$), s-cycle-free graphs and s-cycle-free planar graphs are also studied in other areas. For example, the chromatic number of 3-cycle-free graphs [22] and 5-cycle-free graphs [23] were investigated. Cherlin and Komjáth [6] showed that there is no universal countable 5-cycle-free graph. The class of 4-cycle-free Taner graphs plays an important role in designing low-density parity-check (LDPC) codes [15]. As for s-cycle-free planar graphs, Madhavan [19] gave an approximation algorithm for finding maximum independent set on 3-cycle-free planar graphs. Borodin et al. [4] obtained upper bounds on the game chromatic number of 4-cycle-free planar graphs. Esperet et al. [8] gave several positive and negative results on the adapted list coloring of s-cycle-free planar graphs.

In this paper we study the problem of obtaining a maximum subgraph without cycles of a given length s by edge deletions. This problem is equivalent to the following edge transversal problem. We say an edge e *covers* a cycle C if e is one of the edges in C.

> s-CYCLE TRANSVERSAL: Given an undirected graph G and an integer k, is there a set S of at most k edges in G such that every cycle in G of length s is covered by at least one edge in S? We shall call S a *transversal set*.

s-CYCLE TRANSVERSAL is known to be NP-complete on general graphs [24]. Krivelevich [18] and Kortsarz et al. [17] studied approximation algorithms for s-CYCLE TRANSVERSAL. Krivelevich [18] presented a linear programming-based 2-approximation algorithm for 3-CYCLE-TRANSVERSAL. Kortsarz et al. [17] showed

B. Chen (Ed.): AAIM 2010, LNCS 6124, pp. 293–303, 2010.

that the approximation ratio 2 is likely the best possible by showing that a $(2-\epsilon)$-approximation algorithm for 3-CYCLE-TRANSVERSAL implies a $(2-\epsilon)$-approximation algorithm for VERTEX COVER, which might be impossible [16]. Kortsarz et al. [17] also gave a generalized $(s-1)$-approximation algorithm for s-CYCLE TRANSVERSAL where s is any odd number. A related problem is the vertex version of s-CYCLE TRANSVERSAL, where one asks for a minimum vertex set to cover all cycles of length s. The vertex version of 3-CYCLE TRANSVERSAL was studied in the literatures. In particular, Abu-Khzam [1] obtained a quadratic kernel for this problem by reducing it to 3-HITTING SET. Fernau [9] showed this problem can be solved in running time $O(|V|^3 + 2.1788^k)$.

s-CYCLE TRANSVERSAL has also been studied in the context of *parameterized complexity*. A *parameterized problem* is a set of instances of the form (x, k), where x is the input instance and k is a nonnegative integer called the *parameter*. A parameterized problem is said to be *fixed parameter tractable* if there is an algorithm that solves the problem in time $f(k)|x|^{O(1)}$, where f is a computable function solely dependent on k, and $|x|$ is the size of the input instance. When dealing with NP-hard problems in practice, *kernelization* is a very useful preprocessing technique. The idea of kernelization is to design data reduction rules to reduce the input instance to an equivalent *kernel* of smaller size. Formally, the *kernelization* of a parameterized problem is a reduction to a *problem kernel*, that is, to apply a polynomial-time algorithm to transform any input instance (x, k) to an equivalent reduced instance (x', k') such that $k' \leq k$ and $|x'| \leq g(k)$ for some function g solely dependent on k. It is known that a parameterized problem is fixed parameter tractable if and only if the problem is kernelizable. We refer interested readers to [7,12] for more details.

The kernelization of s-CYCLE-TRANSVERSAL was first studied by Brügmann et al. [5]. Brügmann et al. [5] designed data reduction rules to obtain a $6k$ kernel for 3-CYCLE-TRANSVERSAL on general graphs. They also proved the NP-completeness of 3-CYCLE-TRANSVERSAL on planar graphs and gave a $11k/3$ kernel for the problem.

We present new kernelization results for s-CYCLE TRANSVERSAL when $s > 3$. We show that on general graphs 4-CYCLE TRANSVERSAL admits a $6k^2$ kernel and s-CYCLE TRANSVERSAL where $s > 4$ admits a $O(k^{s-1})$ kernel. We generalize the NP-completeness proof of 3-CYCLE-TRANSVERSAL on planar graphs [5] to cases where $s \geq 3$. We then use the *region-decomposition* framework developed in Guo and Niedermeier [13] to show that 4-CYCLE TRANSVERSAL admits a $74k$ kernel on planar graphs.

We also study a related problem small cycle transversal, referred to as $(\leq s)$-CYCLE TRANSVERSAL, where one asks for a minimum edge set to cover all cycles of length $\leq s$ for a given s. Our NP-completeness proof for s-CYCLE TRANSVERSAL on planar graphs extends naturally to $(\leq s)$-CYCLE TRANSVERSAL on planar graphs. The approximation algorithms for $(\leq s)$-CYCLE TRANSVERSAL has been studied in [17]. In this paper we present a $32k$ kernel for (≤ 4)-CYCLE TRANSVERSAL on planar graphs and a $226k$ kernel for (≤ 5)-CYCLE TRANSVERSAL on planar graphs.

Recently Bodlaender et al. [3] proved that all problems having *finite integer index* and satisfying a compactness condition admit linear kernels on planar graphs. They claim that their results ([3] Theorem 2 and Corollary 2) unify and generalize *all* known linear kernels on planar graph problems. However, their results do not generalize our linear kernel results because both s-CYCLE TRANSVERSAL and ($\leq s$)-CYCLE TRANSVERSAL are subclasses of the EDGE-$\mathcal{S}$-COVERING problem ([3] Corollary 3) which is not known to have finite integer index.

Next, we present several necessary definitions and some backgrounds.

We only consider simple and undirected graphs. All paths and cycles considered in this paper are simple. A path P in a graph G is a sequence of vertices $P = (v_0, v_1, \cdots, v_l)$ such that v_{i-1} and v_i are adjacent for all $1 \leq i \leq l$. The length of P is the number of edges in P. A path Q is called a *sub-path* of P if Q is a subsequence of P; or equivalently, we say that P *contains* Q. A cycle is a closed path where the first vertex is the same as the last vertex in the sequence. For example $C = (a, b, c, d, a)$ is a cycle of length 4 containing edges (a, b), (b, c), (c, d), and (d, a). For a set of edges T, if both cycles C_1 and C_2 contain T, we say T is *shared* by C_1 and C_2. If C_1 and C_2 have no other common edges, we say C_1 and C_2 are *edge-disjoint-sharing* T. Let W be a set of cycles in G, we use $E(W)$ to represent the set of edges in cycles in W, and $G[W]$ to represent the subgraph induced by vertices in cycles in W.

For all the s-CYCLE TRANSVERSAL problems with various s values discussed in this paper, we assume the input graph has been preprocessed by removing all vertices and edges that are not contained in any s-cycle. It is clear that this preprocessing is correct and can be done in polynomial time.

2 Kernels on General Graphs

First we show that 4-CYCLE TRANSVERSAL on general graphs admits a $6k^2$ kernel. Let $G = (V, E)$ be the input graph after the above mentioned preprocessing. Enumerate all 4-cycles in G and compute in polynomial time a maximal set W of 4-cycles that share pairwise at most one edge. We call W a *witness* of G.

Reduction Rule 1: For any edge e in a 4-cycle in W, if e is shared by k other 4-cycles in W, then delete e from G, remove all 4-cycles containing e from W, and decrease k by 1.

Lemma 1. *Reduction Rule 1 is correct.*

Proof. Since W is a maximal set of 4-cycles that share pairwise at most one edge, the $k+1$ 4-cycles are pairwise edge-disjoint-sharing e. If G has a transversal set S of size k, then e must be in S, since otherwise at least $k+1$ edges are needed to cover the $k+1$ 4-cycles. Thus we can safely delete e and decrease k by 1. □

It is clear that Reduction Rule 1 can be applied in polynomial time. Let G' be the reduced graph that cannot be further reduced by Reduction Rule 1. Let

$Q := G' - G[W]$. We show that G' has at most $6k^2$ vertices, by bounding the sizes of both W and Q. Similar approaches have been used in Abu-Khzam [1] and Moser [20].

Theorem 1. 4-CYCLE TRANSVERSAL *admits a* $6k^2$ *problem kernel.*

Proof. We show that if G' has more than $6k^2$ vertices, then G does not have a transversal set of size k. First observe that if W contains more than k^2 4-cycles, then G' does not have a transversal set of size k, since by Reduction Rule 1 any edge in G' covers at most k 4-cycles in W. This implies that $G[W]$ has no more than $4k^2$ vertices, because otherwise W contains more than k^2 4-cycles. In the rest of the proof, we show that Q has no more than $2k^2$ vertices. The total number of vertices in G' is at most $4k^2 + 2k^2 = 6k^2$.

First observe that Q induces an independent set. Suppose that there is an edge e in Q. e must belong to a 4-cycle C. Since the both end points of e are not in W, at most two vertices of C are in W, which implies that C shares at most one edge with any 4-cycle in W. Thus C should be included in W due to the maximality of W.

Let v be a vertex in Q and let C_v be a 4-cycle containing v. There exists a 4-cycle $C \in W$ that shares at least two edges with C_v. We call v a *Q-neighbor* of C in this case. Note that if v is a Q-neighbor of a 4-cycle $C = (a, b, c, d, a) \in W$, then v must be connected to either both a and c, or to both b and d. If C has more than two Q-neighbors, then at least two of them, denoted by x and y, are connected to the same vertex pair $\{a, c\}$ or $\{b, d\}$. Without loss of generality, assume that both x and y are connected to both a and c. Then (x, a, y, c, x) is a 4-cycle that does not share an edge with 4-cycles in W and thus should be included in W. This contradicts the that x and y are not in W. Since every 4-cycle C in W has at most two Q-neighbors, and each vertex in Q is a Q-neighbor of some 4-cycle in W, the number of vertices in Q is at most $2|W| \le 2k^2$. □

In the following, we present a nontrivial generalization of the above techniques to achieve a $O(k^{s-1})$ kernel for s-CYCLE TRANSVERSAL where $s > 4$.

Reduction Rule 2.1: Repeatedly apply **reduce**(G) until the graph cannot be further reduced. Let G be the reduced graph G and let W be the witness of G.

The algorithm **reduce**(G) will compute the witness W for G and check repeatedly whether there are edges in $E(W)$ that can be safely deleted. Once it finds such edges, the algorithm will delete them and start over again. Let H be the set of all s-cycles in G. H can be enumerated in $O(|E|^s)$ time. It is clear that Reduction Rule 2 can be applied in polynomial time.

Lemma 2. *Reduction rule 2.1 is correct.*

Proof. We will prove the lemma by an induction on i. In **reduce**(G) edge deletions occur in the for loop with the loop index i taking values from 0 to $s - 4$. For the base case where $i = 0$, let G and G' be the graphs before and after the edge deletion is applied. We will show that G has a transversal set of size k if

```
Algorithm reduce(G):

compute the set H of all s-cycles in G
compute from H a maximal set W of s-cycles sharing at most s − 3 edges pairwise
for i = 0 to s − 4
    R = ∅
    for each cycle C ∈ W
        for each edge set T of size s − 3 − i in C
            compute the set M ⊆ W of s-cycles that contain T
            if (|M| > s^i(k + 1)^{i+1})
                compute any M' ⊆ M of size |M| − s^i(k + 1)^{i+1}
                W = W − M'; R = R ∪ M'
    if E(R) − E(W) is not empty
        delete the edges in E(R) − E(W) from G and stop
```

and only if G' has a transversal set of size k. One direction is trivial. Suppose G' has a transversal set S of size k, then S is also a transversal set for G. Suppose this is not true and there is a s-cycle C in G that is not covered by S, then C must contain an edge $e \in E(R) - E(W)$. From the way the set R is constructed, C contains a set T of $s-3$ edges where T is shared by C and k other s-cycles in W. Since W is a maximal set of s-cycles sharing at most $s-3$ edges pairwise, the $k+1$ s-cycles sharing T have no other common edges. This implies that one of the edges in T must be in S, since otherwise $k+1$ edges will be needed to cover the $k+1$ s-cycles edge-disjoint-sharing T. This contradicts the fact that C is not covered by S.

Using similar notations for the inductive steps where $0 < i \leq s-4$, we suppose that there is an uncovered s-cycle C containing an edge $e \in E(R) - E(W)$, then C must contain an edge set T of size $s-3-i$ which is shared by a set A of $s^i(k+1)^{i+1}$ many s-cycles in W. By the inductive hypothesis, any edge superset of T with size $s-3-(i-1)$ can be shared by at most $s^{i-1}(k+1)^i$ s-cycles in W. Therefore for any s-cycle $C_1 \in A$, C_1 can intersect with at most $s \cdot s^{i-1}(k+1)^i = s^i(k+1)^i$ s-cycles in A. Since A has $s^i(k+1)^{i+1}$ s-cycles, we can always find $k+1$ edge-disjoint s-cycles in A sharing T. This implies that one of the edges in T must be also in S, contradicting the fact that C is not covered by S. □

Let $Q := G - G[W]$. For a vertex $v \in Q$, we say v is a *Q-neighbor* of a s-cycle $C \in W$ if v forms a s-cycle with a path of length $s-2$ in C. By an argument similar to that in the proof of Theorem 1, Q is an independent set and every vertex $v \in Q$ has to be a Q-neighbor of some s-cycle in W. Construct an auxiliary bipartite graph $A = \{V_1, V_2, E'\}$ as follows. For every path of length $s-2$ in a s-cycle in W, create a vertex in V_1. For every vertex of Q, create a vertex in V_2. There is an edge between $v_1 \in V_1$ and $v_2 \in V_2$, if the vertex corresponding to v_1 and the path corresponding to v_2 form a s-cycle in G.

Reduction Rule 2.2: Compute a maximum set of edges M in A such that every vertex in V_1 is incident to at most $k+1$ edges in M. Delete all vertices in Q that corresponds to a vertex in V_2 that is not incident to an edge in M.

Reduction Rule 2.2 can be applied in polynomial time, since M can be computed in polynomial time by making $k+1$ copies of each vertex in V_1 and computing a maximum matching of the resulting bipartite graph.

Lemma 3. *Reduction Rule 2.2 is correct.*

Proof. Let G and G' be the graph before and after Reduction Rule 2.2 is applied. If there is a transversal set S of size k for G', then S is also a transversal set for G. Suppose this is not true and there is an uncovered s-cycle C in G, C must contain a deleted vertex $v_2 \in Q$, and a path P corresponding to a vertex $v_1 \in V_1$. By the way A is constructed, P is shared by $k+1$ s-cycles edge-disjoint-sharing P in G'. Therefore one of the edges in P must be also in S, contradicting the fact that C is not covered. □

Theorem 2. *For any $s > 4$,* S-CYCLE TRANSVERSAL *admits a $O(k^{s-1})$ problem kernel.*

Proof. Based on the algorithm **reduce(G)**, any single edge in s-cycles in W can cover at most $s^{s-4}(k+1)^{s-3}$ s-cycles. If W has more than $s^{s-4}(k+1)^{s-2}$ s-cycles, then W cannot be covered by any edge set of size k. Therefore, $G[W]$ has no more than $s^{s-3}(k+1)^{s-2}$ vertices. By Reduction Rule 2.2, every s-cycle in W has at most $s(k+1)$ Q-neighbors in Q, so the number of vertices in Q is bounded by $s(k+1) \cdot s^{s-4}(k+1)^{s-2} = s^{s-3}(k+1)^{s-1}$. Overall, the total number of vertices in the reduced graph is $O(k^{s-1})$. □

3 Linear Kernels on Planar Graphs

S-CYCLE TRANSVERSAL for any fixed $s \geq 3$ is known to be NP-complete on general graphs [24]. Brügmann et al. [5] showed it is NP-complete on planar graphs when $s = 3$. Using a similar technique, we are able to prove the NP-completeness of S-CYCLE TRANSVERSAL and ($\leq s$)-CYCLE TRANSVERSAL for any fixed $s \geq 3$ on planar graphs. The proofs are omitted due to lack of space.

Theorem 3. S-CYCLE TRANSVERSAL *and* ($\leq s$)-CYCLE TRANSVERSAL *are NP-complete for any fixed $s \geq 3$ on planar graphs of maximum degree seven.*

We first present a $74k$ kernel for 4-CYCLE TRANSVERSAL on planar graphs. Given an input graph G, we say that a 4-cycle C in G is *dangling* if only one edge of C is shared with other 4-cycles in G.

Reduction Rule 3: (the Dangling Rule) If there is a dangling 4-cycle C in G, then delete all four edges in C and decrease k by 1.

Lemma 4. *Reduction Rule 3 is correct and can be applied in polynomial time.*

Let the reduced graph be $G' = (V, E)$. To bound the size of G', we will use the *region-decomposition* framework developed by Guo and Niedermeier [13]. Suppose that G' has a transversal set S of size k. Let $V(S)$ be the set of endpoints of the edges in S. $V(S)$ have at most $2k$ vertices. With respect to $V(S)$, 4-CYCLE TRANSVERSAL admits the distance property ($C_V = 1, C_E = 1$) which is required to apply the region-decomposition framework. Therefore we can decompose G' into a set of regions.

Definition 1 ([13]). *A region $R(u, v)$ between two distinct vertices $u, v \in V(S)$ is a closed subset of the plane with the following properties:*

1. *The boundary of $R(u, v)$ is formed by two paths between u and v of length at most 3. The two paths do not need to be disjoint or simple. A vertex is said to be* inside *$R(u, v)$ if it lies either on the boundary or strictly inside $R(u, v)$.*
2. *All vertices inside $R(u, v)$ have distance at most 1 to at least one of the vertices u and v. Similarly, all edges whose both endpoints are inside $R(u, v)$ have distance at most 1 to at least one of the vertices u and v.*
3. *With the exception of u and v, none of the vertices inside $R(u, v)$ are from $V(S)$.*

An $V(S)$-region decomposition of G' is a set $\mathcal{R}$ of regions such that no vertex lie strictly inside more than one region from $\mathcal{R}$. The following Lemma directly follows from Lemma 1 in [13].

Lemma 5. *There is a maximal $V(S)$-region decomposition $\mathcal{R}$ for the graph G that consists of at most $6k - 6$ regions.*

Then we show that there are constant number of vertices inside each region in $\mathcal{R}$.

Lemma 6. *Every region $R(u, v)$ in $\mathcal{R}$ contains at most 12 vertices which are not in $V(S)$.*

Proof (Proof of Lemma 6). Consider a region $R(u, v)$. We distinguish two cases.
case 1: $(u, v) \notin E$, or $(u, v) \in E$ but $(u, v) \notin S$. There may not be an edge between u and v. If there is an edge between u and v, then this edge is not in

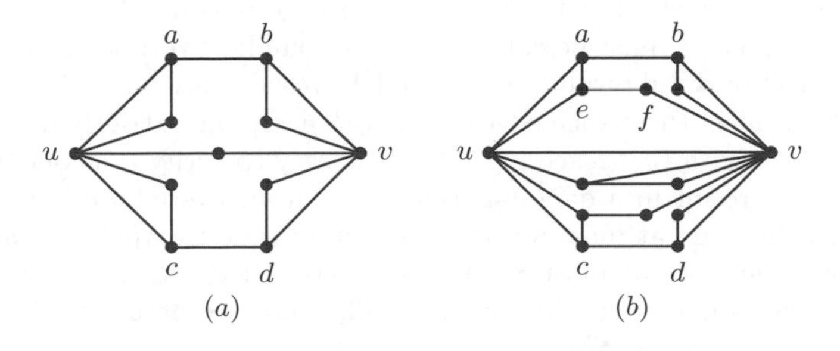

Fig. 1. Two cases in the region decomposition

S, which means u and v are endpoints of two edges $e, e' \in S$, separately, where both e and e' are outside $R(u,v)$. This implies that no edge from S is inside $R(u,v)$.

First, there are no degree-one vertices in $R(u,v)$ since they are not involved in any 4-cycle. Second, there are no edges with both end points strictly inside $R(u,v)$. Suppose there is such an edge e with both end points strictly inside $R(u,v)$, e must form a 4-cycle with two other vertices inside $R(u,v)$, thus all the edges in this 4-cycle must be inside $R(u,v)$. This contradicts the fact that no edge from S is inside $R(u,v)$. Therefore all vertices strictly inside $R(u,v)$ must have degree at least two and must connect to boundary vertices. As shown in Fig 1(a), $R(u,v)$ has at most 6 boundary vertices including u and v. For every pair of boundary vertices, there can be only one vertex strictly inside $R(u,v)$ connecting to both of them, otherwise they will form an uncovered 4-cycle strictly inside $R(u,v)$. Out of the 15 pair of boundary vertices, each of the pairs — $\{u,b\}$, $\{u,d\}$, $\{a,v\}$, $\{c,v\}$ — cannot share a common neighbor vertex strictly inside $R(u,v)$ since these will result in uncovered 4-cycles; each of the pairs — $\{a,b\}$, $\{c,d\}$ — cannot share common neighbor vertices since the neighbor vertices will have distance larger than 1 from u and v; at most one of the pairs — $\{u,v\}$, $\{a,d\}$, $\{b,c\}$ — can have a common neighbor due to planarity of the region. Therefore, there can be at most 5 vertices strictly inside $R(u,v)$, as shown in Fig 1(a).

case 2: $(u,v) \in S$. The edge (u,v) is either inside $R(u,v)$ or a different region between u and v. The analysis for case 1 still applies except that there can be edges with both end points strictly inside $R(u,v)$, such as (e,f) in Fig 1(b) which is contained in a 4-cycle (u,e,f,v,u). Since G' is reduced by Reduction Rule 3, this 4-cycle cannot be dangling, therefore at least one of the edges (e,u), (e,f), and (f,v) has to be involved in another 4-cycle. First, it is not possible for (e,f) to be involved in another 4-cycle. If (e,f) forms another 4-cycle (u,f,e,v,u) with (u,v), then we have an uncovered 4-cycle (u,e,v,f,u) inside $R(u,v)$. If (e,f) forms another 4-cycle with an edge other than (u,v) in $R(u,v)$, again this 4-cycle will be uncovered. Without loss of generality, assume that (e,u) is involved in another 4-cycle. It is not possible for (e,u) to be involved in another 4-cycle that contains (u,v). If there is a 4-cycle (u,e,w,v,u) where w is a vertex inside $R(u,v)$, then (e,w,v,f,e) will be uncovered in $R(u,v)$. So (e,u) has to be involved in a 4-cycle that is covered by an edge outside $R(u,v)$. In order to form such a 4-cycle, e must be connected to a boundary vertex, either a, c, or v (connecting to b or d will result in uncovered 4-cycles). Based on this observation, there can be at most three such edges with both end points strictly inside $R(u,v)$ forming 4-cycles with (u,v), see Fig 1(b). It is easy to verify that one more such edge will either result in a dangling 4-cycle or an uncovered 4-cycle in $R(u,v)$. In this case, there are at most 8 vertices strictly inside the region $R(u,v)$.

Therefore, there are at most 8 vertices strictly inside $R(u,v)$ and at most 4 vertices on the boundary of $R(u,v)$. Overall, there are at most 12 vertices in $R(u,v)$ that are not in $V(S)$. □

Theorem 4. 4-CYCLE TRANSVERSAL *admits a* $74k$ *kernel on planar graphs.*

Proof. Let w be a vertex outside any region in $\mathcal{R}$. w has to be involved in some 4-cycle C. Let $(u, v) \in S$ be the edge that covers C. Without loss of generality, assume that w is adjacent to u, i.e., there exists another vertex w' such that $C = (u, w, w', v, u)$. If w' is also outside any region, then the path (u, w, w', v) either forms a new region between u and v or should be included in an existing region between u and v, contradicting the maximality of $\mathcal{R}$. Therefore w' must be a boundary vertex for some region $R(u, v)$ between u and v. Let (u, a, b, v) and (u, c, d, v) be the boundary paths of $R(u, v)$. w' cannot be b or d since otherwise (u, w, w', a, u) or (u, w, w', c, u) are uncovered 4-cycles. Therefore w' has to be either a or c. This implies that either $(a, v) \in E$ or $(c, v) \in E$. Note that in this case, w is the only vertex outside all regions that is adjacent to boundary vertices of $R(u, v)$. If there is another such vertex w'', then this will result in either uncovered 4-cycles or violation of the planarity of $R(u, v)$.

From the above analysis, a vertex w outside the region $R(u, v)$ will always be adjacent to either u and one of the boundary vertices adjacent to u, or v and one of the boundary vertices adjacent to v. If such a vertex w exists, we cannot have a vertex strictly inside $R(u, v)$ that is adjacent to the same two vertices since this will result in an uncovered 4-cycle. When we count the total number of vertices, we only need to count them once. This implies that we don't need to count the vertices outside all regions in $\mathcal{R}$ in order to bound the total number of vertices, .

There are at most $2k$ vertices in $V(S)$. By Lemma 5 and Lemma 6, there are at most $6k - 6$ regions and each region has at most 12 vertices not in $V(S)$. Therefore the total number of vertices in G' is at most $(6k - 6) \cdot 12 + 2k \leq 74k$. □

Finally, we consider the related problem of small cycle transversal. For the lack of space the details are omitted. We assume the input graph has been preprocessed such that all edges and vertices which are not involved in any small cycle are deleted.

First, the Reduction Rule 3 can be easily adapted for the kernelization of (≤ 4)-CYCLE TRANSVERSAL by removing both dangling 3-cycles and 4-cycles.

Corollary 1. (≤ 4)-CYCLE TRANSVERSAL *admits a* $32k$ *kernel on planar graphs.*

Proof. (Sketch) Similar to the proof of Lemma 6, however there is only one vertex strictly inside a region. There are at most $2k$ vertices in $V(S)$ and each region has at most 5 vertices not in $V(S)$. The number of regions is at most 6k-6. Therefore the total number of vertices in G' is at most $(6k - 6) \cdot 5 + 2k \leq 32k$. □

Using similar techniques and a more involved analysis, we are able to obtain a $226k$ kernel for (≤ 5)-CYCLE TRANSVERSAL. We present the reduction rules in the following. However, the proofs are omitted for the lack of space. We call a simple path from u to v a *chain* if the path has length at least 2 and all vertices in the path except u and v are of degree two. The following reduction rules are used.

Reduction Rule 4.1: (The chain rule) For any two vertices u and v, if there are two or more chains between u and v, delete all chains except the shortest one.

Reduction Rule 4.2: (The generalized dangling rule) For an edge (u, v), if there is a cycle C of length ≤ 5 containing (u, v) such that for any edge e on C, all cycles of length ≤ 5 containing e must go through both u and v, then delete (u, v) and decrease k by 1.

References

1. Abu-Khzam, F.N.: Kernelization algorithms for d-hitting set problems. In: WADS, pp. 434–445 (2007)
2. Alon, N.: Bipartite subgraphs. Combinatorica 16, 301–311 (1996)
3. Bodlaender, H.L., Fomin, F.V., Lokshtanov, D., Penninkx, E., Saurabh, S., Thilikos, D.M.: (meta) kernelization. In: Proceedings of the 50th Annual IEEE Symposium on Foundations of Computer Science (2009)
4. Borodin, O.V., Kostochka, A.V., Sheikh, N.N., Yu, G.: M-degrees of quadrangle-free planar graphs. J. Graph Theory 60(1), 80–85 (2009)
5. Brügmann, D., Komusiewicz, C., Moser, H.: On generating triangle-free graphs. Electronic Notes in Discrete Mathematics 32, 51–58 (2009)
6. Cherlin, G., Komjáth, P.: There is no universal countable pentagon-free graph. J. Graph Theory 18(4), 337–341 (1994)
7. Downey, R., Fellows, M.: Parameterized Complexity. Springer, Heidelberg (1999)
8. Esperet, L., Montassier, M., Zhu, X.: Adapted list coloring of planar graphs. J. Graph Theory 62(2), 127–138 (2009)
9. Fernau, H.: Speeding up exact algorithms with high probability. Electronic Notes in Discrete Mathematics 25, 57–59 (2006)
10. Füredi, Z.: On the number of edges of quadrilateral-free graphs. J. Comb. Theory Ser. B 68(1), 1–6 (1996)
11. Füredi, Z., Naor, A., Verstraete, J.: On the turan number for the hexagon. Advances in Mathematics 203(2), 476–496 (2006)
12. Guo, J., Niedermeier, R.: Invitation to data reduction and problem kernelization. SIGACT News 38(1), 31–45 (2007)
13. Guo, J., Niedermeier, R.: Linear problem kernels for NP-hard problems on planar graphs. In: Arge, L., Cachin, C., Jurdziński, T., Tarlecki, A. (eds.) ICALP 2007. LNCS, vol. 4596, pp. 375–386. Springer, Heidelberg (2007)
14. Györi, E., Lemons, N.: Hypergraphs with no odd cycle of given length. Electronic Notes in Discrete Mathematics 34, 359–362 (2009)
15. Halford, T.R., Grant, A.J., Chugg, K.M.: Which codes have 4-cycle-free tanner graphs? IEEE Transactions on Information Theory 52(9), 4219–4223 (2006)
16. Khot, S., Regev, O.: Vertex cover might be hard to approximate to within 2-epsilon. J. Comput. Syst. Sci. 74(3), 335–349 (2008)
17. Kortsarz, G., Langberg, M., Nutov, Z.: Approximating maximum subgraphs without short cycles. In: Goel, A., Jansen, K., Rolim, J.D.P., Rubinfeld, R. (eds.) APPROX and RANDOM 2008. LNCS, vol. 5171, pp. 118–131. Springer, Heidelberg (2008)
18. Krivelevich, M.: On a conjecture of tuza about packing and covering of triangles. Discrete Math. 142(1-3), 281–286 (1995)
19. Veni Madhavan, C.E.: Foundations of Software Technology and Theoretical Computer Science, vol. 181, pp. 381–392. Springer, Heidelberg (1984)
20. Moser, H.: A problem kernelization for graph packing. In: Nielsen, M., Kucera, A., Miltersen, P.B., Palamidessi, C., Tuma, P., Valencia, F.D. (eds.) SOFSEM 2009. LNCS, vol. 5404, pp. 401–412. Springer, Heidelberg (2009)

21. Naor, A., Verstraëte, J.: A note on bipartite graphs without 2k-cycles. Comb. Probab. Comput. 14(5-6), 845–849 (2005)
22. Thomassen, C.: On the chromatic number of triangle-free graphs of large minimum degree. Combinatorica 22(4), 591–596 (2002)
23. Thomassen, C.: On the chromatic number of pentagon-free graphs of large minimum degree. Combinatorica 27(2), 241–243 (2007)
24. Yannakakis, M.: Node-and edge-deletion NP-complete problems. In: STOC, pp. 253–264 (1978)

Online Splitting Interval Scheduling on m Identical Machines*

Feifeng Zheng[1,2,3,4], Bo Liu[1], Yinfeng Xu[1,2,3,4], and E. Zhang[5]

[1] School of Management, Xi'an JiaoTong University, Xi'an, 710049, China
[2] Ministry of Education Key Lab for Intelligent Networks and Network Security, Xi'an, 710049, China
[3] Ministry of Education Key Lab for Process Control and Efficiency Engineering, Xi'an, 710049, China
[4] State Key Lab for Manufacturing Systems Engineering, Xi'an, 710049, China
[5] School of Information Management and Engineering, Shanghai University of Finance and Economics, Shanghai, 200433, China
zhengff@mail.xjtu.edu.cn

Abstract. This paper investigates online scheduling on m identical machines with *splitting intervals*, i.e., intervals can be split into pieces arbitrarily and processed simultaneously on different machines. The objective is to maximize the throughput, i.e., the total length of satisfied intervals. Intervals arrive over time and the knowledge of them becomes known upon their arrivals. The decision on splitting and assignment for each interval is made irrecoverably upon its arrival. We first show that any non-split online algorithms cannot have bounded competitive ratios if the ratio of longest to shortest interval length is unbounded. Our main result is giving an online algorithm ES (for Equivalent Split) which has competitive ratio of 2 and $\frac{2m-1}{m-1}$ for $m = 2$ and $m \geq 3$, respectively. We further present a lower bound of $\frac{m}{m-1}$, implying that ES is optimal as $m = 2$.

1 Introduction

Interval scheduling problem has a lot of applications such as crew or vehicle scheduling, telecommunication, hotel rental problem, etc (refer to Kolen et al.[1]), in which requests may show up over time and their information is known upon their arrivals. To deal with such problems with dynamic requests, Lipton and Tomkins [2] were the first to introduce online interval scheduling. Consider one machine and intervals with fixed start and end times arrive over time. Upon the arrival of each interval, online algorithms must decide whether or not to irrecoverably schedule it without overlapping, aiming to maximize the total length of satisfied intervals. They gave an optimal 2-competitive algorithm for the special case with two lengths. For the general case where intervals may

* This work was supported by NSF of China under Grants 70525004, 70702030, 70602031, and 60736027, and Doctoral Fund of Ministry of Education of China no. 20070698053.

B. Chen (Ed.): AAIM 2010, LNCS 6124, pp. 304–313, 2010.

be of arbitrary lengths, they gave a randomized algorithm with competitive ratio $O((\log \Delta)^{1+\epsilon})$ as well as a lower bound of $\Omega(\log \Delta)$ where Δ is the ratio between the longest and shortest intervals. Faigle et al. [3] showed that no deterministic algorithm has competitive ratio less than Δ, and a greedy algorithm which schedules an arriving interval if the machine is idle can reach this bound. They also proved a greedy algorithm to be 2-competitive for the case where the machine is provided with a buffer to store one interval.

For interval scheduling on m identical machines, Faigle and Nawijn [4], and independently Carlisle and Lloyd [5] considered the preemptive model with the objective of minimizing the number of missing intervals. They presented a greedy deterministic algorithm which always outputs an optimal solution. Thibault and Laforest [6] investigated a preemptive model with processing time windows and the objective is to maximize the number of satisfied intervals. For the case where there are at most β different interval lengths in an input sequence, they gave a $(4\min\{\beta, \lceil\log_2(\Delta)\rceil + 1\})$-competitive deterministic algorithm. In this paper, we consider interval scheduling on m identical machines to maximize the total length of satisfied intervals. We focus on the case where interval may be split into at most m pieces and processed simultaneously. This case is reasonable in some manufacturing activities such that a manufacturer may divide a production job into some parts and process them in several workshops simultaneously.

1.1 Related Work

One quite related area is online weighted interval scheduling to maximize the total weight of completed intervals. Woeginger [7] introduced a preemptive online interval scheduling on a single machine with arbitrary interval weight. He showed that no deterministic algorithms have finite competitive ratios for general case where there is no relationship between weight and length of interval, and gave a matching upper and lower bound of 4 for deterministic algorithms for the special case of uniform length, leaving it open whether randomization can be used to break this bound. A lot of work followed this open problem. Canetti and Irani [8] showed that even randomized algorithm cannot have finite competitive ratio for the general case. Seiden [9] gave a randomized 3.732-competitive algorithm for the special case where the weight of interval is a continuous convex non-negative function f of interval length with $f(0) = 0$. Fung et al. [10] improved the upper bound to 3.5822 for the case of unit length and gave a lower bound of 4/3. Fung et al. [11] further improved the upper bound to 2, and a matching lower bound for a subset of randomized algorithms.

Another important line is job scheduling where each job has a deadline equal to or larger than the arrival time plus job length. Thus, interval scheduling may be viewed as a special case of job scheduling with tight deadlines. The offline job scheduling with splitting jobs on m identical machines has been extensively studied. See Xing and Zhang [12], Kim and Shim et al. [13], and Shim and Kim [14]. A splitting job can be split into several pieces arbitrarily and processed independently on different machines. Xing and Zhang [12] studied the model to minimize makespan. They proposed two polynomial algorithms for the case without setup time, and a heuristic algorithm with worst-case ratio of $\frac{7}{4}-\frac{1}{m}(m \geq 2)$ for the case where each job has an independent setup time. Both Kim et al. [13] and Shim and Kim [14] investigated the problem to minimize the total tardiness. To the best of our knowledge, there are no results on online case with splitting jobs.

The rest of this paper is organized as follows. Section 2 gives a formal description of the problem to be investigated, and some definitions on competitive ratio. In Section 3 we prove the competitiveness of non-split algorithms which processes each interval in an integral time duration on one machine. We then propose ES algorithm and prove its competitiveness in Section 4. Section 5 gives a lower bound of competitive ratio for deterministic algorithms. Finally, Section 6 concludes this work.

Below are basic notations used in this paper.

m: the number of machines;

M_u: the uth machine, $1 \leq u \leq m$;

J_i: an interval indexed by i;

$r(J_i)$ or r_i: the arrival time of J_i;

$p(J_i)$ or p_i: the processing time or length of J_i;

$d(J_i)$ or d_i: the deadline of J_i, by which J_i must be satisfied. $d_i = r_i + p_i$.

2 Problem Description and Basic Definitions

2.1 Problem Description

The problem of online splitting interval scheduling on m identical machines is formally described as follows. There are m identical machines denoted by $M_1, M_2, \ldots, M_m$, where m is known beforehand. Each interval J_i with length or processing time p_i and deadline $d_i = r_i + p_i$ arrives at time r_i, upon which p_i and d_i become known to online algorithms. J_i may or may not be split into k $(2 \leq k \leq m)$ pieces without any cost and be processed simultaneously on k machines. It is not allowed to assign two or more pieces on one machine. With the power of splitting, although the irrecoverable decision on splitting and assignment for interval J_i must be made upon its arrival, the start time of processing J_i may actually be larger than its arrival time. If all the pieces of J_i are assigned to different machines within time duration $[r_i, d_i]$, we say the assignment of J_i as well as the interval itself is *feasible*. So, J_i has to be rejected if it is not feasible or at least one piece of J_i cannot meet the deadline d_i. The objective is to maximize the total length of satisfied intervals. Let $Prof = \sum_{J_i \in S} p_i$ where S is the set of satisfied intervals. Applying the concept of Three-Field Notation (refer to Graham et al. [15]), we denote the model by $Pm \mid online, r_i, split, d_i = r_i + p_i \mid Prof$.

2.2 Competitive Ratio

For online problems, to gauge the performance of an online algorithm $\mathcal{A}$, the competitive ratio analysis (refer to Borodin and El-yaniv [16]) is often used. For the case with objective of maximizing the profit, the *competitive ratio* of $\mathcal{A}$ is defined as

$$c_{\mathcal{A}} = \sup_{I} \frac{|OPT(I)|}{|\mathcal{A}(I)|}$$

where $|\mathcal{A}(I)|$ and $|OPT(I)|$ denote the total profit gained by $\mathcal{A}$ and an offline optimal algorithm OPT from an arbitrary input instance I , respectively. If $c_{\mathcal{A}}$ is infinite, then $\mathcal{A}$ is *not competitive*; if there does exist an instance I such that $|OPT(I)|/|\mathcal{A}(I)| = c_{\mathcal{A}}$, then we say the competitive analysis of $\mathcal{A}$ is *tight.*

For an online scheduling problem, the *lower bound* of competitive ratio for online algorithms is defined as

$$c^* = \inf_{\mathcal{A}\in\pi} \sup_{I} \frac{|OPT(I)|}{|\mathcal{A}(I)|}$$

where π is the set of all online algorithms for the online problem. If $c_{\mathcal{A}} = c^*$, then algorithm $\mathcal{A}$ is *optimal* in competitiveness.

3 Competitiveness of Non-split Algorithms

Non-split algorithms are defined as a class of algorithms that assign each feasible interval to a single machine on its arrival. For this class of algorithms, we show that their competitive ratios are at least $\sqrt[m]{\Delta}$ where Δ is the ratio of longest to shortest interval length.

Theorem 1. *In the model $Pm|online, r_i, split, d_i = r_i + p_i|Prof$, any non-split algorithm cannot be better than $\sqrt[m]{\Delta}$-competitive.*

Proof. It suffices to present an interval input instance σ to make any non-split algorithm $\mathcal{A}$ behave poorly and be at least $\sqrt[m]{\Delta}$-competitive. Let $\sigma = (J_0, J_1, \ldots, J_n)$ where the value of n depends on the behavior of $\mathcal{A}$. Interval J_k $(0 \leq k \leq n)$ with length x^k arrives at time $k\varepsilon$ where $\varepsilon \in (0, \frac{1}{m})$ and $x = \sqrt[m]{\Delta}$. First, J_0 arrives at time 0. $\mathcal{A}$ has two selections as follows.

Case 1. $\mathcal{A}$ rejects J_0. No more intervals arrive later and σ terminates. $\mathcal{A}$ satisfies no intervals and $|\mathcal{A}(\sigma)|$=0, while OPT will complete J_0 with $|OPT(\sigma)| =$ 1, implying that $\mathcal{A}$ loses.

Case 2. $\mathcal{A}$ accepts J_0. In this case, intervals J_i for $1 \leq i < m$ arrive one by one until $\mathcal{A}$ rejects some J_i. There are two subcases.

Case 2.1. $\mathcal{A}$ accepts J_{i-1} but rejects J_i. No more intervals arrive later. $\mathcal{A}$ completes $\{J_0, \ldots, J_{i-1}\}$ with total profit $|\mathcal{A}(\sigma)|$=$1 + x + \ldots + x^{i-1}$. OPT completes all the $i + 1$ $(\leq m)$ intervals with total profit $|OPT(\sigma)|$=$1 + x + \ldots + x^i$. $|OPT(\sigma)|/|\mathcal{A}(\sigma)|$ $>$$x$ in this subcase.

Case 2.2. $\mathcal{A}$ accepts all the first m intervals $\{J_0, \ldots, J_{m-1}\}$. In this case, the last interval J_m with length x^m arrives at time $m\varepsilon$. Since $m\varepsilon < 1$, all the m machines are busy at time $m\varepsilon$ and then $\mathcal{A}$ has to reject J_m, implying that $|\mathcal{A}(\sigma)|$=$1 + x + \ldots + x^{m-1}$. OPT will reject the first interval J_0 and accept the rest m intervals, gaining a total profit $|OPT(\sigma)|$=$x(1 + x + \ldots + x^{m-1})$. In this case, $|OPT(\sigma)|/|\mathcal{A}(\sigma)|$ =x.

In either case, the ratio of what OPT gains to that of $\mathcal{A}$ is at least $x = \sqrt[m]{\Delta}$. The theorem follows. □

Note that as $m = 1$, the above lower bound reduces to Δ as proved in Faigle et al. [3] for the case of single machine. When Δ is unbounded, Theorem 1 implies that any deterministic algorithms cannot be competitive. Moreover, it is not hard to see that the optimal greedy algorithm with competitive ratio Δ for single machine scheduling in Faigle et al. [3] is still Δ-competitive for the model $Pm|online, r_i, split, d_i = r_i + p_i|Prof$. We may present m copies of intervals with unit length at time 0 and m copies of intervals with uniform length Δ at time 1/2. The greedy algorithm will accept the first m intervals and reject the last m intervals, implying the ratio of Δ.

4 *ES* Algorithm and Its Competitive Analysis

4.1 Algorithm Description

We already know from the previous section that if Δ is infinite, non-split algorithms cannot be competitive. Thus we propose an online algorithm ES whose competitive ratio has no relationship with the value of Δ. ES always splits each interval into m equivalent pieces and assigns them on the m machines in the same time duration, that is, all the m pieces are processed within the same time section on m machines. Thus the status of the m machines, either idle or busy, is always the same. ES algorithm is formally described in the following.

ES **algorithm:** When a new interval J_i arrives at time r_i, ES accepts or rejects J_i according to whether J_i is feasible or not. Ties are broken by selecting the shortest interval arriving at time r_i. Assume that the m machines are supposed to complete the currently assigned tasks at some time $t \geq r_i$. If $t + p_i/m \leq d_i$, then J_i is feasible and ES accepts the interval. ES splits J_i into m equivalent pieces each of which is assigned to one of the m machines, and will process J_i during time period $[t, t + p_i/m)$. Otherwise if $t + p_i/m > d_i$, ES rejects J_i.

ES is simple and easy to operate while it performs well by the power of splitting intervals. It will not miss a very long interval that arrives when all the m machines are busy since a long interval has a long deadline as well. However, ES may actually miss such an interval J_i that arrives when all the machines are busy and p_i together with d_i is not large enough for ES to satisfy the interval. That is, ES may be induced to start too many relatively shorter intervals that arrive earlier than J_i. Hence, ES might not be optimal in competitiveness for general case.

4.2 Competitive Analysis

We first define the concepts of *busy* and *idle* sections. If all the m machines change their status from idle to busy at time t_1 and return idle at time $t_2 (> t_1)$, then $[t_1, t_2)$ is called a *busy* section of ES. And if all the machines become busy again at time $t_3 > t_2$, then $[t_2, t_3)$ is called an *idle* section.

We first consider a special case where $m = 2$, and have the following theorem.

Theorem 2. *In the model $P2|online, r_i, split, d_i = r_i + p_i|Prof$, ES is 2-competitive.*

Proof. We prove this theorem by drawing a contradiction for a constructed *smallest counter-sequence*. Suppose the theorem is not true. Then there must exist an interval input sequence σ with the smallest number of intervals that makes the ratio between the profit of OPT, denoted by $|OPT(\sigma)|$, and that of ES, denoted by $|ES(\sigma)|$, strictly larger than 2. That is, $|OPT(\sigma)|/|ES(\sigma)| > 2$. We call σ the smallest counter-sequence. We prove the theorem below by showing that such σ does not exist at all.

We observe that for the smallest counter-sequence σ, the processing sequence I produced by ES includes exactly one busy section. Otherwise suppose I contains at least two busy section $[t_1, t_2)$ and $[t_3, t_4)$, where $0 \leq t_1 < t_2 < t_3 < t_4$. According to construction of ES, no intervals in σ arrive during $[t_2, t_3)$. Thus we can divide the intervals in σ into two sets such that one set contains the intervals arriving during $[t_1, t_2)$ and the other set contains those arriving during $[t_3, t_4)$. At least one of the two sets makes the ratio of the profit of OPT to that of ES larger than 2. This contradicts that σ includes the smallest number of intervals. So I contains exactly one busy section, denoted by $[t_1, t_2)$. Note that all the intervals in σ arrive during $[t_1, t_2)$ by construction of ES. Let J_n be the last interval processed by ES before time t_2. Then we have $t_1 \leq t_2 - p_n/2$.

For OPT, it is also idle before time t_1 while it may keep the machines busy after time t_2. Assume that OPT keeps at least one machine busy until time t' ($\geq t_2$) while the other machines become idle on or before the time. If $t' - t_2 \leq t_2 - t_1$, then $t' - t_1 \leq 2(t_2 - t_1)$, implying $|OPT(\sigma)|/|ES(\sigma)| \leq 2$ and then the nonexistence of σ. In the following we prove $t' - t_2 \leq t_2 - t_1$ by drawing contradiction.

Assume otherwise that $t' - t_2 > t_2 - t_1$. Let J_k be the interval completed by OPT at time t', implying at least one machine completes J_k at the time. Note that J_k arrives on or after time t_1, i.e., $r_k = d_k - p_k \geq t_1$. Moreover, since OPT completes J_k at time t', $d_k \geq t'$. So, $d_k - \frac{p_k}{2} = \frac{d_k}{2} + \frac{d_k - p_k}{2} \geq \frac{t'}{2} + \frac{t_1}{2} > t_2$, i.e. $d_k > t_2 + \frac{p_k}{2}$. By ES, it will start J_k at time t_2 and satisfy the interval before time d_k, contradicting that ES is idle after time t_2. Hence, $t' - t_2 \leq t_2 - t_1$. The theorem follows. □

Now we consider the general case with $m > 2$. Consider an arbitrary interval input instance Γ. Let I be the corresponding sequence produced by ES. Assume without loss of generality that ES starts the first interval in I at time 0. By ES, I consists of a series of alternant busy and idle sections. Assume there are n busy sections in I, denoted by $B = (B_1, B_2, \cdots, B_n)$ ($n \geq 1$). Between every two consecutive busy sections B_i and B_{i+1}, there is an idle section G_i, and thus there are totally $n-1$ idle sections $G = (G_1, G_2, \cdots, G_{n-1})$. Let $r(B_i)$ and $d(B_i)$ be the start time and end time of each B_i respectively. Then $B_i = [r(B_i), d(B_i))$ and $G_i = [d(B_i), r(B_{i+1}))$.

Let $E = \{J_i\}$ be the set of intervals in B which are completed by ES, and $O = \{o_j\}$ be the set of intervals completed by OPT in Γ. Let $\overline{O} = O/E$ be the interval set that contains the intervals satisfied by OPT but not ES. Denote by $|E|$, $|O|$ and $|\overline{O}|$ the total length of intervals in set E, O and $\overline{O}$, respectively. Since E may contain some intervals out of O, $|O| \leq |\overline{O}| + |E|$. If $\overline{O} = \phi$, then $|O| = |E|$ or $|O|/|E| = 1$. So, we assume that $\overline{O}$ is not empty. For any interval $\overline{o}_j$ in $\overline{O}$, we observe that it arrives in one of the n busy sections.

Lemma 1. *For every $\overline{o}_j \in \overline{O}$, there exists an index $i(1 \leq i \leq n)$ satisfying $r(B_i) \leq r(\overline{o}_j) < d(B_i)$.*

Proof. The lemma directly follows by construction of ES. □

By Lemma 1, we divide all the intervals in $\overline{O}$ into n subsets $\overline{O} = \{S_1, S_2, \cdots, S_n\}$ such that $S_i = \{\overline{o}_j | r(B_i) \leq r(\overline{o}_j) < d(B_i)\}$ $(1 \leq i \leq n)$. Note that some S_i may be empty. For any S_i, we have the following lemma.

Lemma 2. *For every $\overline{o}_j \in S_i$, we have $r(B_i) \leq r(\overline{o}_j)$ and $d(\overline{o}_j) < \frac{md(B_i)-r(\overline{o}_j)}{m-1}$.*

Proof. The first conclusion is straightforward by the definition of S_i. For the second conclusion, assume otherwise that $d(\overline{o}_j) \geq \frac{md(B_i)-r(\overline{o}_j)}{m-1}$, which is equivalent to $d(B_i) \leq d(\overline{o}_j) - \frac{d(\overline{o}_j)-r(\overline{o}_j)}{m} = d(\overline{o}_j) - \frac{p(\overline{o}_j)}{m}$. This implies that ES can satisfy $\overline{o}_j$ within $[d(B_i), d(\overline{o}_j))$, contradicting that $\overline{o}_j \in \overline{O} = O/E$. Hence, it is a contradiction to the assumption. The Lemma follows. □

Let $|S_i|$ be the total length of the intervals in S_i, i.e. $|S_i| = \sum_{\overline{o}_j \in S_i} p(\overline{o}_j)$, $|B_i|$ be that of the intervals processed by ES within B_i. By construction of ES, we have $|B_i| = m(d(B_i) - r(B_i))$.

$$\begin{aligned}
|S_i| &= \sum_{\overline{o}_j \in S_i} (d(\overline{o}_j) - r(\overline{o}_j)) \\
&\leq m\left(\max_{\overline{o}_j \in S_i} d(\overline{o}_j) - \min_{\overline{o}_j \in S_i} r(\overline{o}_j)\right) \\
&< m\left(\frac{m}{m-1} d(B_i) - \frac{r(\overline{o}_j)}{m-1} - r(B_i)\right) \\
&\leq \frac{m^2}{m-1}\left(d(B_i) - r(B_i)\right) \\
&= \frac{m}{m-1}|B_i|
\end{aligned}$$

where the first inequality holds since there are at most m machines being busy during $[\min_{\overline{o}_j \in S_i} r(\overline{o}_j), \max_{\overline{o}_j \in S_i} d(\overline{o}_j))$, and both the second and third inequalities are due to Lemma 2. Hence, the ratio between the profit of OPT and that of ES for Γ can be bounded below.

$$\frac{|O|}{|E|} \leq \frac{|E| + |\overline{O}|}{|E|} = 1 + \frac{\sum_{i=1}^{n} |S_i|}{\sum_{i=1}^{n} |B_i|} < 1 + \frac{m}{m-1} = \frac{2m-1}{m-1}.$$

Based on the above analysis, we have the following theorem.

Theorem 3. *In the model $Pm|online, r_i, split, d_i = r_i + p_i|Prof$, ES has a competitive ratio of $\frac{2m-1}{m-1}$ where $m \geq 3$.*

In the rest of this section, we present an instance $\sigma = (J_0, J_1, \ldots, J_m)$ to show that for the case with $m \geq 3$, the competitive analysis of ES is tight. That is, there exists an instance σ such that $\frac{|OPT(\sigma)|}{|ES(\sigma)|} = \frac{2m-1}{m-1}$.

The instance is illustrated in Fig.1(a). Let $t_3 = 1$ and t_1 be an arbitrarily small positive real number. Let $t_2 = \frac{t_3 - t_1 - m\epsilon}{m-1}$ and $t' = \frac{t_3}{m} - \epsilon$ where $\epsilon \to 0^+$. The first interval J_0 with length $p_0 = t_3$ arrives at time 0, and the rest m intervals $J_1, \ldots, J_m$ with uniform length $p_1 = t_2 - t_1$ arrive at time t_1. It can be verified that $t' = t_2 - \frac{p_1}{m}$. Since $p_1 = t_2 - t_1 = (\frac{p_0 + p_1}{m} - \epsilon) - t_1$, we have by algebraic calculation that

$$p_1 = \frac{1}{m-1} p_0 - \frac{m}{m-1}(\epsilon + t_1).$$

ES starts J_0 at time 0 and completes it at time $\frac{p_0}{m} = t' + \epsilon > t' = t_2 - \frac{p_1}{m}$. Thus ES cannot catch the deadline of any other intervals after completing J_0 (See Fig.1(b)), implying that $|ES(\sigma)| = p_0$. OPT will first start $J_1, \ldots, J_m$ at time t_1 on the m machines respectively, and then satisfy J_0 within $[t_2, t_3)$ since $t_2 + p_0/m \leq t_3$ due to theorem condition $m \geq 3$ (See Fig.1(c)). That is, OPT complete all the $m + 1$ intervals with a total profit $|OPT(\sigma)| = p_0 + mp_1$.

$$\frac{|OPT(\sigma)|}{|ES(\sigma)|} = \frac{p_0 + mp_1}{p_0} = \frac{p_0 + m[\frac{1}{m-1}p_0 - \frac{m}{m-1}(\epsilon + t_1)]}{p_0} = \frac{2m-1}{m-1} - \frac{m^2(\epsilon + t_1)}{(m-1)p_0}$$

As $(\epsilon + t_1) \to 0$, $\frac{|OPT(\sigma)|}{|ES(\sigma)|} \to \frac{2m-1}{m-1}$. So, the competitive analysis of ES in Theorem 3 is tight.

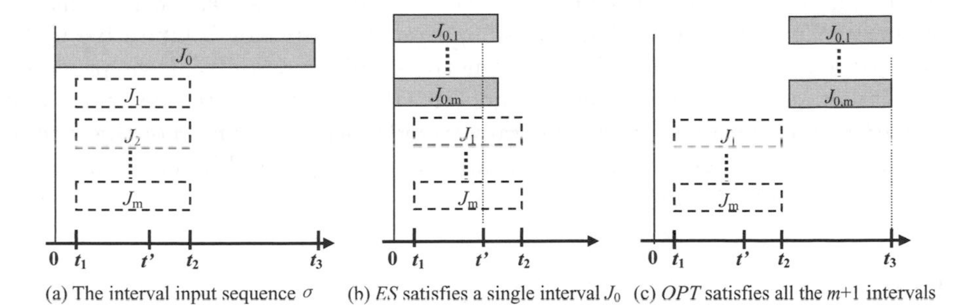

Fig. 1. An illusive instance for tight analysis of ES algorithm

5 A Lower Bound

Theorem 4. *In the model $Pm|online, r_i, split, d_i = r_i + p_i|Prof$ with $m \geq 2$, no deterministic algorithm has a competitive ratio less than $\frac{m}{m-1}$.*

Proof. Similar to the proof of Theorem 1, it is sufficient to construct an interval input instance $\sigma = \{J_0, J_1, \ldots\}$ to make any online deterministic algorithm $\mathcal{A}$ behave poorly and be at best $\frac{m}{m-1}$-competitive. The number of intervals in σ depends on the behavior of $\mathcal{A}$. The first interval J_0 with length $p_0 = 1$ arrives at time 0. $\mathcal{A}$ may have two selections.

Case1. $\mathcal{A}$ rejects J_0. No more intervals arrive later. OPT will complete J_0 with a profit of p_0 while $\mathcal{A}$ gains no profit, implying that $\mathcal{A}$ loses.

Case2. $\mathcal{A}$ accepts J_0. At time $r_1 < 1/m$, there come m uniform intervals with sufficiently large length p_1. For J_0, $\mathcal{A}$ may split the interval into several pieces and arrange those pieces on some machines respectively. Let $C_i(J_0)$ be the completion time of the piece processed on the ith machine M_i. If some M_i does not process J_i, set $C_i(J_0) = 0$. Let $C_0 = \max C_i(J_0)$ $(1 \leq i \leq m)$. We observe that $C_0 \geq 1/m$ otherwise J_0 cannot be satisfied. Since $r_1 < 1/m \leq C_0$, $\mathcal{A}$ completes at most $m-1$ out of the m intervals arriving at time r_1. So, $|\mathcal{A}(\sigma)| = p_0 + (m-1)p_1 = 1 + (m-1)p_1$. OPT will reject J_0 and complete all the last m intervals, implying a profit of $|OPT(\sigma)| = mp_1$. As $p_1 \to \infty$,

$$\frac{|OPT(\sigma)|}{|\mathcal{A}(\sigma)|} \geq \frac{mp_1}{1 + (m-1)p_1} \to \frac{m}{m-1}.$$

The theorem follows. □

By Theorems 2 and 4, ES is optimal in competitiveness for the case $m = 2$.

6 Conclusion

This work investigated interval scheduling on m identical machines to maximize the total length of satisfied intervals. We focused on the case with splitting intervals. We first showed that non-split algorithms are not competitive if the ratio of longest to shortest interval length is unbounded. We then proposed a splitting algorithm which has a competitive ratio related to the number of machines and is optimal as there are only two identical machines. For the general case with $m \geq 3$ machines, it is left a gap between upper and lower bounds that needs to be closed in further work.

References

1. Kolen, A.W.J., Lenstra, J.K., Papadimitriou, C.H., Spieksma, F.C.R.: Interval Scheduling: A survey. Naval Research Logistics 54, 530–543 (2007)
2. Lipton, R.J., Tomkins, A.: Online interval scheduling. In: Proceedings of Fifth Annual ACM-SIAM Symposium on Discrete Algorithms (SODA 1994), pp. 302–311 (1994)
3. Faigle, U., Garbe, R., Kern, W.: Randomized online algorithms for maximizing busy time interval scheduling. Computing, 95–104 (1996)
4. Faigle, U., Nawijn, W.M.: Note on scheduling intervals on-line. Discrete Applied Mathmatics 58, 13–17 (1995)

5. Carlisle, M.C., Lloyd, E.L.: On the k-coloring of intervals. Discrete Applied Mathematics 59, 225–235 (1995)
6. Thibault, N., Laforest, C.: On-line time-constrained scheduling problem for the size on k machines. In: The 8th International Symposium on Parallel Architectures, Algorithms and Networks, pp. 20–24 (2005)
7. Woeginger, G.J.: On-line scheduling of jobs with fixed start and end times. Theoretical Computer Science 130, 5–16 (1994)
8. Canetti, R., Irani, S.: Bounding the power of preemption in randomized scheduling. SIAM Journal of Computing 27, 993–1015 (1998)
9. Seiden, S.S.: Randomized online interval scheduling. Operations Research Letters, 171–177 (1998)
10. Fung, S.P.Y., Poon, C.K., Zheng, F.F.: Online Interval Scheduling: Randomized and Multiprocessor Cases. Journal of Combinatorial Optimization 16, 248–262 (2008)
11. Fung, S.P.Y., Poon, C.K., Zheng, F.F.: Improved Randomized Online Scheduling of Unit Length Intervals and Jobs. In: Bampis, E., Skutella, M. (eds.) WAOA 2008. LNCS, vol. 5426, pp. 53–66. Springer, Heidelberg (2009)
12. Xing, W., Zhang, J.: Parallel machine scheduling with splitting jobs. Discrete Applied Mathematics 103, 259–269 (2000)
13. Kim, Y.D., Shim, S.O., Kim, S.B., Choi, Y.C., Yoon, H.: Parallel machine scheduling considering a job splitting property. International Journal of Production Research 42, 4531–4546 (2004)
14. Shim, S.O., Kim, Y.D.: A branch and bound algorithm for an identical parallel machine scheduling problem with a job splitting property. Computers and Operations Research 35, 863–875 (2008)
15. Graham, R.L., Lawler, E.L., Lenstra, J.K., Rinnooy Kan, A.H.G.: Optimization and approximation in deterministic sequencing and scheduling: A survey. Annals of Discrete Mathematics 5, 287–326 (1979)
16. Borodin, A., El-yaniv, R.: Online computation and competitive analysis. Cambridge University Press, Cambridge (1998)

Extended Tabu Search on Fuzzy Traveling Salesman Problem in Multi-criteria Analysis

Yujun Zheng

Institute of Software, Chinese Academy of Sciences, Beijing 100080, China
yujun.zheng@computer.org

Abstract. The paper proposes an extended tabu search algorithm for the traveling salesman problem (TSP) with fuzzy edge weights. The algorithm considers three important fuzzy ranking criteria including expected value, optimistic value and pessimistic value, and performs a three-stage search towards the Pareto front, involving a preferred criterion at each stage. Simulations demonstrate that our approach can produce a set of near optimal solutions for fuzzy TSP instances with up to 750 uniformly randomly generated nodes.

Keywords: Tabu search, traveling salesman problem, fuzzy optimization, multi-criteria decision making.

1 Introduction

The traveling salesman problem (TSP) is one of the most well-known combinatorial optimization problems, and many seemingly different managerial problems such as vehicle routing, job scheduling, network design, etc., can be modeled as TSP and its variants [1]. Given a weighted graph $G = \langle V, E, w \rangle$ with n nodes, the problem is to find a Hamiltonian cycle (i.e., a cycle that starts from a node and passes through every other node exactly once) of minimum weight, which can be mathematically stated as follows:

$$\min \sum_{i,j} w_{ij} x_{ij} \tag{1}$$

$$\text{s.t.} \sum_{i=1}^{n} x_{ij} = 1, \quad j = 1, 2, ..., n \tag{2}$$

$$\sum_{j=1}^{n} x_{ij} = 1, \quad i = 1, 2, ..., n \tag{3}$$

$$\sum_{i,j \in S} x_{ij} \leq |S| - 1, \quad 2 \leq |S| \leq n - 2, \quad S \subset \{1, 2, ..., n\} \tag{4}$$

$$x_{ij} \in \{0, 1\}, \quad i, j = 1, 2, ..., n, i \neq j \tag{5}$$

In the classical TSP the edge weight of the graph is considered to be exact. However, in practice, the weight may represent distance, time, cost, transmission

B. Chen (Ed.): AAIM 2010, LNCS 6124, pp. 314–324, 2010.

power, etc, which are always referred to as a fuzzy set. Since fuzzy numbers are represented by possibility distributions, it is difficult to determine clearly whether one fuzzy number is larger or smaller than the other [2], which improves the difficulty of problem-solving. For example, given two fuzzy path weights w_1 and w_2, the minimum of them does not need to be either w_1 or w_2. A straightforward idea is to transform the fuzzy numbers into real numbers, but there is a variety of measures or ranking methods for fuzzy numbers, and in most cases, a single measure is insufficient to support comprehensive decision-making. When multiple criteria are involved, there is no natural notion of a distinct optimal solution and we have to be content with non-dominated solutions, i.e., solutions that are not dominated by any other solution in all objectives.

The paper presents a novel approach to the fuzzy traveling salesman problem (FTSP) base on tabu search [3,4], a meta-heuristic search that repeatedly moves from a current solution to the best of neighboring solutions while avoiding being trapped in local optima by keeping a tabu list of forbidden moves. In this paper we take into consideration three important fuzzy measures, namely expected value, optimistic value and pessimistic value, and extend the tabu search method to generate a set of near optimal solutions with a distinct preference ranking of the objectives. Our approach has the following advantages:

- It guides the search towards the Pareto-optimal front, giving preference to the expected objective functions and at the same time providing optimistic and pessimistic alternatives for the decision maker.
- The algorithm is a simple extension of the basic tabu search algorithm and thus is simple to implement, but more advanced features of tabu search can be easily integrated.
- Although dealing with a multi-objective version of a typical NP-hard problem, the algorithm performs well for reasonably large-size problem instances, as suggested by the experimental result.
- The proposed metaheuristic is general-purpose and can be applied to a large variety of other fuzzy optimization problems.

1.1 Related Work

The general TSP is known to be NP-hard and APX-hard and thus requires effective heuristics and metaheuristics (e.g., genetic algorithm [5,6], simulated annealing [7], tabu search [8,9]) to obtain good solutions. In their survey conducted in 2008, Basu and Ghosh [10] found that tabu search is possibly the most widely-used and successful metaheuristic procedure for the TSP and its related problems in the literature, but most of the research deals only with problem instances with less than 500 nodes and only three papers address the instances with more than 1000 nodes (although few complicated extensions of tabu search, e.g., [11], have been tested on instances with up to 85,900 nodes).

In the early literature, uncertainties have been introduced into TSP with probability distribution functions and stochastic models [12,13,14], but static formulations only considering expected values of parameters have inherent limitations, while dynamic formulations such as Markov Process incurs high computational costs. In real-world decision analysis, fuzzy numbers are much more convenient to describe the performance of alternatives and thus more frequently employed in modeling combinatorial optimization problems.

Fuzzy optimization also appears in literature with multiple criteria, and a multi-objective Pareto front can be interpreted as the solution for a fuzzy problem [15]. Therefore, Ulungu and Teghem [16] suggested to use metaheuristics including evolutionary algorithms and tabu search to handle multi-objective problems more effectively than traditional programming methods. During the last decades, evolutionary algorithms have been a primary focus of multi-objective TSP approaches [17,18,19,20], but the experimental results reported restrict to relatively small size problems. On the other hand, most tabu search methods for multi-objective optimization are based on the combination of objectives, which cannot identify all points in a trade-off surface of non-convex solution spaces [21]. The first tabu search approach that evolves a set of solutions in parallel was developed by Hansen [22], which selects an active objective in each iteration based on the Tchebycheff metric. Under a similar scheme, Kulturel-Konak *et al* [23] employed a multinomial probability mass function to eliminate the requirements of weighting and scaling, while Alves and Climaco [24] use an interactive method that requires the decision-maker to specify the regions of major interest.

2 Basic Tabu Search

In this section we introduce a basic tabu search procedure for the TSP, the notation of which is given below.

Notation

x	current solution
x^*	the best solution seen so far
$N(x)$	neighborhood of solution x
$w(x)$	objective value of solution x
T	tabu list
λ	tabu tenure
$\overline{k}$	upper limit of the number of iterations
$\overline{l}$	upper limit of the number of non-improving iterations

The search starts from an arbitrary Hamiltonian cycle x. A neighborhood to x is defined as any solution obtained by a pairwise exchange of any two nodes in x. At each step of the iterative procedure, the move to the best neighboring solution which is not tabu is chosen. However, if a move results in a solution better than the current best one, it is allowed even if it is tabu (which is called as the aspiration criterion). In order to prevent from cycling in a small set of recently visited solutions, the tabu list keeps the number for which a given pair of nodes is prohibited from exchange: after a pairwise exchange of node i and j, the corresponding number $T(i, j)$ is set to λ, while any other $T(i', j')$ decreases one. The search terminates if the number of iterations reaches $\overline{k}$, or the solution is not improved for $\overline{l}$ iterations. The pseudo-code of the algorithm is as follows.

ALGORITHM: TS
[input: $G = \langle V, E, w \rangle, \lambda, \overline{k}, \overline{l}$]
[output: $x*$]
//*Initialization*:
let x be an arbitrary feasible solution;
$x^* \leftarrow x, k \leftarrow 0, l \leftarrow 0$;
Initialize the tabu list T;
//*Iterative Moves*:
while($k < \overline{k}$) do
  $N \leftarrow \{y \in N(x) | T(x, y) \leq 0 \vee w(y) < w(x^*)\}$;
  if $N = \emptyset$ then break;
  $x' \leftarrow min(N)$;
  if $(w(x') < w(x^*))$ then $x^* \leftarrow x', l \leftarrow 0$;
  else $l \leftarrow l + 1$;
  if $(l = \overline{l})$ then break;
  Update T with (x, x');
  $x \leftarrow x', k \leftarrow k + 1$;
endwhile;
return x^*;

Algorithm 1. The basic tabu search algorithm

3 An Extended Tabu Search for FTSP

3.1 Fuzzy Ranking Methods Used

For fuzzy numbers, there is a variety of measures developed ranging from the trivial to the complex. Zadeh [25] first proposed the concept of possibility measure, but it is not self-dual; In [26] Liu and Liu defined a self-dual credibility measure and refined it in [27]; In recent years the concepts of centroid have been widely used for developing ranking index (e.g. [28,29,30,31]). Chen and Hwang [32] classified the ranking methods of fuzzy numbers into four major classes including preference relation, fuzzy mean and spread, fuzzy scoring, and linguistic expression. In this paper, we use the credibility measure from [27] that satisfies normality, monotonicity, self-duality, and maximality. Let ξ be a fuzzy variable with membership function μ; then for any set B of real numbers, we have:

$$Cr\{\xi \in B\} = (\sup_{x \in B} \mu(x) + 1 - \sup_{x \in B^C} \mu(x))/2 \tag{6}$$

Thus the expected value, α-optimistic value and α-pessimistic value ($\alpha \in (0, 1]$) of ξ are respectively defined by Equation (7)~(9):

$$E(\xi) = \int_0^{\infty} Cr\{\xi \geq r\}dr - \int_{-\infty}^{0} Cr\{\xi \leq r\}dr \tag{7}$$

$$\xi_{\sup}(\alpha) = \sup\{r | Cr\{\xi \geq r\} \geq \alpha\} \tag{8}$$

$$\xi_{\inf}(\alpha) = \inf\{r | Cr\{\xi \leq r\} \geq \alpha\} \tag{9}$$

For example, let $\xi = (l, m, u)$ be a triangular fuzzy variable, then we have:

$$E(\xi) = (l + 2m + u)/4 \tag{10}$$

$$\xi_{\sup}(\alpha) = \begin{cases} 2\alpha m + (1 - 2\alpha)u & \text{if } \alpha \leq 0.5 \\ (2\alpha - 1)l + (2 - 2\alpha)m & \text{else} \end{cases} \tag{11}$$

$$\xi_{\inf}(\alpha) = \begin{cases} (1 - 2\alpha)l + 2\alpha m & \text{if } \alpha \leq 0.5 \\ (2 - 2\alpha)m + (2\alpha - 1)u & \text{clsc} \end{cases} \tag{12}$$

3.2 Algorithmic Framework

For the problem with fuzzy weights, it is difficult to exactly determine the optimum from a set of candidate solutions, and hence a comprehensive approach should take more than one ranking criteria into consideration. Given a set of ranking method $f_1, f_2, ...f_m$, a fuzzy number x is said to be dominated by another number y if $f_i(y) \leq f_i(x)$ for all i and $f_i(y) < f_i(x)$ for at least one i. For an optimization problem with fuzzy objectives, if a solution's objective value is not dominated by that of any other solution, it is called a non-dominated solution. Thus, the essential goal becomes to find a set of non-dominated solutions.

Our extended tabu search algorithm for FTSP considers three ranking criteria: expected value, optimistic valuc and pessimistic value. The algorithm keeps a set Q of non-dominated solutions seen so far and performs a three-stage search, each dealing with a different ranking criterion:

Stage 1. Starting from a feasible solution, repeatedly searches the neighborhood $N(x)$, updates Q with non-dominated solutions, and moves to a solution with the best expected value in $N(x)$.

Stage 2. Turns to a solution with the best optimistic value seen so far, restarts the neighborhood search and non-dominated solution updates, moving to a solution with the best optimistic value in $N(x)$ at each step.

Stage 3. Turns to a solution with the best pessimistic value seen so far, restarts the neighborhood search and non-dominated solution updates, moving to a solution with the best pessimistic value in $N(x)$ at each step.

The proposed stage arrangement is for general decision-making, but not mandatory. In practice, the most concerned ranking criterion is selected in the first stage, and the least concerned one in the last stage. For example, for a pessimistic decision-maker the ranking criteria can be used in the order of pessimistic, expected, and optimistic value.

For the sake of simplicity, we use $E(x)$ rather than $E(w(x))$ to denote the expect value of the objective value of solution x, and so are $O_\alpha(x)$ and $P_\alpha(x)$ for the optimistic value and pessimistic value at confident level α respectively. Other notation used is given below, and the extended tabu search procedure is shown in Algorithm 2.

Notation

$x \prec y$	x dominates y
$x \succ y$	x is dominated by y
x_E^*, x_O^*, x_P^*	known best expected, optimistic, pessimistic solutions, resp.
Q	set of non-dominated (ND) solutions
$NDUpdate(Q, y)$	remove any $x \succ y$ from Q, and add y to Q if $\nexists x \in Q : y' \prec y$
α	confidence level

3.3 Parameter Settings and Data Structure Implementation

The settings of tabu tenure λ, upper limit $\overline{k}$ of total iterations and upper limit $\overline{l}$ of non-improving iterations have great impacts on the algorithm efficiency and the solution qualities. Obviously, large values of $\overline{k}$ and $\overline{l}$ contribute to the solution qualities but degrade the algorithm efficiency. As previous comparative studies suggest, tabu tenures varying based on problem instance sizes generally give better solutions than fixed tabu tenures, and most of them range from $n/32$ to $3n/2$ for TSP with n nodes [10,33]. In our framework, λ is suggested to be about $\sqrt{n}$ and $\overline{l}$ to be $\sqrt{n/2}$, whereas $\overline{k}$ can be selected from the range $10n \sim 100n$ based on the user's preference.

The data structure implementation for solution set Q also has a great impact on algorithm performance, especially for large-size problems. The most straightforward way is to use a linear list, for which a candidate solution x has to be tested against each $y \in Q$ in the worst case. Habenicht [34] employed the Quad-tree structure for identifying non-dominated criterion vectors with much fewer comparisons, and Sun and Steuer [35,36] improved the structure to achieve more computational and storage savings. Here we implement our algorithm with the Quad-tree structure in [35], which is mostly suitable for problems with small sets of criteria and probably large sets of non-dominated solutions.

Thus in each iteration of the search, there are $n(n-1)/2$ neighboring solutions to be tested on three criteria, and the run time complexity for non-dominated solution update is $O(\frac{3}{2}n^2|Q|)$. However, previous computational results show that the average number of pairwise comparisons on Quad tree can be less than one tenth of linear list.

4 Experimental Results

The presented algorithm *ExTS* has been test on a set randomly generated symmetric FTSP instances, in which the weight are represented as fuzzy triangular numbers. Three single objective optimization algorithms, namely *TS-exp*, *TS-opt* and *TS-pes*, which perform basic tabu search for optimal expected value, optimistic value and pessimistic value respectively based on Algorithm 1, are used for evaluating the quality of non-dominated solutions. For comparison, we also implement Hansen's *MOTS* algorithm [22], where FTSP is considered as a three objective optimization problems. The basic parameter values are given in Table 1.

```
ALGORITHM: ExTS
[input: G = ⟨V, E, w⟩, α, λ, k̄, l̄]
[output: Q]
//Initialization:
let x be an arbitrary feasible solution;
x*_E ← x, x*_O ← x, x*_P ← x;
Q ← ∅, N ← ∅, k ← 0, l ← 0;
Initialize the tabu list T;
//Stage 1:
while(k < k̄) do
  N ← {y ∈ N(x)|T(x, y) ≤ 0 ∨ (∄y' ∈ Q : y' ≺ y)};
  if N = ∅ then break;
  x' ← min_E(N);
  if (E(x') < E(x*_E)) then x*_E ← x';
  else if (E(x) ≤ E(x') then l ← l + 1;
  if (l = l̄) then break;
  foreach y ∈ N : NDUpdate(Q, y);
  x*_O ← min_O(x*_O, min_O(N)), x*_P ← min_P(x*_P, min_P(N));
  Update T with (x, x');
  x ← x', k ← k + 1;
endwhile;
//Stage 2:
x ← x*_O, k ← 0, l ← 0;
let T be the tab list with x*_O;
while(k < k̄) do
  N ← {y ∈ N(x)|T(x, y) ≤ 0 ∨ (∄y' ∈ Q : y' ≺ y)};
  if N = ∅ then break;
  x' ← min_O(N);
  if (O_α(x') < O_α(x*_O)) then x*_O ← x';
  else if (O_α(x) ≤ O_α(x') then l ← l + 1;
  if (l = l̄) then break;
  foreach y ∈ N : NDUpdate(Q, y);
  x*_P ← min_P(x*_P, min_P(N));
  Update T with (x, x');
  x ← x', k ← k + 1;
endwhile;
//Stage 3:
x ← x*_P, k ← 0, l ← 0;
let T be the tab list with x*_P;
while(k < k̄) do
  N ← {y ∈ N(x)|T(x, y) ≤ 0 ∨ (∄y' ∈ Q : y' ≺ y)};
  if N = ∅ then break;
  x' ← min_P(N);
  if (P_α(x') < P_α(x*_P)) then x*_P ← x';
  else if (P_α(x) ≤ P_α(x') then l ← l + 1;
  if (l = l̄) then break;
  foreach y ∈ N : NDUpdate(Q, y);
  Update T with (x, x');
  x ← x', k ← k + 1;
endwhile;
return Q;
```

Algorithm 2. The extended tabu search algorithm for FTSP

In addition to the CPU time, the performance measures of the *MOTS* and *ExTS* algorithms also include:

- $|Q|$: the number of non-dominated solutions found.
- $\delta_E = (E_b - \overline{E})/E_b \times 100\%$: where E_b is the best expected value found by *TS-exp* and $\overline{E}$ is that found by *MOTS/ExTS*.
- $\delta_O = (O_b - \overline{O})/O_b \times 100\%$: where O_b is the best optimistic value found by *TS-opt* and $\overline{O}$ is that found by *MOTS/ExTS*.
- $\delta_P = (P_b - \overline{E})/P_b \times 100\%$: where P_b is the best pessimistic value found by *TS-pes* and $\overline{P}$ is that found by *MOTS/ExTS*.

The experiments are conducted on a 2.6 GHz AMD Athlon64 X2 Computer. The summary of computational costs are presented in Table 2, which demonstrates that, although generating a set of non-dominated FTSP solutions is rather harder than obtaining a single solution, our extended algorithm can effectively tackle the FTSP instances with up to 750 nodes. In comparison, *MOTS* takes much more CPU time on the same instances and is hard to deal with the instances with more than 300 nodes.

The comparison of solution qualities between *MOTS* and *ExTS* are presented in Table 3. For both the algorithms, the number of non-dominated solutions $|Q|$ is about one tenth of the problem size. However, for large size problems $|Q|$ usually reaches its max size limit, which indicates that the size of the real Pareto set may be greater than the limit.

Table 1. Parameter values used in the algorithms

Algorithm	λ	α	$\overline{k}$	$\overline{l}$	max size of Q
TS-exp/ TS-opt/ TS-pes	$\sqrt{n}$	0.8	$60n$	$\sqrt{n/2}$	-
MOTS/ExTS	$\sqrt{n}$	0.8	$20n$	$\sqrt{n/2}$	$n/10$ (10 if $n < 100$)

Table 2. The CPU time (in seconds) consumed by the algorithms

n	*TS-exp*	*TS-opt*	*TS-pes*	*MOTS*	*ExTS*
10	0.00	0.00	0.00	0.02	0.01
20	0.01	0.01	0.01	0.03	0.03
50	0.10	0.12	0.10	0.65	0.42
100	0.79	0.76	0.71	7.31	2.53
150	2.26	2.60	2.69	137.30	12.15
300	20.70	20.10	22.45	10108.33	112.32
450	84.23	97.07	96.52	--	736.42
600	139.87	151.57	145.16	--	2881.97
750	510.71	547.25	554.68	--	12392.08

Table 3. The comparison of solution qualities between *MOTS* and *ExTS*

n	*MOTS*				*ExTS*			
	$\|Q\|$	δ_E	δ_O	δ_P	$\|Q\|$	δ_E	δ_O	δ_P
10	1	0	0	0	1	0	0	0
20	2	0	0	0	2	0	1.37%	0
50	6	1.12%	0.50%	0	5	1.76%	0.50%	2.52%
100	**10**	3.67%	2.54%	3.10%	**10**	1.72%	-0.07%	2.65%
150	**15**	3.88%	4.86%	-0.85%	9	1.45%	-1.50%	-2.11%
300	**30**	2.72%	-2.03%	3.79%	26	1.00%	-1.50%	0.79%
450	**45**	--	--	--	**45**	0.84%	-7.82%	-5.23%
600	**60**	--	--	--	**60**	0.66%	-13.83%	0.80%
750	**75**	--	--	--	**75**	0.97%	-5.09%	-20.25%

5 Concluding Remarks

Tabu search has been demonstrated to be a successful optimization method for a wide range of single-objective optimization problems, but its applications in fuzzy optimization and multi-criteria analysis has not been deeply studied. The paper presents an FTSP algorithm that performs a three-stage tabu search, involving a preferred fuzzy ranking criterion at each stage. Simulations demonstrate that our approach effectively tackles the FTSP instances with up to 750 randomly generated nodes.

Our algorithm is conceptually simple and can be easily extended by integrating more tabu search features studied in the literature, e.g., k-opt moves [8], dynamic tabu tenures [37], adaptive aspiration levels [38], to improve the algorithm performance and solution quality. More importantly, the proposed metaheuristic is general-purpose, which we believe can be applied to a large variety of fuzzy optimization problems. Future work also includes the parallel implementation of the extended tabu search.

Acknowledgement

The author would like to thank the AAIM's reviewers for their suggestions to improve the algorithm description and comparison.

References

1. Johnson, D.S., Gutin, G., McGeoch, L.A., Yeo, A., Zhang, W., Zverovich, A.: The Traveling Salesman Problem and Its Variations. Kluwer Academic Publishers, Dordrecht (2002)
2. Kwang, H.L., Lee, J.H.: A method for ranking fuzzy numbers and its application to decision making. IEEE Trans. Fuzzy Sys. 7, 677–685 (1999)
3. Glover, F.: Tabu search, part I. ORSA J. Comput. 1, 190–206 (1989)

4. Glover, F.: Tabu search, part II. ORSA J. Comput. 2, 4–32 (1990)
5. Homaifar, A., Guan, S., Liepins, G.E.: A new approach on the traveling salesman problem by genetic algorithms. In: 5th Int'l Conf. Genetic Algorithms, San Francisco, USA, pp. 460–466 (1993)
6. Snyder, L.V., Daskin, M.S.: A random-key genetic algorithm for the generalized traveling salesman problem. European J. Oper. Res. 174, 38–53 (2006)
7. Pepper, J., Golden, B., Wasil, E.: Solving the traveling salesman problem with annealing-based heuristics: a computational study. IEEE Trans. Sys. Man. Cyber., Part A 32, 72–77 (2002)
8. Knox, J.: Tabu search performance on the symmetric traveling salesman problem. Comput. Oper. Res. 21, 867–876 (1994)
9. Misevicius, A.: Using iterated tabu search for the traveling salesman problem. Inform. Tech. Control 32, 29–40 (2004)
10. Basu, S., Ghosh, D.: A review of the tabu search literature on traveling salesman problems. In: IIMA Working Papers 2008-10-01, Indian Institute of Management Ahmedabad (2008)
11. Hasegawa, M., Ikeguchi, T., Aihara, K.: Solving large scale traveling salesman problems by chaotic neurodynamics. Neural Networks 15, 271–283 (2002)
12. Leipala, T.: On the solutions of stochastic traveling salesman problems. European J. Oper. Res. 2, 291–297 (1978)
13. Gendreau, M., Laporte, G., Seguin, R.: A tabu search heuristic for the vehicle routing problem with stochastic demands and customers. Oper. Res. 44, 469–477 (1996)
14. Choi, J., Lee, J.H., Realff, M.J.: An algorithmic framework for improving heuristic solutions: Part II. A new version of the stochastic traveling salesman problem. Comput. Chem. Eng. 28, 1297–1307 (2004)
15. Jimenez, F., Cadenas, J.M., Sanchez, G., Gomez-Skarmeta, A.F., Verdegay, J.L.: Multi-objective evolutionary computation and fuzzy optimization. Int. J. Approx. Reasoning 43, 59–75 (2006)
16. Ulungu, E.L., Teghem, J.: Multi-objective combinatorial optimization problems: A survey. J. Multi-Criteria Decision Anal. 3, 83–104 (1994)
17. Yan, Z., Zhang, L., Kang, L., Lin, G.: A new MOEA for multi-objective TSP and its convergence property analysis. In: Fonseca, C.M., Fleming, P.J., Zitzler, E., Deb, K., Thiele, L. (eds.) EMO 2003. LNCS, vol. 2632, pp. 342–354. Springer, Heidelberg (2003)
18. Gaspar-Cunha, A.: A multi-objective evolutionary algorithm for solving traveling salesman problems: Application to the design of polymer extruders. In: 7th Int'l Conf. Adaptive & Natural Computing Algorithms, Coimbra, Portugal, pp. 189–193 (2005)
19. Jozefowiez, N., Glover, F., Laguna, M.: Multi-objective meta-heuristics for the traveling salesman problem with profits. J. Math. Model. Algor. 7, 177–195 (2008)
20. Samanlioglu, F., Ferrell, W.G., Kurz, M.E.: A memetic random-key genetic algorithm for a symmetric multi-objective traveling salesman problem. Comput. Ind. Eng. 55, 439–449 (2008)
21. Fonseca, C.M., Fleming, P.J.: An overview of evolutionary algorithms in multiobjective optimization. Evolutionary Computation 3, 1–16 (1995)
22. Hansen, M.P.: Tabu search for multiobjective optimization: MOTS. In: 13th Int'l Conf. Multi Criteria Decision Making, Cape Town, South Africa (1997)
23. Kulturel-Konak, S., Smith, A.E., Norman, B.A.: Multi-objective tabu search using a multinomial probability mass function. Eur. J. Oper. Res. 169, 915–931 (2006)

24. Alves, M.J., Clomaco, J.: An interactive method for 0-1 multiobjective problems using simulated annealing and tabu search. J. of Heuristics 6, 385–403 (2000)
25. Zadeh, L.A.: Fuzzy sets as a basis for a theory of possibility. Fuzzy Set. Sys. 1, 3–28 (1978)
26. Liu, B., Liu, Y.K.: Expected value of fuzzy variable and fuzzy expected value models. IEEE Tran. Fuzzy Sys. 10, 445–450 (2002)
27. Liu, B.: Uncertainty Theory, 2nd edn. Springer, Berlin (2007)
28. Cheng, C.H.: A new approach for ranking fuzzy numbers by distance method. Fuzzy Set. Sys. 95, 307–317 (1998)
29. Chu, T.C., Tsao, C.T.: Ranking fuzzy numbers with an area between the centroid point and original point. Comput. Math. Appl. 43, 111–117 (2002)
30. Chen, S.J., Chen, S.M.: Fuzzy risk analysis based on the ranking of generalized trapezoidal fuzzy numbers. Appl. Intell. 26, 1–11 (2007)
31. Ramli, N., Mohamad, D.: A comparative analysis of centroid methods in ranking fuzzy numbers. European J. Sci. Res. 28, 492–501 (2009)
32. Chen, S.J., Hwang, C.L.: Fuzzy multiple attribute decision making methods and applications. Springer, New York (1992)
33. Tsubakitani, S., Evans, J.R.: Optimizing tabu list size for the traveling salesman problem. Comput. Oper. Res. 25, 91–97 (1998)
34. Habenicht, W.: Quad trees, a data structure for discrete vector optimization problems. Lect. Note. Econom. Math. Sys. 209, 136–145 (1983)
35. Sun, M., Steuer, R.E.: Quad trees and linear list for identifying nondominated criterion vectors. INFORM J. Computing 8, 367–375 (1996)
36. Sun, M.: A primogenitary linked quad tree data structure and its application to discrete multiple criteria optimization. Ann. Oper. Res. 147, 87–107 (2006)
37. Chiang, W., Russell, R.: A reactive tabu search metaheuristic for the vehicle routing problem with time windows. INFORMS J. Computing 9, 417–430 (1997)
38. Cordeau, J., Gendreau, M., Laporte, G.: A tabu search heuristic for periodic and multi-depot vehicle routing problems. Networks 30, 105–119 (1997)

Efficient Exact and Approximate Algorithms for the Complement of Maximal Strip Recovery

Binhai Zhu

Department of Computer Science, Montana State University,
Bozeman, MT 59717-3880, USA
bhz@cs.montana.edu

Abstract. Given two genomic maps G and H represented by a sequence of n gene markers, a *strip* (syntenic block) is a sequence of distinct markers of length at least two which appear as subsequences in the input maps, either directly or in reversed and negated form. The problem *Maximal Strip Recovery* (MSR) is to find two subsequences G' and H' of G and H, respectively, such that the total length of disjoint strips in G' and H' is maximized (i.e., conversely, the complement of the problem CMSR is to minimize the number of markers deleted to have a feasible solution). Recently, both MSR and its complement are shown to be NP-complete. A factor-4 approximation is known for the MSR problem and an FPT algorithm is known for the CMSR problem which runs in $O(2^{3.61k}n+n^2)$ time (where k is the minimum number of markers deleted). We show in this paper that there is a factor-3 asymptotic approximation for CMSR and there is an FPT algorithm which runs in $O(3^k n+n^2)$ time for CMSR, significantly improving the previous bound.

1 Introduction

In comparative genomics, one of the first steps is to decompose two given genomes into syntenic blocks—segments of chromosomes which are deemed to be homologous in the two input genomes. In the past, many methods have been proposed, but they are very vulnerable to ambiguities and errors. In the past several years, a method was proposed to eliminate noise and ambiguities in genomic maps, through handling a problem called Maximal Strip Recovery (MSR) (see below for the formal definition) [5,12]. In [4], a factor-4 polynomial-time approximation algorithm was proposed for the problem, and several close variants of the problem were shown to be intractable. In [11], both MSR and its complement CMSR was shown to be NP-complete, and an $O(2^{3.61k}n + n^2)$ time (where k is the minimum number of markers deleted) FPT algorithm was also proposed for CMSR. Most recently, MSR was shown to be APX-hard [2,8] and CMSR was also shown to be APX-hard [9].

In this paper, we focus on solving the CMSR problem with both exact and approximate solutions. We show, with a bounded search tree method, that CMSR can be solved in $O(3^k n + n^2)$ time. For the approximation part, we present the first (asymptotic) approximation algorithm for CMSR, with an asymptotic

B. Chen (Ed.): AAIM 2010, LNCS 6124, pp. 325–333, 2010.

approximation factor of 3. In the following, we formally define the problems and give a more detailed sketch of the development of the problems.

In comparative genomics, a genomic map is represented by a sequence of gene markers. A gene marker can appear in several different genomic maps, in either positive or negative form. A *strip* (syntenic block) is a sequence of distinct markers that appears as subsequences in two or more maps, either directly or in reversed and negated form. Given two genomic maps G and H, the problem *Maximal Strip Recovery* (MSR) [5,12] is to find two subsequences G' and H' of G and H, respectively, such that the total length of disjoint strips in G' and H' is maximized. Intuitively, those gene markers not included in G' and H' are noise and ambiguities. The problem of deleting the minimum number of noise and ambiguous markers to have a feasible solution (i.e., every remaining marker must be in some strip) is exactly the *complement of MSR*, which will be abbreviated as CMSR.

We give a precise formulation of the generalized problem MSR: Given two signed permutations (genomic maps) G_i of $\langle 1, \ldots, n \rangle$, $1 \leq i \leq 2$, find q sequences (strips) S_j of length at least two, and find two signed permutations π_i of $\langle 1, \ldots, q \rangle$, such that each sequence $G_i^\star = S_{\pi_i(1)} \cdots S_{\pi_i(q)}$ (here S_{-j} denotes the reversed and negated sequence of S_j) is a subsequence of G_i, and the total length of the strips S_j is maximized. Note that we can easily generalize the problem to handle $d > 2$ input permutations, or to MSR-d, as in [4,11]. In this paper, we will focus only on the complement of MSR, or the CMSR problem. We refer to Fig. 1 for an example. In this example, each integer represents a marker.

$$
\begin{aligned}
G_1 &= \langle 1, 2, 3, 4, 5, 6, 7, 8, 9, 10, 11, 12 \rangle \\
G_2 &= \langle -9, -4, -7, -6, 8, 1, 3, 2, -12, -11, -10, -5 \rangle \\
S_1 &= \langle 1, 2 \rangle \\
S_2 &= \langle 6, 7, 9 \rangle \\
S_3 &= \langle 10, 11, 12 \rangle \\
\pi_1 &= \langle 1, 2, 3 \rangle \\
\pi_2 &= \langle -2, 1, -3 \rangle \\
G_1^\star &= \langle 1, 2, 6, 7, 9, 10, 11, 12 \rangle \\
G_2^\star &= \langle -9, -7, -6, 1, 2, -12, -11, -10 \rangle
\end{aligned}
$$

Fig. 1. An example for the problem MSR and CMSR. MSR has a solution size of eight. CMSR has a solution size of four: the deleted markers are 3,4,5 and 8.

In 2007, a heuristic based on Maximum Clique (and its complement Maximum Independent Set) was proposed for the problem MSR [5,12], which does not guarantee finding the optimal solution. In [4], this heuristic was modified to achieve a factor-4 approximation for MSR. This was done by converting the problem to computing the maximal independent set in t-interval graphs, which admit a factor-$2t$ approximation [1]. Not surprisingly, recently MSR was shown to be NP-complete (which implies that CMSR is also NP-complete) [11] and APX-hard [2,8]. Bulteau et al. also introduced a parameter called *gap*, which is the number of non-selected markers between two markers in a valid strip [2]. They showed that δ-gap-MSR is NP-complete for any $\delta \geq 1$ and APX-hard for any $\delta \geq 2$. They also presented a factor-1.8 approximation for $\delta = 1$ and raised several interesting questions regarding the approximability of the problem for different values of δ.

For CMSR, an $O(2^{3.61k}n + n^2)$ time (where k is the minimum number of markers deleted) FPT algorithm was known [11]. In this paper, we improve this running time to $O(3^k n + n^2)$. Before this work, there has been no approximation result known for CMSR. In this paper, we present the first polynomial-time approximation algorithm for CMSR, with a factor of 3. Jiang proved recently that CMSR is APX-hard [9], indicating unlikely a PTAS for CMSR.

This paper is organized as follows. In Section 2, we present a new fixed-parameter algorithm for CMSR. In Section 3, we present a factor-3 asymptotic approximation for CMSR. Finally in Section 4, we conclude the paper with a few open questions.

2 A Bounded Search Tree Algorithm for CMSR

In this section, we consider solving CMSR with an FPT algorithm. Basically, an FPT algorithm for an optimization problem Π with optimal solution value k is an algorithm which solves the problem in $O(f(k)n^c)$ time, where f is any function only on k, n is the input size and c is some fixed constant not related to k. More details on FPT algorithms can be found in the monograph by Downey and Fellows [6].

We first review the following lemma which was proved (and revised) in [11,7].

Lemma 1. *Before any marker is deleted, if $xyzw$ or $-w-z-y-x$ appears in both G_1 and G_2 as maximal common substrings (or, if $xyzw$ appears in G_1 and $-w-z-y-x$ appears in G_2, and vice versa), then there is an optimal solution for MSR which has $xyzw$ or $-w-z-y-x$ as a strip.*

An example for the above lemma is as follows: $G_1 = cdaxyzwbef$ and $G_2 = e-w-z-y-xfcdab$. $xyzw$ appears in G_1, $-w-z-y-x$ appears in G_2 (in signed reversal order). So we have one optimal solution $G_1^\star = cdxyzw$ and $G_2^\star = -w-z-y-xcd$. On the other hand, the optimal solution is not unique as we can select $G_1^+ = cdabef$ and $G_2^+ = efcdab$. It should be noted that the above lemma also holds when a strip is of length greater than four.

Let Σ be the alphabet for the input maps G_1 and G_2. In [11], the above lemma was applied to obtain a simple weak kernelization procedure, which is incomplete [7]. For completeness, we sketch the complete version below. Note that the weak kernel definition is new. For a problem Π in NP, Π has a kernel iff it has a weak kernel [7]. (In short, weak kernel refers to the search space where a solution can be directly or indirectly searched or drawn.)

Before any marker is deleted, we can identify all maximal common substrings of length at least one (possibly in negated and reversed form, which will also be called maximal common substrings for convenience) of G_1 and G_2. We also call a length-1 maximal common substring (which is a letter) an *isolate*. Two substrings are called *neighbors* if there is no other string in between them.

Lemma 1 holds for maximal common substrings of length greater than 4. In fact, similar to that, we can show that a length-3 maximal common substring of G_1 and G_2 which has at most 3 isolated neighbors in G_1 and G_2 can be a strip in some optimal solution of MSR, etc. The weak kernelization procedure is as follows.

1. Without deleting any gene marker in G_1 and G_2, identify a set of maximal common substrings of length at least four, a set of maximal common substrings of length three which has at most 3 isolated neighbors, and a set of maximal common substrings of length two which has at most 2 isolated neighbors, from the two sequences G_1 and G_2.
2. For each common substring identified, change it to a new letter in Σ_1, with $\Sigma_1 \cap \Sigma = \emptyset$. Let the resulting sequences be G'_1, G'_2.

The correctness of this procedure follows from the fact that if a maximal common substring S of length-s in G_1 and G_2 has t isolated neighbors and $t \leq s$, then S is a strip in some optimal solution of MSR. (If not, then we could delete the t isolated neighbors of S, making S a strip and hence obtaining a solution at least as good as the previous one.) This implies that we can focus on a special kind of solution for CMSR in which the maximum number of isolated letters are deleted.

Let Σ_1 be the set of new letters used in the weak kernelization process, with $\Sigma_1 \cap \Sigma = \emptyset$. The two lemmas for obtaining the final result are: (1) There is an optimal CMSR solution of size k for G_1 and G_2 if and only if the solution can be obtained by deleting k markers in Σ from G'_1 and G'_2 respectively. (2) In G'_1 (resp. G'_2), there are at most $5k$ letters (markers) in Σ [11]. To see a slightly revised proof of the last lemma (due to the revised weak kernelization procedure), let k_i be number of length-i common substrings deleted in the optimal CMSR solution. Consequently we have

$$k = k_1 + 2k_2 + 3k_3.$$

The size of the weak kernel is the number of letters that can possibly be deleted, i.e., the number of letters in Σ. Let S be a length-s maximal common substring to be deleted. If $s = 3$, then S has at most 4 isolated neighbors and we have 7 associated letters for S. If $s = 2$, then S has 3 or 4 isolated neighbors; and we can have 6 associated letters for S (when we have 4 isolated neighbors), 7 or 8 (three isolated neighbors and another neighbor of length-2 or length-3). Now let us consider the remaining letters which must be all isolates. Let x be a marker to be deleted, let it appear in G'_1 as $\cdots axb \cdots cd \cdots$ and let it appear in G'_2 as $\cdots cxd \cdots ab \cdots$. Clearly, in this example x is associated with $\{x, a, b, c, d\}$. In other words, for each isolate, we have 5 associated letters. Putting these together, the number of letters in Σ is

$$5k_1 + 8k_2 + 7k_3 \leq 5k_1 + 10k_2 + 15k_3 = 5k.$$

Let Σ_1 be the set of new letters used in the weak kernelization process, with $\Sigma_1 \cap \Sigma = \emptyset$. It is easily seen that there is an optimal CMSR solution of size k for G_1 and G_2 if and only if the solution can be obtained by deleting k markers in Σ from G_1' and G_2' respectively. Then we can easily find the exact solution or report that such a solution does not exist by checking $\binom{5k}{k} \approx 2^{3.61k}$ possible solutions. This presents an FPT algorithm for CMSR which runs in $O(2^{3.61k}n + n^2)$ time [11].

To improve the running time of this algorithm, we first try to avoid enumerating $\binom{5k}{k} \approx 2^{3.61k}$ solutions. We need a few new lemmas. In G_1', G_2' we call any maximal continuous block (or substring) only made of letters in Σ a *pseudo-block*.

An example regarding the pseudo-block concept is as follows: $G_1 = abcde - f - w - z - y - x$ and $G_2 = bcdef - axyzw$. After the weak kernelization step, $G_1' = a\boxed{\text{bcde}} - f\boxed{\text{-w-z-y-x}}$ and $G_2' = \boxed{\text{bcde}}f - a\boxed{\text{xyzw}}$. There are two pseudo-blocks in G_1': a and $-f$. There is one pseudo-block in G_2': $f - a$. We now proceed to prove the following lemmas.

Lemma 2. *In G_1' (resp. G_2'), there are at most k pseudo-blocks. If there is a pseudo-block of length one (i.e., an isolated letter z in Σ), then z must be deleted to obtain an optimal solution for the original CMSR problem for G_1 and G_2.*

Lemma 3. *In any pseudo-block in G_1', if any 2-substring xy (i.e., two consecutive letters xy in Σ) is retained for an optimal solution then all the markers in Σ, between x and y, and within the same pseudo-block in G_2', must be deleted to obtain an optimal solution for the original CMSR problem for G_1 and G_2.*

Proof. In G_2', if some marker in Σ between x and y is not deleted, then xy cannot be a strip in the final solution. A contradiction.

On the other hand, notice that if x and y in G_2' belong to different pseudo-blocks, then by a similar argument for Lemma 1, we can simply delete x, y in both G_1', G_2' to obtain an optimal solution. □

The above lemmas imply a bounded search tree algorithm for CMSR. The idea is that one can start searching at any 2-substring xy within a pseudo-block. Either both x, y are deleted, or one of x and y is deleted, or xy is kept (then following Lemma 3, at least one other marker would be deleted). We have the following bounded search tree algorithm.

1. Without deleting any gene marker in G_1 and G_2, identify a set of maximal common substrings of length at least four, a set of maximal common substrings of length three which has at most 3 isolated neighbors, and a set

of maximal common substrings of length two which has at most 2 isolated neighbors, from the two sequences G_1 and G_2.
2. For each substring identified, change it to a new letter in Σ_1, with $\Sigma_1 \cap \Sigma = \emptyset$. Let the resulting sequences be G_1', G_2'.
3. $G_1" \leftarrow G_1', G_2" \leftarrow G_2'$. Loop over the following steps until exactly k letters in Σ have been deleted with a valid CMSR solution computed, or report that no such solution exists.
4. Identify all pseudo-blocks in $G_1", G_2"$. Delete any pseudo-block with size one.
5. Start from the leftmost pseudo-block in $G_1"$ and let the first two letters of the pseudo-block be xy. Perform an exhaustive search with 4 possibilities: (I) only x is deleted, (II) only y is deleted, (III) both x and y are deleted, and (IV) xy is kept as a strip but all letters in Σ between x and y and within the same pseudo-block in $G_2"$ are deleted. Delete correspondingly x, y, x and y, and the markers between x and y in $G_2"$ respectively (and also delete the corresponding markers in $G_1"$ for case IV) and solve the resulting sub-problem recursively by repeating Step (4)-(5).

Let $f(k)$ be the size of the search tree. It is easy to see the following recurrence relation

$$f(k) = \begin{cases} 0 & \text{if } k = 0, \\ 1 & \text{if } k = 1, \\ \leq 3f(k-1) + f(k-2) & \text{if } k > 1. \end{cases}$$

Solving this recurrence relation, we have

$$f(k) = \frac{\sqrt{13}}{13}\left(\frac{3+\sqrt{13}}{2}\right)^k - \frac{\sqrt{13}}{13}\left(\frac{3-\sqrt{13}}{2}\right)^k,$$

which implies $f(k) \leq \frac{\sqrt{13}}{13}(\frac{\sqrt{13}+3}{2})^k + \frac{\sqrt{13}}{13}(\frac{\sqrt{13}-3}{2})^k = \frac{\sqrt{13}}{13}3.302^k + O(1) \approx \frac{\sqrt{13}}{13}2^{1.73k}$.

When a possible solution is obtained (i.e., after up to k markers are deleted), it takes $O(n)$ time to check whether the solution is valid (i.e., whether each of the remaining markers is within some strip). Of course, we need to spend $O(n^2)$ time to build the correspondence between the markers in G_1 and G_2. Hence, CMSR can be solved in $O(2^{1.73k}n + n^2)$ time.

It is obvious that the algorithm can be improved: for case (III), when both x and y need to be deleted, we can delete only one of x and y and delay the deletion of the other to the next level of search. Hence the main recurrence becomes $f(k) \leq 3f(k-1)$ and $f(k) \leq 3^k$. Consequently, we can slightly improve the above result to have the following theorem.

Theorem 1. *CMSR can be solved in $O(3^k n + n^2)$ time.*

3 An Approximation Algorithm for CMSR

In this section, we consider solving CMSR with an approximation algorithm. A polynomial time algorithm $\mathcal{A}$ for a minimization problem Π with optimal solution value O^* is a factor-α symptotic approximation if for all instances of Π the solution value returned by $\mathcal{A}$, APP, satisfies $APP \leq \alpha \times O^* + \beta$, where β is some constant. More details on approximation algorithms can be found in a standard textbook, e.g., the one by Vazirani [10].

In [4], a factor-4 approximation was proposed for the MSR problem. The algorithm directly makes use of the approximation algorithm for MWIS in 2-interval graphs [1]; unfortunately, it does not make use of the property of MSR. In the following, we present a factor-3 asymptotic approximation for CMSR, by exploiting its property.

We first assume that a weak kernelization step has been performed, i.e., G'_1, G'_2 and the resulting (up to) K pseudo-blocks in G'_1 or G'_2 have been obtained. If $K = 0$, the instance is easily solvable (as no marker needs to be deleted); so, without loss of generality, we assume from now on that $K \neq 0$.

We define a special version of the CMSR problem in which one has to delete the minimum number of markers so that each strip hence obtained is of length exactly two. We call the resulting minimization problem CMSR$_2$, with optimal solution value OPT$_2$. Let the optimal solution value for CMSR be OPT. Obviously, we have OPT $\geq K$. We first prove the following lemma.

Lemma 4. *$OPT_2 \leq 2 \times OPT + 1$.*

Proof. In an optimal solution for any CMSR instance, each strip is of length at least two. So feeding the CMSR instance to CMSR$_2$, i.e., only using length-2 strips, one would delete at most one extra letter for each strip (of odd length). Assume that in an optimal solution for any CMSR instance we have a total of L strips of odd length. When running an optimal solution for CMSR$_2$ for this instance, we have

$$OPT_2 \leq OPT + L.$$

It is obvious that L is bounded by the (maximum) number of pseudo-blocks in G'_1 and G'_2 plus one, i.e., $L \leq K + 1$. Therefore, $L \leq K + 1 \leq OPT + 1$. Putting all together, we have

$$OPT_2 \leq 2 \times OPT + 1.$$ □

All we need to do from now on is to have a polynomial-time approximation algorithm for CMSR$_2$. We show below that CMSR$_2$ is exactly a vertex cover problem on 2-interval graphs.

After the weak kernelization step (i.e., after G'_1, G'_2 are obtained), we identify all possible length-2 candidate strips, for letters in Σ, in all the pseudo-blocks in G'_1 and G'_2. Notice that due to Lemma 1, we do not have to consider any length-2 candidate strip crossing some letter in Σ_1. Any letter not appearing in any of the length-2 candidate strip will be in the optimal solution for CMSR$_2$, hence must be deleted. We form a 2-interval graph $\mathcal{G}$ for these length-2 candidate strips: each node corresponds to a length-2 candidate strip, two such strips form an edge if they are in conflict, i.e., if the corresponding intervals in either G'_1 or G'_2 have an intersection. For example, $G'_1 = \cdots acbd \cdots$, $G'_2 = \cdots abcd \cdots$, then the two candidate strips $[a, b], [c, d]$ form an edge in $\mathcal{G}$.

Now one can see that this is exactly a vertex cover problem on 2-interval graphs: we need to delete the minimum number of nodes in $\mathcal{G}$ so that there is no edge (conflict) left. Using the known approximation algorithm by Butman et al. [3], we can obtain a factor-1.5 approximation for this problem. Let the optimal vertex cover in $\mathcal{G}$ have value OPT_1 and let the resulting approximation solution value be APP_1. We have $\text{APP}_1 \leq 1.5 \times \text{OPT}_1$. We now proceed to prove the following theorem.

Theorem 2. *There is a factor-3 polynomial time asymptotic approximation for CMSR.*

Proof. Our approximation algorithm is exactly what we have just been described. We delete all letters in Σ in G_1', G_2' which are not associated with any length-2 candidate strip. Let OPT_0 be the number of such letters deleted.

We then build the 2-interval graph $\mathcal{G}$ and run the approximate vertex cover algorithm by Butman et al. [3]. Let the optimal vertex cover in $\mathcal{G}$ have value OPT_1 and let the resulting approximation solution value be APP_1. Following the result by Butman et al. [3], we have $\text{APP}_1 \leq 1.5 \times \text{OPT}_1$.

Now let APP be the solution value of the whole approximation algorithm, combined with Lemma 4, we have

$$\begin{aligned} APP &= OPT_0 + APP_1 \\ &\leq OPT_0 + 1.5 \times OPT_1 \\ &\leq 1.5 \times OPT_2 \\ &\leq 3 \times OPT + 1.5. \end{aligned}$$

Note that $OPT_2 = OPT_0 + OPT_1$, so $OPT_0 + 1.5 \times OPT_1 \leq 1.5 \times OPT_2$. Also, if there is no valid solution found for CMSR_2 then it implies that there is no valid solution for the corresponding CMSR instance. □

4 Concluding Remarks

We summarize the current FPT, hardness and approximation results for MSR and CMSR in the following table.

Table 1. Summary of the current status on MSR and CMSR

	MSR	CMSR
FPT	———	$O(3^k n + n^2)$ — this paper
Approximation	factor 4 [4]	factor 3 (asymptotic) — this paper
Hardness	APX-hard [2,8]	APX-hard [9]

It would be interesting to know whether our FPT algorithm for CMSR can be further improved. The running time of the FPT algorithm we have obtained in this paper for CMSR, though much faster than its predecessor in [11], is still not efficient enough to make them truely useful in practice. To make such an FPT algorithm practical for MSR datasets, which usually has k between 50 to 150 (with n usually at least 600 or bigger), it must be more efficient. The current FPT algorithm can probably only handle the cases when $k \leq 30$. Even though whether an exact vertex cover FPT algorithm can be applied to solve CMSR directly is still unknown, CMSR is inherently a vertex cover problem. So it might be possible to improve the $O(3^k n + n^2)$ time bound for CMSR. Moreover, to make the algorithm applicable some tuning is necessary so that the algorithm works for the multichoromosal genomes or genomic maps. Another interesting theoretical question is to decide whether the corresponding asymptotic approximation factor 3 can be further improved.

Acknowledgments

I thank Henning Fernau for his valuable comment on the search tree algorithm in Section 2, i.e., on 'delayed' search. I also thank anonymous reviewers for several insightful comments. This research is partially supported by NSF grant DMS-0918034 and NSF of China under project 60928006.

References

1. Bar-Yehuda, R., Halldórsson, M.M., Naor, J.(S.), Shachnai, H., Shapira, I.: Scheduling split intervals. SIAM Journal on Computing 36, 1–15 (2006)
2. Bulteau, L., Fertin, G., Rusu, I.: Maximal strip recovery problem with gaps: hardness and approximation algorithms. In: Dong, Y., Du, D.-Z., Ibarra, O. (eds.) ISAAC 2009. LNCS, vol. 5878, pp. 710–719. Springer, Heidelberg (2009)
3. Butman, A., Hermelin, D., Lewenstein, M., Rawitz, D.: Optimization problems in multiple-interval graphs. In: Proceedings of the 18th ACM-SIAM Symposium on Discrete Algorithms (SODA 2007), pp. 268–277 (2007)
4. Chen, Z., Fu, B., Jiang, M., Zhu, B.: On recovering syntenic blocks from comparative maps. Journal of Combinatorial Optimization 18, 307–318 (2009)
5. Choi, V., Zheng, C., Zhu, Q., Sankoff, D.: Algorithms for the extraction of synteny blocks from comparative maps. In: Giancarlo, R., Hannenhalli, S. (eds.) WABI 2007. LNCS (LNBI), vol. 4645, pp. 277–288. Springer, Heidelberg (2007)
6. Downey, R., Fellows, M.: Parameterized Complexity. Springer, Heidelberg (1999)
7. Jiang, H., Zhu, B.: Weak kernels. ECCC Report, TR10-005 (May 2010)
8. Jiang, M.: Inapproximability of maximal strip recovery. In: Dong, Y., Du, D.-Z., Ibarra, O. (eds.) ISAAC 2009. LNCS, vol. 5878, pp. 616–625. Springer, Heidelberg (2009)
9. Jiang, M.: Inapproximability of maximal strip recovery, II. In: Proceedings of the 4th Annual Frontiers of Algorithmics Workshop, FAW 2010 (to appear 2010)
10. Vazirani, V.: Approximation Algorithms. Springer, Heidelberg (2001)
11. Wang, L., Zhu, B.: On the tractability of maximal strip recovery. J. of Computational Biology (to appear 2010); An earlier version appeared in TAMC 2009
12. Zheng, C., Zhu, Q., Sankoff, D.: Removing noise and ambiguities from comparative maps in rearrangement analysis. IEEE/ACM Transactions on Computational Biology and Bioinformatics 4, 515–522 (2007)

Author Index

Printing: Mercedes-Druck, Berlin
Binding: Stein+Lehmann, Berlin